The *Crystal Field Handbook* is based on the modern conceptual understanding of crystal fields. It clarifies several issues that have historically produced confusion, particularly the effects of covalency and ligand polarization on the energy spectra of magnetic ions.

Clear instructions and a set of computer programs are provided for the phenomenological analysis of energy spectra of magnetic ions in solids. Readers are shown how to employ a hierarchy of parametrized models to extract as much information as possible from observed spectra. Special attention is given to the superposition model, the parametrization of transition intensities and methods to be used when the standard (one-electron) crystal field parametrization fails.

This is the first book on crystal field theory which brings together the whole range of phenomenological models, together with the conceptual and computational tools necessary for their use. It will be of particular interest to graduate students and researchers working on the development of opto-electronic systems and magnetic materials.

BETTY NG is presently Head of the Air Quality Modelling and Assessment Unit in the Environment Agency. She taught undergraduate courses in quantum mechanics and group theory. She has carried out research work in several areas of computational physics, including crystal field theory, superconducting phase transitions and the dispersion of pollutants in tidal waters. Her doctoral research pioneered the use of many-body techniques in the *ab initio* calculation of crystal field and correlation crystal field parameters. She is author and co-author of 23 refereed papers on various aspects of crystal field theory.

DOUG NEWMAN has carried out research on crystal field theory since 1963. His work with Margaret Curtis on the *ab initio* calculation of lanthanide crystal fields provided the basis for the introduction of several parametrized models into the literature: the superposition model (1967), its application to $S$-state ions (1972), the correlation crystal field model (1968), the spin-correlated crystal field model (1970) and the vector crystal field model (for transition intensities) in 1975. He is author and co-author of more than 150 refereed papers on crystal field theory and related topics in mathematical physics. He is joint author of *How to use Groups*, with J. W. Leech, published by Methuen in 1969 and *Computational Techniques in Physics*, with P. K. McKeown, published by Adam Hilger in 1987.

# Crystal Field Handbook

edited by

## D. J. Newman

*University of Southampton*

and

## Betty Ng

*Environment Agency Wales*

CAMBRIDGE
UNIVERSITY PRESS

CAMBRIDGE UNIVERSITY PRESS
Cambridge, New York, Melbourne, Madrid, Cape Town, Singapore, São Paulo

Cambridge University Press
The Edinburgh Building, Cambridge CB2 8RU, UK

Published in the United States of America by Cambridge University Press, New York

www.cambridge.org
Information on this title: www.cambridge.org/9780521591249

First published 2000
This digitally printed version 2007

*A catalogue record for this publication is available from the British Library*

*Library of Congress Cataloguing in Publication data*
Crystal field handbook / edited by D.J. Newman and Betty Ng.
    p.  cm.
  Includes bibliographical references.
  ISBN 0 521 59124 4
  1. Crystal field theory.  I. Newman, D.J. (Douglas John)  II. Ng, Betty, 1959–

QD475.C76. 2000
538'.43–dc21    99-052178

ISBN 978-0-521-59124-9 hardback
ISBN 978-0-521-03936-9 paperback

To our parents. . .

# Contents

# Contributors

Chan Kwok Sum
*Department of Physics and Materials Science, City University of Hong Kong, Kowloon, Hong Kong*

Guokui Liu
*Chemistry Division, Argonne National Laboratory, Argonne, USA*

Douglas J. Newman
*Department of Physics and Astronomy, University of Southampton, UK*

Betty K. C. Ng
*The Environment Agency Wales, Cardiff, UK*

Michael F. Reid
*Department of Physics and Astronomy, University of Canterbury, Christchurch, New Zealand*

Czeslaw Rudowicz
*Department of Physics and Materials Science, City University of Hong Kong, Kowloon, Hong Kong*

Yau Yuen Yeung
*Department of Science, Hong Kong Institute of Education, New Territories, Hong Kong*

# Preface

Our aim in producing this book is to make a wide range of up-to-date techniques for the analysis of crystal field splittings accessible to non-specialists and specialists alike. All of these techniques are based on the *phenomenological* approach, in which the aims are to

(i) parametrize observed crystal field splittings using models which provide the maximum predictive power,

(ii) test model assumptions directly by means of the quality of parameter fits and the accuracy of their predictions,

(iii) use model parameters as a convenient interface between first principles calculations and experiment.

It is expected that readers will be mainly interested in the first two of these aims. Hence relatively little space is devoted to the description of first principles calculations. While most of the book is concerned with practical applications of the phenomenological approach, some space is given to providing a critical appraisal of relationships between the this approach and other approaches to crystal field theory.

A series of examples, backed with a set of QBASIC programs, is provided to enable readers to rapidly attain proficiency in the basic techniques of phenomenological analysis. Those who wish to embark on more sophisticated analyses are directed to other available program packages. In addition to the procedures covered in the QBASIC programs, the book covers three topics which have barely been touched upon elsewhere. These are the method of crystal field invariants (Chapter 8), the semiclassical model (Chapter 9) and the use of parametrized models in the analysis of transition intensities between crystal field split levels (Chapter 10).

In the process of constructing the bibliography, a file (cfh_all.bib) containing a far more comprehensive list of references on crystal fields and related topics was generated. This is formatted as an input file to BibTeX, making it useful in the construction of all types of reference list. The editors are grateful to Dr M. F. Reid for his assistance in generating this file, and for making it available on the website

`http://www.phys.canterbury.ac.nz/crystalfield`

It is hoped that readers will assist Dr Reid in keeping this file accurate and up-to-date.

One editor (DJN) would like to express his gratitude to all his former colleagues and students who, over the last 40 years, have participated in the development of the phenomenological approach ot crystal fields. Thanks are also due to the staff and students at the University of Southampton who contributed to the initiation and the development of this book project. We are especially indebted to the contributing authors, who have respected deadlines and shown considerable forbearance with the fussiness of the editors in their attempts to integrate the notation and conceptual approach in different parts of the book.

The editors would also like to express their gratitude to the editorial staff at Cambridge University Press for their patient help in eliminating typographical errors and improving the layout. We are particularly grateful to the respondents in 'texline' for helping with our TeX problems and adapting the standard 'cupplain' format to our requirements.

*Doug Newman and Betty Ng*

# Introduction

D. J. NEWMAN

*University of Southampton*

BETTY NG

*Environment Agency*

The energy spectrum of an isolated magnetic ion in a crystal carries information about the magnetic ion itself, its crystalline host and the interaction between these two components of the system. Crystal field theory comprises a range of techniques for extracting as much of this information as possible from observed spectra. The aim is to express the information in a form that can be used to predict the energy spectra of related systems.

Magnetic ions in crystals have many useful physical properties. In particular, there is an ongoing search for new types of laser crystal [Kam95, Kam96] and new magnetic materials. In order to design the new systems required for specific applications, it is necessary to be able to predict energy level structures and transition intensities for any magnetic ion in any crystalline environment. The techniques described in this book go a considerable way towards the achievement of this goal.

The term 'crystal field theory' has been applied to describe two quite different approaches. One of these, the so-called *phenomenological approach*, which involves the use of linear parametrized operator expressions to fit experimental results, provides a highly successful predictive tool. The alternative, so-called *ab initio* approach, in which energy levels and transition intensities are calculated from first principles, has proved far less useful. This book is mainly concerned with showing how the phenomenological approach can be used to obtain information about the physical properties of magnetic ions in crystals directly from observed spectra.

Our aim in this book is to make modern techniques for analysing and predicting spectra of isolated magnetic ions in crystals more widely known and more easily accessible. Physical assumptions inherent in the various parametrized Hamiltonian and transition intensity models are clarified, and conceptual and computational tools are provided for the analysis of experimental results. The emphasis is on extracting as much information as possi-

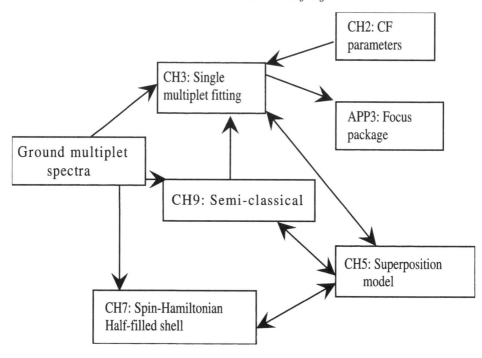

Fig. 0.1 Routes for the analysis of ground multiplet spectra.

ble from observed spectra by using a range of tried and tested parametrized models as tools. Many examples from the recent literature on crystal field theory are quoted, and an extensive bibliography is included. Nevertheless, the book is *not* intended to provide a comprehensive review of the literature.

Readers of this book either have access to experimental results that they wish to analyse, or wish to predict the energy level structure or transition intensities of a system which has yet to be studied experimentally. The criteria for choosing an appropriate parametrized model are the same in both cases, and depend on the type of spectrum that is to be analysed or predicted. Optical spectra contain information about energy levels and transition intensities for many multiplets, while spectra obtained from inelastic neutron scattering, electron spin resonance or far-infrared spectroscopy normally only provide information about the lowest energy or ground multiplets. We distinguish these as the 'many-multiplet' and 'ground multiplet' problems in the following discussion.

The analysis of a given set of experimental results may involve the use of several parametrized models, as shown in Figures 0.1 and 0.2. Figure 0.1 shows three possible routes for handling the ground multiplet problem.

(i) If the experimental spectrum is for a magnetic ion with a half-filled shell, it is appropriate to fit the spectrum to spin-Hamiltonian parameters as described in Chapter 7.

(ii) Complete ground multiplet spectra can be analysed using the parameter fitting procedure described in Chapter 3. One of the tables in Chapter 2 will often provide a suitable set of starting parameters.

(iii) The 'superposition model' parametrization, described in Chapter 5, may be employed if relevant starting parameters are not available or when experimental results are insufficient to carry out the procedures described in Chapter 3. A more sophisticated procedure, designed specially for the analysis of neutron scattering results, is mentioned in Appendix 3.

(iv) When a ground multiplet spectrum is only partially known, or the lines are broad, it may be possible to use the semiclassical crystal field model described in Chapter 9. This is particularly appropriate for ground multiplets corresponding to large angular momenta.

Figure 0.2 (on the following page) shows four possible routes for handling the 'many-multiplet' problem.

(i) If complete sets of energy levels are available for only a few multiplets it may be appropriate to analyse them one at a time using the methods described in Chapter 3, possibly using a set of starting parameters provided in Chapter 2. When information about crystal structure is available, further checks and analyses can be carried out using the 'superposition model', as described in Chapter 5. Separate multiplet analyses can also be used to provide information about correlation effects, as described in Chapter 7.

(ii) When the observed spectrum provides energy levels for many multiplets, the most productive method of analysis is provided by specially developed computer packages. One of these is described in detail in Chapter 4, and another is described briefly in Appendix 3.

(iii) When quantitative information about transition intensities is available, the parametrization techniques described in Chapter 10 should be employed. This parametrization is particularly relevant for predicting the properties of crystal lasers.

(iv) In cases when it is difficult to carry out the fitting procedures described in Chapter 4, e.g. for reasons such as low site symmetry, multiplet moments can be used to provide a starting point for the techniques described in Chapter 8.

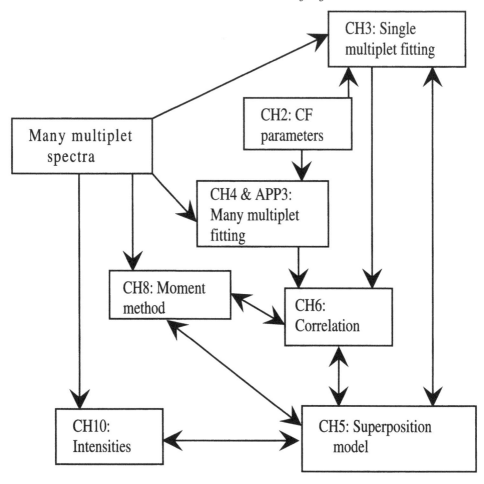

Fig. 0.2 Routes for the analysis of many-multiplet spectra.

It is important to start with a clear idea of the purpose of any proposed analysis of experimental results. While this is generally not relevant to the choice of route, it will determine just how far it is sensible to go with a specific analysis. The intention may be to:

(i) verify the interpretation of an observed set of energy levels;

(ii) determine the single-ion part of a parametrized Hamiltonian as the starting point of a more general investigation of the dynamic interactions between magnetic ions, or between magnetic ions and phonons;

(iii) provide information which will assist in predicting properties of related systems, e.g. as the first step in designing systems which are intended to have specific opto-electronic properties;

(iv) investigate the various types of parametrized model, e.g. study their reliability and predictive power;

(v) use parametrized models as a route to obtain an improved understanding of the electronic structure of magnetic ions in crystals.

The theoretical justification of simple parametrized models, given in Chapter 1, is particularly relevant to the last two intentions.

The parametrized models used in this book for the analysis of energy levels and transition intensities are all based on 'effective operators' which act solely on the many-electron states of the open-shell electrons of the magnetic ions. The structure of these operators depends upon a few general, but nevertheless clearly defined, physical assumptions. None of the parameters has predetermined values. Given sufficient experimental results, an attempt can be made to fit all the model parameters. This phenomenological approach is quite rare in physics because it is unusual to have the large amounts of information that are provided by energy spectra [New78].

The most important magnetic ions are all within the transition metal series (with partially filled $3d$ or $4d$ shells), the lanthanide (or rare-earth) series (with partially filled $4f$ shells) and the actinide series (with partially filled $5f$ shells). Most of the examples in this book are concerned with the analysis of observed energy spectra of ions in the lanthanide and actinide series. Some excellent treatises are already available on the transition metal series, notably the books by Gerloch and Slade [GS73] and Ballhausen [Bal62]. While there are some differences in detail arising from the different relative importance of the crystal field and other interactions involved in generating the spectra, *the phenomenological models discussed in this book, and their justifications, are essentially the same for all magnetic ions.*

It is worth noting the close parallels that exist between the phenomenological models used for the analysis of energy level structures and those used in the analysis of transition intensities. Similar parallels also relate the conceptual analysis of contributions to the transition intensities, developed in Chapter 10, and the analysis contributions to the crystal field, developed in Chapter 1.

A package of QBASIC programs, described in Appendix 2, has been designed to provide a simple 'tool kit', designed to illustrate the various techniques of analysing spectra which are discussed in this book. The use of these programs is described in Chapters 3 and 5. More specialized computer packages, which the reader can access, are briefly described in Chapter 4, and Appendices 3 and 4.

# 1

# Crystal field splitting mechanisms

D. J. NEWMAN

*University of Southampton*

BETTY NG

*Environment Agency*

In order to interpret the information obtained from magnetic ion spectra using parametrized crystal field models it is necessary to have a qualitative understanding of the physical mechanisms through which the crystalline environment induces energy level splittings. The discussion of mechanisms given in this chapter should be seen in this light. It is not intended to provide a practical basis for quantitative *ab initio* calculations of crystal field splittings. Instead, it provides a conceptual description of the various mechanisms that contribute to crystal field splittings.

The first three sections give a qualitative description of the most important mechanisms which contribute to crystal field splittings. These comprise the electrostatic, charge penetration, screening, exchange, overlap and covalency contributions. Sections 1.4 and 1.5 express these contributions in terms of a simple *algebraic formalism* related to the 'tight-binding' model in solid state physics. Several other formal approaches to crystal field theory, including the phenomenological approach, are described in the subsequent sections. Numerical results for a particular system, based on the formalism developed in Sections 1.4 and 1.5, are given in Section 1.9. The significance of these results to the phenomenological methods used in this book is summarized in Section 1.10.

## 1.1 The crystal field as a perturbation of free-ion open-shell states

Crystals can be regarded as assemblies of free ions, bound together in several possible ways. Ionic crystals are held together by a balance between long-range electrostatic attractive forces and short-range repulsions, the latter being largely attributable to the exclusion principle acting between the outer electrons on neighbouring ions. Most crystals also have some degree of covalent bonding, in which the total energy is reduced by the sharing of electrons between orbitals on neighbouring ions. In conductors, bonding is effected through the coupling between the ions and a sea of conducting electrons.

Free magnetic ions are characterized by a Hamiltonian with spherical symmetry. In a crystalline environment, the site symmetry of magnetic ions is lowered, thus breaking the degeneracy of the free-ion $J$ or $L$ multiplets. The crystal field can therefore be thought of as a perturbation, generated by the same mechanisms that produce crystal bonding, acting on 'free' magnetic ions. The qualitative form of the splitting (i.e. the effect of the crystal field on the degeneracies) can be predicted from a knowledge of the site symmetry of the magnetic ion, as will be explained in Chapter 2 and Appendix 1.

An electrostatic potential at points $\mathbf{r}$ near to a magnetic ion, due to any external charge distribution, can be expressed as the multipolar expansion centred on that ion, viz.

$$V_{\mathrm{CF}}(\mathbf{r}) = \sum_k V^{(k)}(\mathbf{r}), \qquad (1.1)$$

where the individual contributions $V^{(k)}(\mathbf{r})$ can be expressed as sums (over $q$) of numerical factors times the functions

$$r^k Y_{k,q}(\theta, \phi).$$

Here $\mathbf{r}$ is expressed in terms of spherical polar coordinates $(r, \theta, \phi)$ centred on the magnetic ion, and $Y_{k,q}$ are spherical harmonic functions.

Whatever the magnitudes of the terms in (1.1), their $r^k$ radial dependence is determined by the requirement that the Laplace equation is satisfied. Given that the one-electron wavefunctions can be factorized as a product of radial and angular parts, it is possible to replace (1.1) by an expansion in which the radial integration has already been carried out, making the potential a function of the angular coordinates $\theta$ and $\phi$ alone. Equation (1.1) can then be rewritten as

$$V_{\mathrm{CF}}(\theta, \phi) = \sum_k V^{(k)}(\theta, \phi). \qquad (1.2)$$

The individual contributions $V^{(k)}(\theta, \phi)$ are now written in terms of

$$\langle r^k \rangle Y_{k,q}(\theta, \phi),$$

where $\langle r^k \rangle$ is the 'expectation' of $r^k$ for an open-shell electron, and can be expressed as a radial integral (e.g. see equation (16.3) of [AB70]). Free-ion radial wavefunctions necessary for evaluating these integrals can be obtained by Hartree–Fock free-ion calculations (e.g. [FW62]). Given the uncertainties in the precise form of the radial wavefunctions, the calculated values of the $\langle r^k \rangle$ are necessarily inaccurate.

In the phenomenological approach used in this book, the radial integrals $\langle r^k \rangle$ are absorbed into the numerical coefficients of equation (1.2). As shown subsequently in this chapter, the resulting expansion is not restricted to the representation of electrostatic potentials and can take into account the full range of mechanisms that contribute to the crystal field splitting. The value of using a multipolar expansion for $V_{CF}$ is that only a few of the multipole moments, i.e. those with even $k \leq 2l$, have non-zero angular factors in the matrix elements between open-shell single-electron states with angular momentum $l$. A discussion of these selection rules is given in Appendix 1. The first term, with $k = 0$, does not produce any crystal field splitting. The second term, with $k = 2$, is frequently referred to as the 'quadrupolar' contribution, while the $k = 4$ term is occasionally referred to as the 'hexadecapolar' contribution. Happily, this nomenclature is seldom extended to the case $k = 6$. It is more convenient to refer to contributions by their rank $k$.

The first step in formulating an *ab initio* calculation of crystal field splittings is to construct the open-shell states for a 'free' magnetic ion. Considerable simplifications will result if these are taken to be the many-electron wavefunctions corresponding to the free-ion configurations $l^n$, where $n$ is the number of open-shell electrons. These many-electron functions can be constructed as determinants of one-electron wavefunctions, which are derived from a Hartree–Fock self-consistent mean field calculation for the ion [LM86]. In addition to the one-electron wavefunctions, the Hartree–Fock calculation also produces an isotropic (or spherically symmetric) electrostatic potential. The crystal field can be regarded as an anisotropic perturbation of this potential.

It must be noted that free-ion interelectronic coulomb interactions (that are not included in the mean field) and spin–orbit interactions produce even greater splittings of the degenerate configurational energies than does the crystal field. The resulting free-ion many-electron wavefunctions are usually expressed as linear combinations of the aforementioned determinantal wavefunctions constructed using the one-electron wavefunctions of a Hartree–Fock calculation. However, considerable progress in calculating crystal field splittings from first principles can be made simply by regarding these interactions as 'other perturbations', which act independently on the Hartree–Fock single-electron states.

## 1.2 The electrostatic crystal field model

Historically, the first approximation to the crystal field in ionic crystals was obtained by assuming that, apart from the magnetic ion under consideration, all the other ions in the crystal may be treated simply as sources of an electrostatic field. In this, so-called *electrostatic crystal field model*, each of the electrons in the open shell of a magnetic ion is supposed to move independently in an electrostatic potential, the anisotropic component of which is produced by the surrounding crystal. It turns out that this model is incorrect, and much of this chapter is devoted to showing why this is the case. However, given that the electrostatic model is so entrenched in the literature of the subject, it provides a useful starting point for the present discussion.

### 1.2.1 Quantitative development of the electrostatic model

The simplest form of electrostatic model is the 'point charge model', in which the electrostatic potential at a magnetic ion is obtained by summing the contributions of all the other ions (regarded as point charges) in the lattice. The physical justification for doing this is that the electrostatic potential external to any spherically symmetric charge distribution can be shown to be the same as that generated by the same net charge concentrated at a point at the centre of that distribution. Hence an approximation inherent in the point charge model is that the open-shell electrons on the magnetic ion lie outside the charge distributions on the neighbouring ions. Although this approximation is poor, the simple point charge model is successful in the sense that, at least in the case of ionic crystals, it almost always predicts crystal field parameters of the correct sign. This corresponds to a repulsion of the open-shell electrons by the negative charges carried by the neighbouring ions.

The rank $k$ point charge contributions to the electrostatic potential can be shown to have the simple functional dependence $1/R^{k+1}$, where $R$ is the distance between the point charge and the centre of the magnetic ion. This distance dependence, especially in the case $k = 2$, leads to the expectation that distant ions in the crystal can make a significant contribution to the electrostatic potential. Because there are strongly cancelling positive and negative contributions in the point charge lattice sums, quite sophisticated methods have to be used to obtain reliable results.

While it is feasible to overcome the difficulties involved in making accurate calculations of wavefunctions and lattice sums, the results obtained still depend on the doubtful assumptions inherent in the point charge model. Experience has shown that, however much effort is spent on improving their accuracy, such calculations never even come close to reproducing the experimentally observed crystal field splittings. Typically, in the case of the lanthanide ions, they overestimate the rank 2 contributions by a factor of 10 and underestimate the rank 6 contributions by a similar factor [New71].

Early attempts to improve the point charge model (e.g. see [HR63]) were focussed on taking account of the lack of exact spherical symmetry of ionic charge distributions in crystals. It was suggested that, apart from point charge contributions, electrostatic contributions from induced (point) dipole and quadrupole moments on the ions should be included in the lattice sums. However, these improvements did not enhance agreement with experiment, the gross errors mentioned above remaining unchanged.

The overestimation of the rank 2 crystal field in electrostatic models of lanthanide and actinide crystal fields can be largely attributed to the neglect of the screening effects of the filled $(N + 1)s^2p^6$ shell, which lies outside the partially filled $Nf$ shell. Virtual excitations of the outer-shell electrons modify the externally generated electrostatic field experienced by an open-shell electron. This screening effect can be incorporated into the electrostatic point charge model by means of multiplicative *screening factors*, denoted $(1 - \sigma_k)$. Several attempts have been made to calculate the screening factors which, in accord with intuition, are found to be less than unity. For the trivalent lanthanide ions, $\sigma_4$ and $\sigma_6$ are small, but $\sigma_2$ has been calculated to be about 0.7 [SBP68]. This clearly has a significant effect on calculated rank 2 electrostatic contributions.

There are, however, considerable difficulties in carrying out calculations of screening factors which are both realistic and accurate. For example, the calculation by Sternheimer *et al.* [SBP68] was carried out for free ions, rather than ions in a crystalline environment. More recent computational [NBCT71] and phenomenological [NP75] analyses suggest that the value of $\sigma_2$ is in fact close to 0.9, allowing only about 10% of an externally generated rank 2 point charge contribution to 'get through' to the $4f$ electrons in the trivalent lanthanides.

Although the point charge approximation may provide an adequate representation of distant ion contributions, it cannot possibly give an accurate description of the contributions from the immediate neighbours (or ligands) of a magnetic ion. One reason for this is that the charge distributions of the ligands significantly overlap the open-shell electronic wavefunctions of a magnetic ion. Allowing for the effect this has on the electrostatic field makes the results much worse, in the sense that the net rank 4 electrostatic contributions become too small and the lanthanide and actinide rank 6 contributions change sign. Moreover, no purely electrostatic contribution has been found which could correct these discrepancies.

In summary, every attempt to improve the results obtained with the simple point charge model introduces further computational difficulties and a widening of the gap between theoretical predictions and experimental observations. This exposes the fundamental deficiencies of the electrostatic model.

The failure of the more sophisticated forms of the electrostatic model has led some authors to attempt to exploit the fact that a simple point charge model usually gives crystal field contributions of the correct sign by postulating different 'effective' magnitudes of the point charges for the contributions of each rank $k$. However, while this fudge may work for a given system, the 'effective' charges have no physical reality, so there is no basis for making comparisons between the effective charges for different systems. In consequence, effective point charge models have no predictive capability.

## 1.2.2 Qualitative features of the electrostatic model

Although the electrostatic model does not provide an adequate *quantitative* description of the crystal field, several of its *qualitative* features carry over to more realistic models. The first of these features is the interaction of the electrostatic potential separately with each of the open-shell electrons. In other words, crystal field splittings can be described in terms of one-electron matrix elements. This feature makes the number of parameters required to describe observed crystal field splittings independent of the number of electrons in the shell. The second feature is that the electrostatic potential is spin independent: the one electron matrix elements do not depend on spin. Thirdly, the electrostatic potential can be expressed as a finite multipolar expansion, with terms spanning very few ranks. These three properties of the electrostatic crystal field model are all expressed by equation (1.2).

A fourth feature of the electrostatic point charge model, not represented in (1.2), is that the electrostatic potentials in the neighbourhood of a magnetic ion can be calculated as a simple sum, or *superposition*, of contributions from all the different sources. It is shown in Chapter 5 that this property provides a useful means for the analysis and interpretation of empirically determined crystal fields.

## 1.3 Other contributions to the effective potential

'Ligand field' models put the emphasis on the construction of an appropriate representation for the open-shell wavefunctions, rather than on determining the form of an effective crystal field potential which acts upon the open-shell electrons. 'Molecular orbitals' are constructed from an appropriate linear combination of open-shell and ligand wavefunctions determined by *overlap* and *covalency* (see Section 1.7). The alternative 'crystal field' approach used in the present analysis, in which an effective potential acts upon the open-shell electrons, must certainly take account of overlap and covalency.

Overlap contributions arise from the non-orthogonality of the one-electron free-ion wavefunctions on neighbouring ions. In particular, overlap between the electronic wavefunctions in the open-shell and the outer-shell wavefunctions on the ligand results in Pauli exclusion. This has the effect of repelling the open-shell electrons from the ligand directions, producing a crystal field contribution of the same sign as that produced by their negative ionic charges.

Covalency arises from a mixing of open-shell and ligand wavefunctions, which contributes to crystal bonding. Because different open-shell electronic states mix with the ligand states differently, covalency also produces a contribution to the crystal field splittings. As will be demonstrated below, the signs of the overlap and covalency contributions to the crystal field are invariably the same, although their origin is different.

The non-local nature of covalency and overlap makes it apparent that their contributions to the crystal field cannot be expressed simply as a function of electron position, as in (1.1). Nevertheless, it can be represented by a one-electron *operator*. It turns out that this makes no difference to the *number* of parameters required to characterize a given one-electron crystal field.

## 1.4 Crystal field energies in terms of many-electron matrix elements

In this section, we formulate a crystal field model in which the open-shell electrons of a magnetic ion interact individually with their crystalline environment. The aim is to demonstrate that the crystal field potential energy of the magnetic ion can be expressed as a sum of these individual electron contributions. Furthermore, as all the electrons other than those in the open shell are supposed to be in filled orbitals, these contributions are independent of the spin of the open-shell electron.

The formulation includes one electron in an open-shell orbital on the magnetic ion and all the outer $s^2p^6$-shell electrons on the ligands. The wavefunction of the open-shell electron is denoted by $\phi_\alpha$. The single-electron wavefunctions corresponding to electrons in the outer (closed) shells of all the ligands, are denoted by $\chi_\nu$. The ground state of the complex is approximated as a single determinant including all $N = 8M + 1$ single-electron wavefunctions, where $M$ is the number of ligands. It is written

$$\Psi_{\alpha 0} = (1/\sqrt{N})|\phi_\alpha \Pi_\nu \chi_\nu|, \tag{1.3}$$

where the Greek subscripts differentiate between both orbitals and spins. Wavefunctions for strongly bound, or 'core', electrons are included with the atomic nucleus as contributing only to the electrostatic field. Hence they do not appear explicitly in the determinantal wavefunction.

To first order, the energy $E_{\alpha 0}$ of the approximate ground state $\Psi_{\alpha 0}$ is given by

$$E_{\alpha 0} = \frac{\langle \Psi_{\alpha 0}|H|\Psi_{\alpha 0}\rangle}{\langle \Psi_{\alpha 0}|\Psi_{\alpha 0}\rangle}. \tag{1.4}$$

Here $H$ is the Hamiltonian containing the coulomb interactions $e^2/r_{ij}$ between electrons and the (one-electron) contributions arising from the charges on the atomic cores of the magnetic ion and ligands. The Hamiltonian $H$ also contains the kinetic energy operator, but spin–orbit coupling is ignored.

The effects of covalency are included in the formalism by including an admixture of many excited states into the ground state. The excited states used in the construction of this admixture are related to the ground state by supposing that ligand electrons are excited into the open shell. They take the form

$$\Psi_{\alpha\beta\tau} = (1/\sqrt{N})|\phi_\alpha \phi_\beta \Pi_{\nu\neq\tau} \chi_\nu|, \tag{1.5}$$

where $\tau$ denotes the ligand state vacated by the excited electron.

The admixture of the excited states into the ground state can be determined using perturbation theory or through the variational principle. It takes the general form

$$\Psi_\alpha = \Psi_{\alpha 0} + \sum_{\beta\tau} \gamma_{\alpha\beta\tau} \Psi_{\alpha\beta\tau}, \tag{1.6}$$

where the admixture coefficients are given by

$$\gamma_{\alpha\beta\tau} = -\frac{N_{\alpha\beta\tau}}{D_{\alpha\beta\tau}}, \tag{1.7}$$

with numerator

$$N_{\alpha\beta\tau} = [\langle\Psi_{\alpha\beta\tau}|H|\Psi_{\alpha 0}\rangle - \langle\Psi_{\alpha\beta\tau}|\Psi_{\alpha 0}\rangle E_{\alpha 0}]$$
$$\times \langle\Psi_{\alpha\beta\tau}|\Psi_{\alpha\beta\tau}\rangle^{-1/2}\langle\Psi_{\alpha 0}|\Psi_{\alpha 0}\rangle^{-1/2} \tag{1.8}$$

and denominator

$$D_{\alpha\beta\tau} = \frac{\langle\Psi_{\alpha\beta\tau}|H|\Psi_{\alpha\beta\tau}\rangle}{\langle\Psi_{\alpha\beta\tau}|\Psi_{\alpha\beta\tau}\rangle} - E_{\alpha 0}. \tag{1.9}$$

The energy corresponding to $\Psi_\alpha$ is given by

$$E_\alpha = E_{\alpha 0} - 2\sum_{\beta,\tau}\frac{|N_{\alpha\beta\tau}|^2}{D_{\alpha\beta\tau}} + \sum_{\tau}\frac{|N_{\alpha\alpha\tau}|^2}{D_{\alpha\alpha\tau}}. \tag{1.10}$$

The last term in (1.10) appears because the open-shell orbital $\phi_\alpha$ under consideration is already occupied, and is therefore not available for electron transfer. The non-orthogonality of the open-shell and ligand orbitals makes it necessary to include overlap integrals, such as $\langle\Psi_{\alpha\beta\tau}|\Psi_{\alpha 0}\rangle$, in (1.8).

When the magnetic ion is at a site of sufficiently high symmetry (e.g. cubic or axial), open-shell wavefunctions can be chosen so as to make the off-diagonal energy matrix elements, $\langle\Psi_\alpha|H|\Psi_\beta\rangle$ (with $\beta \neq \alpha$) identically zero. The crystal field is then determined entirely in terms of the energy differences $E_\alpha - E_\beta$. While there are many real systems with cubic symmetry, axial symmetry ($C_{\infty v}$) occurs only when there is one ligand. Although it does not correspond to any real (crystalline) system, one-ligand systems provide a convenient way to determine the relative importance of the various crystal field contributions.

Further simplifications result from noting that the energy denominators $D_{\alpha\beta\tau}$ are effectively independent of the suffices $\alpha$ and $\beta$, and the many-electron matrix elements $N_{\alpha\beta\tau}$ are effectively independent of the suffix $\alpha$ (see [New71]). We can therefore write

$$D_{\alpha\beta\tau} = D_\tau(\text{all } \alpha, \beta) \tag{1.11}$$

and

$$N_{\alpha\beta\tau} = N_{\beta\tau}(\text{all } \alpha). \tag{1.12}$$

Cancellation between terms then produces a remarkably simple expression for the crystal field splitting, viz.

$$E_\alpha - E_\gamma = \left(E_{\alpha 0} + \sum_\tau\frac{|N_{\alpha\tau}|^2}{D_\tau}\right) - \left(E_{\gamma 0} + \sum_\tau\frac{|N_{\gamma\tau}|^2}{D_\tau}\right). \tag{1.13}$$

Alternatively, the energy of an electron in orbital $\alpha$ can be written as

$$E_\alpha = E_{\alpha 0} + \sum_\tau \frac{|N_{\alpha\tau}|^2}{D_\tau} + E_0, \qquad (1.14)$$

where the energy $E_0$ is independent of $\alpha$. It is now clear that the approximations made in (1.11) and (1.12) correspond to making the covalency, or charge transfer, contributions dependent only on a single open-shell orbital. In other words, these approximations produce covalency contributions to a *one-electron crystal field*.

## 1.5  Crystal field energies in terms of one-electron matrix elements

The formulae derived in Section 1.4 were expressed in terms of matrix elements between many-electron states. The analysis required to express the previous results explicitly in terms of one-electron wavefunctions is rather tedious and can be found elsewhere [New71]: only results are given here. For simplicity, the system is taken to have an axial symmetry, with only one ligand in the $+z$ coordinate direction. In this case the one-electron orbitals on the magnetic ion which give only diagonal energy matrix elements are the spherical harmonic functions, labelled $l, m$. In the following discussion, therefore, magnetic quantum number labels $(m)$ will replace the Greek labels previously used for the one electron orbitals on the magnetic ion.

The energy $E_m$ of an electron in the orbital $l, m$ is composed of contributions from four distinct processes: *electrostatic*, including point and distributed sources; *overlap*; *exchange* and *covalency*. Symbolically

$$E_m = E_m(\text{electrostatic}) + E_m(\text{overlap}) + E_m(\text{exchange}) + E_m(\text{covalency}). \qquad (1.15)$$

In the present formulation, screening effects are omitted from the consideration but will be included in the numerical example given in Section 1.9. Electrostatic contributions to the crystal field energies are given by

$$E_m(\text{electrostatic}) = \langle m|1/r_T|m\rangle + \left(2\sum_\tau \langle m\tau||m\tau\rangle - 8\langle m|1/r_T|m\rangle\right). \qquad (1.16)$$

Here $\langle m\tau||m\tau\rangle$ stands for the coulomb matrix element $\langle \phi_m \chi_\tau|1/r_{12}|\phi_m \chi_\tau\rangle$ between electrons 1 and 2 in the open shell and on the ligand, respectively. The term $\langle m|1/r_T|m\rangle$ abbreviates the matrix element describing the potential energy of an open-shell electron (in orbital $\phi_m$) in the electrostatic field produced by a point charge at ligand $T$. The distance between the electron and ligand $T$ is denoted by $r_T$. Similar abbreviations are used in the following equations.

The bracketed expression on the right hand side of (1.16) is the difference in electrostatic energy for eight electronic point charges at the origin of ligand $T$ and that for the same eight charges in the outer ligand $s$ and $p$ orbitals. This expression measures the so-called *charge penetration* contribution arising from the distributed charge of the eight ligand outer-shell electrons. In evaluating the full determinantal expressions for $E_{\alpha 0}$ in equation (1.14) an exchange contribution also appears, viz.

$$E_m(\text{exchange}) = -\sum_{\tau}\langle m\tau||\tau m\rangle. \tag{1.17}$$

The fact that the ligand and open-shell orbitals are non-orthogonal produces the *overlap* contribution in (1.15), given by

$$E_m(\text{overlap}) = -\sum_{\tau}\langle \tau|m\rangle(N_{m\tau 1} + 2N_{m\tau 2}), \tag{1.18}$$

where the expressions for $N_{m\tau i}$ are given in (1.21) and (1.22). The covalency contribution to the energy matrix has the closely related form

$$E_m(\text{covalency}) = -\sum_{\tau}\left|\sum_{i=1}^{3} N_{m\tau i}\right|^2/D_\tau, \tag{1.19}$$

where the denominator is given by

$$D_\tau = \epsilon_0 - \epsilon_\tau + U^+ + U^- + \langle \phi\phi||\phi\phi\rangle - \langle \phi\tau||\phi\tau\rangle. \tag{1.20}$$

Here $\phi$ refers to an arbitrary open-shell orbital; $\epsilon_0$ and $\epsilon_\tau$ are, respectively, the Hartree–Fock energies of the open-shell and ligand electrons. The numerators in (1.19) have the three contributions

$$N_{m\tau 1} = \langle m|\tau\rangle(\epsilon_0 - \epsilon_\tau + U^+ + U^- + 1/R), \tag{1.21}$$

$$N_{m\tau 2} = \left(2\sum_{\tau'}\langle m\tau'||\tau\tau'\rangle - 8\langle m|1/r_T|\tau\rangle\right) - \langle m\tau||\tau\tau\rangle, \tag{1.22}$$

and

$$N_{m\tau 3} = \langle m\phi||\tau\phi\rangle - \langle m|\tau\rangle\langle \phi\tau||\phi\tau\rangle. \tag{1.23}$$

In these expressions $\tau'$ and $\tau$ are ligand orbitals. Note that $\phi$ without a suffix is used to denote the radial part of open-shell electronic wavefunctions in all cases where the inclusion of the angular part has a negligible effect on the value of the integrals. In the above equations $R$ is the distance between the magnetic ion and ligand centres, $U^+$, $U^-$ are, respectively, the Madelung potentials at the magnetic ion and the ligand. The values chosen for these quantities must, of course, reflect a specific crystalline environment.

Equations (1.16)–(1.23) involve only the outer-shell orbitals labelled $\tau$ on a single ligand. What happens if several ligands contribute to the crystal field? According to the superposition principle for electrostatic potentials, both the point charge and the charge penetration contributions in $E_m$(electrostatic) generalize to a sum over the ligands. $E_m$(exchange) involves only one ligand orbital and again generalizes to a sum over individual ligand contributions. The ligand orbital $\tau$ which appears in $E_m$(overlap) and $E_m$(covalency) refers to a specific ligand. Hence the only possibility for involving more than one ligand in these contributions arises from the bracketed expression for $N_{m\tau 2}$. In this expression it is possible for both $T$ and $\tau'$ to refer to a ligand different to that which the orbitals $\tau$ belong, giving rise to three-centre contributions involving inter-ligand charge penetration. However, such contributions are expected to be small because of the strong electrostatic repulsion between ligands. It is found, in practice, that the most important three-centre contributions occur in the next order of perturbation theory, and these are an order of magnitude smaller than the corresponding two-centre contributions [CN70]. Hence, it is a reasonable approximation to assume that *all the crystal field contributions can be built from a superposition of contributions from the individual ligands.* This provides the theoretical underpinning of the *superposition model*, which expresses the crystal field as a sum of single ligand contributions. Experimental evidence for the effectiveness of this model, and its development as a practical tool for the analysis of crystal field splittings will be given in Chapter 5.

## 1.6 Many-body approach to crystal field calculations

The aim of the simple *ab initio* formalism described in Sections 1.4 and 1.5 was to identify the dominant contributions to the one-electron crystal field. If it is desired to obtain more accurate results, together with magnitudes of the two-electron and spin-dependent contributions, both the neglected terms in the Hamiltonian (e.g. spin–orbit coupling) and the unfilled higher energy states located on both the magnetic ion and the ligands must be taken into account.

Diagrammatic many-body theory [LM86] provides a comprehensive approach for including the contributions of all perturbations and excited states in an *ab initio* calculation of the energy matrix. This approach also provides an appropriate formalism for the calculation of transition intensities (discussed in Chapter 10). However, the multicentre nature of the crystal field problem introduces special difficulties in applying many-body theory to the calculation of crystal field parameters. In particular, the standard formalism needs to be generalized because realistic orbital basis sets contain occupied states on each of the centres, which are non-orthogonal. There are also problems in determining the appropriate excitation energies when electrons are transferred between ions, and in carrying out the summations of energy contributions over higher order perturbations. These issues have been discussed, and some calculations for the system $Pr^{3+}$–$Cl^-$ have been carried out [NN87b, NN87a].

## 1.7 Relation to other formalisms

The analysis in Section 1.4 showed that the mechanisms considered all contribute predominantly to a one-electron crystal field. This is in broad agreement with the experimental evidence. It is exploited in the phenomenological approach by combining all the contributions into a single effective energy operator, denoted $W_{CF}$, which acts on one-electron open-shell states. The energies $E_m$ defined in equation (1.15) can then be expressed in terms of the diagonal one-electron matrix elements of this operator as

$$e_m = E_m = \langle m|W_{CF}|m\rangle. \tag{1.24}$$

The lower case '*e*' is introduced in this equation to emphasize the fact that $W_{CF}$ operates on one-electron states and defines one-electron energies.

The exchange, overlap and covalency contributions to $W_{CF}$ have a similar form to the 'pseudo-potential' operator, which has been employed in the theory of metals [Har66]. Attempts have been made to calculate the crystal field pseudo-potential directly from density functional theory. However, such calculations have required the introduction of a new type of (fitted) parameters. We take the view that 'hybrid' theories of this type, which mix *ab initio* calculations and fitted parameters, do not advance our understanding of crystal field splitting mechanisms.

In the simplified theory of crystal field splittings developed in Sections 1.4 and 1.5 it has been assumed that the symmetry of the system is sufficiently high that the open-shell orbitals can be chosen to give off-diagonal matrix elements of $W_{CF}$ which are identically zero. When this is not the case, the orbital energies alone cannot provide a complete description of the crystal field. It is then necessary to introduce a parametrization of the complete $(2l + 1)$ by $(2l + 1)$ energy matrix.

### *1.7.1 Crystal field parameters*

The numerical coefficients of the terms $\langle r^k \rangle Y_{k,q}(\theta, \phi)$ in equation (1.2) are termed 'crystal field parameters'. In order to obtain a clear understanding of the relationship between the energy matrix, the energy level splittings and the crystal field parameters it is best to start with the simplest model system: a single open-shell electron for which the spin–orbit coupling is neglected. Let us suppose that this electron has an orbital angular momentum $l$. The standard approach is to use a linear expansion of the operator $W_{CF}$, which generalizes the multipolar expansion of (1.2) by removing its explicit functional dependence on angular coordinates, viz.

$$\langle l m_1 | W_{CF} | l m_2 \rangle = \sum_{k,q} \langle l m_1 | t_q^{(k)} | l m_2 \rangle \hat{B}_q^k. \tag{1.25}$$

Here the $\hat{B}_q^k$ are the *crystal field parameters*, $m_i$ are the magnetic quantum numbers with values between $-l$ and $+l$, and $\langle l m_1 | t_q^{(k)} | l m_2 \rangle$ are matrix elements of the one-electron *tensor operators* $t_q^{(k)}$ (defined in Appendix 1). The crystal field parameters $\hat{B}_q^k$ are, in general, complex quantities.

Tensor operators are employed in the crystal field expansion because of their orthogonality properties and the factorizability of their matrix elements, viz.

$$\langle l m_1 | t_q^{(k)} | l m_2 \rangle = (-1)^{l-m_1} \begin{pmatrix} l & k & l \\ -m_1 & q & m_2 \end{pmatrix} (l \| k \| l). \tag{1.26}$$

Here the $2 \times 3$ array is a $3j$ symbol and $(l \| k \| l)$ denotes a so-called 'reduced' matrix element, which defines the normalization of the tensor operator $t_q^{(k)}$. As the notation implies, the reduced matrix elements are independent of the labels $m_1$, $m_2$ and $q$. The definition and properties of tensor operators, together with the normalization adopted in this work and the allowed values of $k$ and $q$, are given in Section A1.1.1 (also see [Jud63] and Chapter 4).

Using (1.25) and (1.26), one-electron energy matrix elements of the crystal field may be written, in terms of crystal field parameters, as

$$\langle l m_1 | W_{CF} | l m_2 \rangle = (-1)^{l-m_1} \sum_{k,q} \hat{B}_q^k \begin{pmatrix} l & k & l \\ -m_1 & q & m_2 \end{pmatrix} (l \| k \| l). \tag{1.27}$$

Given values of the crystal field parameters, the crystal field energies for a single open-shell electron, with no spin–orbit coupling, can be obtained from the eigenvalues of this energy matrix.

The orthogonality properties of $3j$ symbols (e.g. see [CO80], p. 180) make it possible to invert equation (1.27) analytically, to obtain an expression for the crystal field parameters in terms of matrix elements, viz.

$$\hat{B}_q^k = \sum_{m_1, m_2} (-1)^{l+m_1} \begin{pmatrix} l & k & l \\ -m_1 & q & m_2 \end{pmatrix} (2k+1)(l||k||l)^{-1} \langle lm_1|W_{CF}|lm_2 \rangle.$$

(1.28)

In the case of axial symmetry, as was assumed in the theoretical discussion of Section 1.5, a very simple relationship exists between the orbital energies and the *axial* crystal field parameters $\hat{B}_0^k$. This is because the energy matrix is diagonal, reducing the relationships between the axial parameters and the $l+1$ distinct single-electron energies $e_m (= e_{-m})$ to simple linear expressions. In order to obtain explicit numerical relationships it is necessary to adopt a specific normalization of the tensor operators. Using the Wybourne normalization, defined by equation (A1.2), the axial parameters $\hat{B}_0^k$ for $f$ electrons can be shown to be given by

$$\begin{aligned}
\hat{B}_0^0 &= (1/7)(e_0 + 2e_1 + 2e_2 + 2e_3), \\
\hat{B}_0^2 &= (5/14)(2e_0 + 3e_1 - 5e_3), \\
\hat{B}_0^4 &= (3/7)(3e_0 + e_1 - 7e_2 + 3e_3), \\
\hat{B}_0^6 &= (13/70)(10e_0 - 15e_1 + 6e_2 - e_3).
\end{aligned}$$

(1.29)

The inverse relations are

$$\begin{aligned}
e_0 &= \hat{B}_0^0 + (4/15)\hat{B}_0^2 + (2/11)\hat{B}_0^4 + (100/429)\hat{B}_0^6, \\
e_1 &= \hat{B}_0^0 + (1/5)\hat{B}_0^2 + (1/33)\hat{B}_0^4 - (25/143)\hat{B}_0^6, \\
e_2 &= \hat{B}_0^0 - (7/33)\hat{B}_0^4 + (10/143)\hat{B}_0^6, \\
e_3 &= \hat{B}_0^0 - (1/3)\hat{B}_0^2 + (1/11)\hat{B}_0^4 - (5/429)\hat{B}_0^6.
\end{aligned}$$

(1.30)

The corresponding relations between $d$-electron energies and axial crystal field parameters are

$$\begin{aligned}
\hat{B}_0^0 &= (1/5)(e_0 + 2e_1 + 2e_2), \\
\hat{B}_0^2 &= (e_0 + e_1 - 2e_2), \\
\hat{B}_0^4 &= (3/5)(3e_0 - 4e_1 + e_2),
\end{aligned}$$

(1.31)

and

$$\begin{aligned}
e_0 &= \hat{B}_0^0 + (2/7)\hat{B}_0^2 + (2/7)\hat{B}_0^4, \\
e_1 &= \hat{B}_0^0 + (1/7)\hat{B}_0^2 + (4/21)\hat{B}_0^4, \\
e_2 &= \hat{B}_0^0 - (2/7)\hat{B}_0^2 + (1/21)\hat{B}_0^4.
\end{aligned}$$

(1.32)

In both cases the number of (necessarily real) axial parameters is equal to the number of diagonal one-electron energy matrix elements. As the mean energy, corresponding to $\hat{B}_0^0$, is unobservable, the crystal field is represented by the energy differences $e_1 - e_0$, etc. It can, in fact, be shown that there are *always* exactly the same number of crystal field parameters as there are *independent* one-electron matrix elements for any site symmetry.

### 1.7.2 Ligand field approach

In the 'ligand field' approach crystal field splittings are seen as the result of overlap and covalent bonding between the open-shell electrons on the magnetic ion and the outer-shell electrons on the ligands. This can be expressed mathematically by constructing *molecular orbitals* from a linear combination of atomic orbitals on the two ions.

Molecular orbitals take a particularly simple form for the axially symmetric one-ligand system. The modified 'open-shell' orbital is written

$$\phi'_m = (\phi_m - \lambda_{m\tau}\chi_\tau)/\sqrt{1 - 2\lambda_{m\tau}\langle m|\tau\rangle + \lambda_{m\tau}^2}. \tag{1.33}$$

In this expression the 'covalent mixing parameter' $\lambda_{m\tau}$ is given by

$$\lambda_{m\tau} = \gamma_{m\tau} + \langle m|\tau\rangle,$$

with

$$\gamma_{m\tau} = -\frac{N_{m\tau}}{D_\tau},$$

the numerator $N_{m\tau} = \sum_i N_{m\tau i}$ and denominator $D_\tau$ being defined in equations (1.20)–(1.23). The corresponding normalized molecular orbitals for the ligand outer-shell electrons are orthogonal to $\phi'_m$, and can be written

$$\chi'_\tau = (\chi_\tau + \gamma_{m\tau}\phi_m)/\sqrt{1 + 2\gamma_{m\tau}\langle m|\tau\rangle + \gamma_{m\tau}^2}. \tag{1.34}$$

Molecular orbitals provide a good approximation to the real one-electron states of the system. In particular, the energy expressions

$$e_m = \langle\phi'_m|h|\phi'_m\rangle$$

can be brought into accord with $E_m$ (equation (1.15)) if an appropriate choice of the one-electron Hamiltonian $h$ is made. Molecular orbitals can also be used to calculate the effects of covalent bonding on other observables, such as $g$-factors. If, as is usually the case, an *ab initio* calculation is impracticable, molecular orbitals can be estimated from the phenomenological parameters together with calculated overlap integrals, as described in Chapter 5.

## 1.8  Relationship with the tight-binding model of band theory

Band structure calculations and *ab initio* crystal field calculations have in common that they both determine the energies of a single electron. While band structure calculations determine the energies of states delocalized over the whole crystal as a function of wave number, crystal field calculations determine the energies of electrons localized in the open-shell of a single magnetic ion in the presence of its ligands. Nevertheless, the short range of the dominant interactions which contribute to crystal field splittings makes it possible to establish simple formal relationships between delocalized and localized single-electron energies. In principle, at least, this enables the use of band structure calculations to determine crystal field splittings. This section outlines the formal relationships required and assesses the practicability of this approach.

It is assumed here that readers have some prior knowledge of the techniques used in band structure calculations. Several of the standard computational band structure techniques can be applied to crystals (e.g. NiO) that incorporate magnetic ions. The states corresponding to the open-shell electron energies determined in these calculations are delocalized, in the sense that they have the same amplitudes (but differing phases) in each unit cell of the crystal. Nevertheless, the so-called 'tight-binding model' can be used to fit these delocalized state energies to the energies of isolated ions and cross matrix elements between the states of neighbouring ions. The matrix elements determined by this fitting procedure can be related to matrix elements which appear in Section 1.5.

The fitting procedure can be bypassed in some cases. Energies of the open-shell electrons in a crystal field can be expressed as an average of band energies over the entire Brillouin zone. The 'decoupling transformation' approach [New73a] exploits the short-range nature of the dominant interactions, and provides a very good approximation to Brillouin zone averages by averaging energies over just a few selected high-symmetry points.

The decoupling transformation approach can, in principle, be used to determine the crystal field energies for any system, whether it be ionic, semiconducting or metallic, and for any site symmetry. However, only the simple case of $d$ electrons in sites of cubic symmetry in ionic crystals has been explored in any detail [New73b, New73a, LN73]. In these systems there are only two localized $d$ energies, conventionally labelled $E(\Gamma_{12})$ and $E(\Gamma_{25'})$. The difference between these energies determines the single cubic crystal field parameter. A discussion of the application of the decoupling transformation to metallic copper appears in [New74].

Given that it is possible to relate delocalized and localized state energies, the main question is just how accurately the band energies need to be determined in order to provide realistic determinations of the crystal field splittings. It would, for example, be particularly difficult to calculate lanthanide crystal fields using this approach, as the $4f$ crystal field splittings are orders of magnitude less than the typical band energy dispersions of the other orbitals. Band structure calculations are most often carried out using some form of the augmented plane wave technique (e.g. see [Sla65]) in which the crystalline potential function is an approximate construction. In particular, this approach often assumes that the electrostatic potential is spherical in the neighbourhood of the magnetic ion. This necessarily underestimates the electrostatic contributions to the crystal field (e.g. see [LN73] for a discussion of the problem in the case of NiO). It may therefore be more appropriate to regard the phenomenological crystal field as a useful way to correct errors in band structure calculations, rather than to use band structure calculations as a means of calculating crystal field parameters.

In recent years several attempts have been made to calculate the crystal field potential directly using band structure techniques. This approach is likely to be of most use in metallic systems where the formalism developed in Sections 1.4 and 1.5 is inapplicable. Some recent calculations along these lines have been carried out by Hummler and Fähnle [HF96a, HF96b] for the crystal field at lanthanide ions (R) in the intermetallics RCo$_5$. Their procedure is to add contributions from a lattice sum and the local charge distribution of the conduction electrons. As no contributions from the hybridization of the conduction and $4f$ electrons are included, the calculated crystal field is of purely electrostatic origin. There is no reason to expect these omitted contributions to be negligible. Much more work is necessary before reliable *ab initio* calculations of crystal field parameters in metals and semiconductors become possible.

## 1.9 Numerical results

The evaluation of the expressions given in Section 1.5 is a major computational task, involving the determination of Hartree–Fock wavefunctions for the open-shell electrons on the magnetic ion and the $s^2p^6$ outer-shell ligand electrons. Some calculations, such as screening, also require the generation of excited state wavefunctions on the magnetic ion. Several of the matrix elements involve wavefunctions on both centres, so that special techniques have to be used to expand wavefunctions determined in relation to one centre with respect to another centre. It is not appropriate here to discuss the

Table 1.1. Contributions to the crystal field parameters $\hat{B}_0^k$ (cm$^{-1}$) for the axial system Pr$^{3+}$–F$^-$, spaced at 4.6 au. Contributions are due to (i) the ligand point charge (allowing for screening), (ii) charge interpenetration, (iii) covalency, (iv) overlap and (v) coulomb exchange.

| Contrib. | $k = 2$ | $k = 4$ | $k = 6$ |
|----------|---------|---------|---------|
| (i)      | 246     | 287     | 69      |
| (ii)     | −302    | −205    | −136    |
| (iii)    | 216     | 300     | 227     |
| (iv)     | 376     | 522     | 399     |
| (v)      | −88     | −118    | −63     |
| Total    | 448     | 786     | 496     |

details of such calculations, but readers will appreciate that, with all these complications, a high degree of accuracy cannot be expected in calculated crystal field parameters. Nevertheless, such calculations are useful in that they provide a means of obtaining firm estimates for the relative importance of the various mechanisms contributing to crystal fields. They have been carried out for a number of $3d$ and $4f$ open-shell systems.

Numerical results are given here for a specific calculation based on the formalism described in Sections 1.4 and 1.5. The example chosen is the single-ligand system Pr$^{3+}$–F$^-$, with a distance between ion centres of 4.6 atomic units. A lanthanide system has been chosen in order to counter the folklore that, because of the localization of the $4f$ open-shell wavefunctions, covalency and overlap contributions to the lanthanide crystal field are negligible. The results quoted are adapted from a calculation by Newman and Curtis [NC69] (see also [New71]). Table 1.1 gives the major contributions to the axial crystal field parameters. In accordance with the discussion in Section 1.2.1, 90% screening of the $k = 2$ point charge contribution has been assumed.

While no experimental results for single-ligand crystal fields exist, results derived from experiment using the methods described in Chapter 5 roughly agree with the total calculated results given in Table 1.1. Given the inaccuracies to be expected in such calculations, it is reasonable to conclude that the formalism described in this chapter covers the most important mechanisms that provide contributions to the crystal field.

The so-called 'exclusion' contributions (combining (iv) and (v)) are responsible for about half the total parameter values, the other half coming mostly from covalency (iii). When the charge penetration contributions (ii) are taken into account, the magnitudes of the net electrostatic contributions (i.e. (i) + (ii)) are only of the order of 10% of the combined contributions from covalency and exclusion. Hence, in the analysis of lanthanide crystal fields, approximations which neglect electrostatic contributions are likely to be far more realistic than those which neglect overlap and covalency. This qualitative conclusion is also valid for systems with $3d$ and $5f$ open-shells.

The calculations used to obtain the results given in Table 1.1 were carried out for several ligand distances $R$, and therefore provide a theoretical prediction of the distance dependence of the single-ligand parameters. Given that the distance dependence of the point charge contributions to $\hat{B}_q^k$ can be expressed in terms of the power law $R^{-(k+1)}$ (see Section 1.2.1), it is convenient to assume that any monotonically reducing function can, for a sufficiently small range of $R$, be fitted to a power law of the form $R^{-t_k}$, with a positive *power law exponent* $t_k$. The values of $t_k$ corresponding to the full range of ligand distances considered in [NC69] have been calculated as $t_2 = 3.4 \pm 0.6$, $t_4 = 5.2 \pm 0.4$ and $t_6 = 5.1 \pm 0.6$. Empirical determinations of power law exponents are discussed in Chapter 5.

## 1.10 Summary

In order to extract useful information from observed crystal field splittings it is necessary to replace the electrostatic model as a conceptual basis of the crystal field parametrization, as described by Hutchings [Hut64] and Abragam and Bleaney [AB70], with a more realistic conceptual approach. In the 'phenomenological' approach used in this book, the crystal field is expressed in terms of parametrized operator expressions, which are derived from very general assumptions.

The most important results of the foregoing analysis are that the total interaction energy between the crystalline environment and the open-shell electrons can be analysed into

(i) a sum of interactions of the environment with single open-shell electrons, and
(ii) a sum of interactions of the open-shell electrons with separate ligands.

The first of these results leads to the use of the 'one-electron model of the crystal field', discussed in the following three chapters, while the second leads to the 'superposition model', discussed in Chapter 5. For further discussion of the relationships between phenomenological models of the crystal field see Chapter 6.

# 2

# Empirical crystal fields

## D. J. NEWMAN
*University of Southampton*

## BETTY NG
*Environment Agency*

In this chapter the standard one-electron crystal field parametrizations are defined, and a selection of empirical crystal field parameters are tabulated. Section 2.1 describes the structure of the optical spectra of magnetic ions with partially filled $3d$, $4f$ and $5f$ shells. Section 2.2 is concerned with the definition and normalization of one-electron crystal field parameters. Section 2.3 provides a brief description of the main experimental techniques currently used to determine crystal field parameters. Selected tabulations of experimentally determined values of lanthanide and actinide crystal field parameters are given in Section 2.4.

## 2.1 Spectra and energy levels of magnetic ions

The spectra of magnetic ions in solids are usually comprised of many sharp lines. This is especially the case when the spectra are obtained at low temperatures. Spectral lines correspond to transitions between the energy levels of the states of the open-shell electrons induced by interactions with an electromagnetic field. For example, in absorption, an incoming photon raises the energy of the open shell from a low- to a higher-lying level. The advantage of working at low temperatures is twofold. A sufficiently low temperature ensures that only the lowest-lying level is occupied, so that the transition energies correspond to the relative positions of higher-lying energy levels. Low temperatures also ensure that the electron–phonon interactions in the solid material are minimized, making the observed transition energies better defined.

Selection rules, discussed in Chapters 3, 4, 10 and Appendix 1, restrict possible transitions, so that not all pairs of energy levels can be linked by a particular type of coupling with an electromagnetic field. Hence, when absorptions from the ground state are observed, not all the higher-lying

Table 2.1. Total numbers of multiplets ($N_m$) in the $f^n$ configurations, and low-lying terms of trivalent lanthanide and actinide ions which are accessible to optical spectroscopy.

| | | $n$ | Low-lying terms | $N_m$ |
|---|---|---|---|---|
| $Ce^{3+}$ | $Th^{3+}$ | 1 | $^2F$ | 2 |
| $Pr^{3+}$ | $Pa^{3+}$ | 2 | $^3H,^3F,^1G,^1D,^3P^1I,^1S$ | 13 |
| $Nd^{3+}$ | $U^{3+}$ | 3 | $^4I,^4G,^4F,^4D,^2L,^2H,^2D,^2P,^2S,^2G$ | 41 |
| $Pm^{3+}$ | $Np^{3+}$ | 4 | $^5I,^5F,^5S,^3K,^5G,^3H,^3L,^3F,^3D,^3M,^5D,^3P,^3I$ | 107 |
| $Sm^{3+}$ | $Pu^{3+}$ | 5 | $^6H,^6F,^4G,^4F,^4I,^4M,^4P,^4L,^4K$ | 198 |
| $Eu^{3+}$ | $Am^{3+}$ | 6 | $^7F,^5D,^5L$ | 295 |
| $Gd^{3+}$ | $Cm^{3+}$ | 7 | $^8S,^6P,^6I,^6D,^6G$ | 327 |
| $Tb^{3+}$ | $Bk^{3+}$ | 8 | $^7F,^5D,^5L,^5G$ | 295 |
| $Dy^{3+}$ | $Cf^{3+}$ | 9 | $^6H,^6F,^4F,^4I,^4G$ | 198 |
| $Ho^{3+}$ | $Es^{3+}$ | 10 | $^5I,^5F,^5S,^3K,^5G,^3H,^3L,^3F,^3D,^3M,^5D,^3P,^3I$ | 107 |
| $Er^{3+}$ | $Fm^{3+}$ | 11 | $^4I,^4F,^4S,^2H,^4G,^2G,^2K,^2P$ | 41 |
| $Tm^{3+}$ | $Md^{3+}$ | 12 | $^3H,^3F,^1G,^1D,^1I,^3P$ | 13 |
| $Yb^{3+}$ | $No^{3+}$ | 13 | $^2F$ | 2 |

energy levels will be observed. Fluorescence and two-photon spectroscopy are sometimes used to fill the gaps. Energy levels determined from transition energies are usually sufficient to determine sets of crystal field parameters.

Apart from the transition energies it is sometimes possible to determine both the intensity and breadth of spectral lines (usually with rather less accuracy). This information, most particularly the intensities of the lines, can sometimes be used to provide further constraints on crystal field parameters.

### 2.1.1 Structure of $f^n$ optical spectra

Table 2.1 provides a summary of the terms of the trivalent lanthanide and actinide ions which are accessible to optical spectroscopy, and gives the total number of $J$ multiplets in each case. Tetravalent actinides with the same number ($n$) of open-shell electrons in the $f$ shell have the same spectral structure as the corresponding trivalent ions. As described in the following chapter, the crystal field splitting of single multiplets is often sufficient to determine a complete set of crystal field parameters. Optical spectra, therefore, are usually sufficient to considerably overdetermine the crystal field parameters.

In addition to the one-electron crystal field, due to the anisotropic crystalline environment, several isotropic processes also contribute to the optical spectrum. The most important of these is the coulomb interaction, which separates terms characterized by a total angular momentum label $L$. These terms are split by spin–orbit coupling into $J$ multiplets. All significant isotropic contributions to the effective Hamiltonian are discussed in Chapter 4. Small anisotropic contributions to the effective Hamiltonian can be produced by modifications of the coulomb interaction and spin–orbit coupling induced by the crystalline environment. These are discussed in Chapter 6. They are difficult to distinguish empirically from the one-electron crystal field.

### 2.1.2  Transition metals

In the case of transition metal ions, with a $3d$ partially filled shell, the magnitude of the spin–orbit coupling is usually comparable to that of the crystal field. Hence, both the spin–orbit and crystal field potentials must be included in the effective Hamiltonian at the same time. Most observed spectra include several terms, with different values of $L$ and $S$ (e.g. see Gerloch and Slade [GS73]). Thus the effective Hamiltonians usually include the parametrized coulomb interaction as well as the crystal field parameters and spin–orbit coupling parameter. For this reason it is usual to tabulate fitted values of the crystal field parameters together with the (simultaneously fitted) coulomb and spin–orbit coupling parameters. Because of these complications, tabulations of crystal field parameters for $3d$ ions are not provided in this book. Interested readers can find examples of tabulations of these fitted parameters in the book by Gerloch and Slade [GS73]. Examples of more recent parameter fits can be found in [NN86b] (for $Mn^{2+}$) and [RDYZ93] (for $Cr^{2+}$).

## 2.2  The phenomenological crystal field parametrization

As was explained in Chapter 1, the predominant contributions to crystal field splittings can be represented by a one-electron spin-independent operator, denoted $W_{CF}$. This acts upon many-electron free-ion open-shell states. An explicit algebraic expression for $W_{CF}$ was obtained for the contributions of a single ligand on the axis of quantization, taken to be the $z$-axis. To avoid confusion between the algebraic expressions for $W_{CF}$ obtained in Chapter 1 and the parametrized, or *phenomenological* crystal field operator discussed in this chapter, it is convenient to introduce the notation $V_{CF}$ for the parametrized form. We shall, however, retain the same symbols for the crystal field parameters. The phenomenological crystal field operator acting on many-electron states will be written

$$V_{CF} = \sum_{k,q} \hat{B}_q^k T_q^{(k)},$$

(2.1)

where the crystal field parameters $\hat{B}_q^k$ are, in general, complex. The tensor operators $T_q^{(k)}$ are sums of one-electron tensor operators $t_q^{(k)}(i)$ acting on the states of single electrons, labelled $i$, in the open ($f$ or $d$) shells, *viz.*

$$T_q^{(k)} = \sum_i t_q^{(k)}(i).$$

(2.2)

### 2.2.1 Hermiticity

Because of the quantum mechanical requirement that energy operators must be Hermitian, it is convenient to express $V_{CF}$ in terms of Hermitian operators, so that real crystal field parameters can be employed. The adjoint, or Hermitian conjugate, of a tensor operator is given by

$$T_q^{(k)\dagger} = (-1)^q T_{-q}^{(k)}.$$

(2.3)

Hence, the Hermiticity of $V_{CF}$ implies that the *crystal field parameters* $\hat{B}_q^k$ must satisfy the relation $(\hat{B}_q^k)^* = (-1)^q \hat{B}_{-q}^k$, where the star denotes complex conjugation.

It follows from (2.3) that the following combinations of tensor operators are Hermitian:

$$\Omega_{k,q} = (T_q^{(k)} + (-1)^q T_{-q}^{(k)}), \text{ for } q > 0$$
$$\Omega_{k,q} = i(T_{-q}^{(k)} - (-1)^q T_q^{(k)}), \text{ for } q < 0$$
$$\Omega_{k,0} = T_0^{(k)}.$$

(2.4)

The operators $\Omega_{k,q}$, with $-k < q < k$, do not transform in the same way as the $T_q^{(k)}$ and hence do *not* form a tensor. The crystal field can now be expressed in terms of *real* parameters $B_q^k$ as

$$V_{CF} = \sum_{k,q} B_q^k \Omega_{k,q},$$

(2.5)

where, for $q = 0$,

$$\hat{B}_0^k = B_0^k$$

and, for $q > 0$,

$$\hat{B}_q^k = B_q^k + iB_{-q}^k, \; \hat{B}_q^k = (-1)^q(B_q^k - iB_{-q}^k).$$

Because of this expression, the parameters $B_q^k$ with negative $q$ are sometimes referred to as *imaginary parameters*, although their definition in (2.5) ensures that they take real values.

### 2.2.2 Normalization of the tensor operators

In order to be able to assign numerical values to $B_q^k$ it is necessary to adopt a specific normalization of the tensor operators $t_q^{(k)}(i)$ (see Chapter 1 and Appendix 1). The most often used convention, which was employed in Chapter 1, is to use tensor operators which have the same normalization as the *functions* $C_q^{(k)}(i) = C_q^{(k)}(\theta_i, \phi_i)$. This is generally known as the 'Wybourne normalization'. The functions $C_q^{(k)}(i)$ are related to the spherical harmonics as follows

$$C_q^{(k)}(i) = \sqrt{\frac{4\pi}{2k+1}} Y_{k,q}(\theta_i, \phi_i). \tag{2.6}$$

Rather than introducing a new notation for the Hermitian (i.e. real) combinations of the complex functions $C_q^{(k)}(i)$, as in (2.4), the crystal field is usually expressed in terms of explicit Hermitian combinations. In terms of the *real* parameters $B_q^k$,

$$V_{\text{CF}} = \sum_{k,q>0} [B_q^k(C_q^{(k)} + (-1)^q C_{-q}^{(k)}) + B_{-q}^k i(C_{-q}^{(k)} - (-1)^q C_q^{(k)})] + \sum_k B_0^k C_0^{(k)}, \tag{2.7}$$

where the tensor operators act on *all* the electrons in the open shell. That is to say,

$$C_q^{(k)} = \sum_i C_q^{(k)}(i). \tag{2.8}$$

Although the crystal field operator $V_{\text{CF}}$ is expressed as an explicit function of the angles $\theta_i$, $\phi_i$, *it cannot be interpreted as an electrostatic potential*, for it is not a function of the radial coordinates $r_i$.

Taking $t_q^{(k)} = C_q^{(k)}(\theta_i, \phi_i)$, the matrix elements of $V_{\text{CF}}$ can be expressed as

$$\langle l\alpha|V_{\text{CF}}|l\beta\rangle = \sum_{k,q}(-1)^{l-m} \begin{pmatrix} l & k & l \\ -\alpha & q & \beta \end{pmatrix} \hat{B}_q^k(l\|C^{(k)}\|l), \tag{2.9}$$

where the so-called *3j symbol*

$$\begin{pmatrix} l & k & l \\ -\alpha & q & \beta \end{pmatrix}$$

and the reduced matrix element $(l||C^{(k)}||l)$ are both defined in Appendix 1.

Several other definitions of the phenomenological crystal field parameters can be found in the literature. The most commonly used alternative was introduced by Stevens [Ste52], who employed a $q$-dependent normalization. The standard notation for crystal field parameters in this case is $A_{kq}\langle r^k \rangle$. This rather odd looking notation was chosen because, as explained in Chapter 1, it was then thought to be possible to factorize the crystal field parameters into products of a lattice sum $A_{kq}$ and a radial integral $\langle r^k \rangle$ determined from the one-electron open-shell wavefunctions. Although this interpretation of the crystal field is not tenable, the original, widely used, notation has been retained in this book.

Ratios of Stevens parameters [Ste52] and Wybourne parameters

$$\lambda_{k,q} = A_{kq}\langle r^k \rangle / B_q^k \tag{2.10}$$

are given in Table 2.2. An obvious property of all the expressions for $C_q^{(k)}$ in Table 2.2 is that a change in sign of all the coordinates leaves them unchanged. That is to say, all the contributions to the crystal field potential are invariant under inversion (see Appendix 1). From a mathematical point of view, this is due to the fact that the crystal field can be expressed entirely in terms of tensor operators with even $k$ values. An important consequence of this is that the *effective* symmetry of the crystal field always contains the inversion, whatever the actual symmetry of the site.

### 2.2.3 Non-vanishing crystal field parameters

Both the site symmetry and matrix element selection rules are used in determining which tensor operator components should be included in the crystal field Hamiltonian. These restrictions can both be understood in terms of group representation theory, which is discussed in Appendix 1. The set of tensor operators for a given $k$ value correspond to an irreducible representation of the full rotation group $O_3$. The site symmetry group of the paramagnetic ion is necessarily a subgroup of $O_3$.

In order to determine the number of crystal field parameters it is only necessary to determine the number of site symmetry invariants for each of the allowed $k$ values. Each of the irreducible representations of $O_3$ generally contain one or more invariant representations of the site symmetry group.

Table 2.2. Ratios $\lambda_{k,q}$ (see (2.10)) of crystal field parameters in Wybourne
and Stevens normalizations, together with algebraic expressions for the
tensor operators $C_q^{(k)}$ with $q \geq 0$. The ratios are the same for both positive
and negative $q$. Expressions for the operators with negative $q$-values are
obtained by replacing $r_+ = x + iy$ with $r_- = x - iy$ and multiplying the
overall expression by $(-1)^q$.

| $k$ | $q$ | $\lambda_{k,q}$ | $r^k C_q^{(k)}$ |
|---|---|---|---|
| 2 | 0 | $\frac{1}{2}$ | $\frac{1}{2}(3z^2 - r^2)$ |
| 2 | 1 | $-\sqrt{6}$ | $-\frac{1}{2}\sqrt{6}\,zr_+$ |
| 2 | 2 | $\frac{1}{2}\sqrt{6}$ | $\frac{1}{4}\sqrt{6}\,r_+^2$ |
| 4 | 0 | $\frac{1}{8}$ | $\frac{1}{8}(35z^4 - 30r^2z^2 + 3r^4)$ |
| 4 | 1 | $-\frac{1}{2}\sqrt{5}$ | $-\frac{1}{4}\sqrt{5}(7z^2 - 3r^2)zr_+$ |
| 4 | 2 | $\frac{1}{4}\sqrt{10}$ | $\frac{1}{8}\sqrt{10}(7z^2 - r^2)r_+^2$ |
| 4 | 3 | $-\frac{1}{2}\sqrt{35}$ | $-\frac{1}{4}\sqrt{35}\,zr_+^3$ |
| 4 | 4 | $\frac{1}{8}\sqrt{70}$ | $\frac{1}{16}\sqrt{70}\,r_+^4$ |
| 6 | 0 | $\frac{1}{16}$ | $\frac{1}{16}(231z^6 - 315r^2z^4 + 105r^4z^2 - 5r^6)$ |
| 6 | 1 | $-\frac{1}{8}\sqrt{42}$ | $-\frac{1}{16}\sqrt{42}(33z^4 - 30r^2z^2 + 5r^4)zr_+$ |
| 6 | 2 | $\frac{1}{16}\sqrt{105}$ | $\frac{1}{32}\sqrt{105}(33z^4 - 18r^2z^2 + r^4)r_+^2$ |
| 6 | 3 | $-\frac{1}{8}\sqrt{105}$ | $-\frac{1}{16}\sqrt{105}(11z^2 - 3r^2)zr_+^3$ |
| 6 | 4 | $\frac{3}{16}\sqrt{14}$ | $\frac{3}{32}\sqrt{14}(11z^2 - r^2)r_+^4$ |
| 6 | 5 | $-\frac{3}{8}\sqrt{77}$ | $-\frac{3}{16}\sqrt{77}\,r_+^5 z$ |
| 6 | 6 | $\frac{1}{16}\sqrt{231}$ | $\frac{1}{32}\sqrt{231}\,r_+^6$ |

The number of invariant representations of the subgroup generated in this
way is equal to the number of independent crystal field parameters. The
method for this calculation is described in Appendix 1. A selection of the
results is given in Table 2.3, which determines the form of the crystal field
operator for all the commonly occurring site symmetries. As has already
been pointed out, the effective symmetry of any site also contains the inversion, so that the results summarized in Table 2.3 will be unchanged if the
inversion operator is added to the operators in the site symmetry group.

## 2.2.4 Implicit coordinate systems

As tensor operator functions are defined in terms of coordinates centred on
the magnetic ions, the use of these operators implies some particular choice

Table 2.3. Non-vanishing crystal field parameters in sites with commonly occurring point symmetries. Non-vanishing parameters for $+q$ are indicated by '+' entries. $\pm$ indicates that both $+q$ and $-q$ parameters are non-zero.

| $k$ | $|q|$ | $C_{3v}$ | $C_{3h}/D_{3h}$ | $C_2$ | $D_{2d}/C_{2v}$ | $D_{2h}$ | S | $D_{4h}$ |
|-----|-------|----------|-----------------|-------|-----------------|----------|-----|----------|
| 2 | 0 | + | + | + | + | + | + | + |
| 2 | 2 |   |   | $\pm$ | + | + | $\pm$ |   |
| 4 | 0 | + | + | + | + | + | + | + |
| 4 | 2 |   |   | $\pm$ | + | + | $\pm$ |   |
| 4 | 3 | + |   |   |   |   |   |   |
| 4 | 4 |   |   | $\pm$ | + | + | $\pm$ | + |
| 6 | 0 | + | + | + | + | + | + | + |
| 6 | 2 |   |   | $\pm$ | + | + | $\pm$ |   |
| 6 | 3 | + |   |   |   |   |   |   |
| 6 | 4 |   |   | $\pm$ | + | + | $\pm$ | + |
| 6 | 6 | + | + | $\pm$ | + | + | $\pm$ |   |

of the coordinate system. While this choice is, in principle, arbitrary, it is constrained by the practical requirement that the number of parameters describing the crystal field should be as small as possible. This ensures that the orientation of the chosen coordinate system is related to the symmetry operators which describe the site symmetry.

In setting up the relationship between coordinate system and symmetry operators certain conventions are adopted. For example, whenever a principal axis exists, with a higher rotational symmetry than all other axes, this is taken to correspond to the $z$-axis. For most site symmetries this fixes the $z$-axis, so the only degree of freedom lies in rotations of the $x-y$ plane. However, in low symmetries, such as $D_2$, there can be as many as three possible choices of the $z$-axis. A consequence of this is that there may be several different, but equally good, sets of crystal field parameters for a given system. While each of these sets of parameters fits the experimentally determined energy levels with the same precision, they correspond to different (implicit) choices of coordinate system. The relationships between such 'equivalent' sets of parameters are generally complex, and the particular set arrived at in the fitting process (as described in Chapter 3) will depend on the starting values that were used. The 'CST' program package described in Appendix 4 can be used to transform parameters from one coordinate system to another.

In some cases the choice of coordinate system can affect which set of parameters are chosen to represent the crystal field. For example, the non-vanishing parameters for the site symmetry $C_2$ shown in Table 2.3 are based on the usual assumption that the $z$-axis is taken to be in the same direction as the $C_2$ symmetry axis. However, this $C_2$ axis could alternatively be taken to point in the $y$-direction. Combined with the effective inversion symmetry, common to all crystal fields, this would produce reflection symmetry in the $x$–$z$ plane. Consequently the crystal field operator has to be invariant under a change of sign of $y$. In this case only crystal field parameters with $q \geq 0$ are non-zero. The adoption of this alternative parametrization makes it possible to use the energy level and fitting programs described in Chapter 3, which are restricted to fitting crystal field parameters with positive values of $q$.

Even when the $z$-axis is uniquely determined by the principal axis, the directions of the $x$- and $y$-axes will not, in general, be fixed. The effect of rotating the axes in the $x$–$y$ plane is to change the sign of certain $q \neq 0$ parameters which contribute to the off-diagonal matrix elements of the crystal field. Hence, in many cases, the signs of the fitted parameters with $q \neq 0$ are not unique.

### 2.2.5 Sites of cubic symmetry

When the site symmetry is cubic only two crystal field parameters, one of rank 4 and one of rank 6, survive. If the $z$-axis is taken in the direction of one of the $C_4$ symmetry axes, the crystal field potential takes the form

$$V_{\mathrm{CF}} = B_0^4 \left[ C_0^{(4)} + \sqrt{\frac{5}{14}} (C_4^{(4)} + C_{-4}^{(4)}) \right] + B_0^6 \left[ C_0^{(6)} - \sqrt{\frac{7}{2}} (C_4^{(6)} + C_{-4}^{(6)}) \right]. \quad (2.11)$$

Note that, although there are only two independent crystal field parameters, six operators occur in this expression. In sites with cubic symmetry the pattern of the crystal field splittings depends only on the ratio of the values of the two independent crystal field parameters, while the magnitude of the splittings depends upon a weighted average of their values. The classic paper by Lea, Leask and Wolf [LLW62] exploits this property to produce tabulations of the eigenstates and diagrams of the energy levels for all the lanthanide ion ground multiplets in cubic symmetry. Their results are also applicable to actinide ions. Lea, Leask and Wolf introduced parameters, denoted $\mathcal{B}_k$ in this book, which are related to the Stevens and Wybourne crystal field parameters as follows:

$$\mathcal{B}_4 = \beta A_{40}\langle r^4 \rangle = \beta B_0^4/8, \ \mathcal{B}_6 = \gamma A_{60}\langle r^6 \rangle = \gamma B_0^6/16. \tag{2.12}$$

In this equation, $\beta$ and $\gamma$ are the so-called Stevens multiplicative factors, which are discussed in Section 2.4. It was found [LLW62] to be convenient to introduce the special parameters $-1 < x < +1$ and $W$ through the relationships

$$\mathcal{B}_4 F(4) = Wx, \ \mathcal{B}_6 F(6) = W(1 - |x|). \tag{2.13}$$

Here the values of the arbitrary numerical factors $F(4)$ and $F(6)$ are chosen for convenience in each calculation. Tabulated values can be found in [LLW62]. Some values of the ratio $\beta F(4)/\gamma F(6)$ are provided in Table 9.1. The parameter ratio can be expressed as

$$\frac{\mathcal{B}_4}{\mathcal{B}_6} = \frac{x}{1 - |x|} \frac{F(6)}{F(4)}. \tag{2.14}$$

Lea, Leask and Wolf [LLW62] express the energy levels (in both numerical and graphical forms) as factors of $W$ over the full range of the parameter $x$. Hence the parameters $W$ and $x$ are a form of crystal field parameters for lanthanide and actinide ions in cubic sites. The Lea, Leask and Wolf energy level diagrams can be generated by the Mathematica program LLWDIAG, which accompanies this book (see Appendix 3).

## 2.3 Experimental determination of crystal field parameters

The development, in the second world war, of techniques to generate electromagnetic energy at microwave frequencies led to the use of electron spin resonance techniques to probe the electronic structure of magnetic ions in crystals. Much of this work was carried out on lanthanide ions in crystalline hosts, such as ethyl sulphates, anhydrous chlorides and double nitrates. Most of the early determinations of crystal field parameters from ground multiplets of lanthanide ions were carried out in this way. The theoretical techniques developed at that time (e.g. see Abragam and Bleaney [AB70]) are still widely used today, although they belong to a pre-computer age.

In the late 1950s, optical spectroscopy began to play an important role, and is now by far the most powerful technique available for the study of magnetic ions in transparent insulating host crystals. Absorption spectroscopy enables the energy levels to be determined from about $1000 \ \mathrm{cm}^{-1}$ up to the ultraviolet. Low, normally liquid helium, temperatures are necessary to obtain sharp lines. Some of the excited energy levels within the ground state multiplet may also be observed, but are more appropriately studied using emission or two-photon techniques. Various techniques are available which enable electric dipole transition intensities to be determined (see Chapter 10). However, a limitation of optical spectroscopy is that many physically significant systems, such as magnetic ions in conducting and superconducting materials, do not have transparent hosts.

Crystal field splittings of magnetic ions in opaque systems are now routinely studied using inelastic neutron scattering. The main limitation of this technique is that only low-lying energy levels are accessible. Nevertheless, as explained in Appendix 3, additional information about magnetic dipole transition intensities still makes it possible to determine unique sets of crystal field parameters for lanthanide and actinide ions.

## 2.4 Lanthanide and actinide crystal field parameters

The tables given in this section provide a broad survey of empirical lanthanide and actinide crystal field parameters. This is not intended to be exhaustive, but provides illustrative material for the discussions in Chapters 3, 5 and 9. In many cases it is convenient to tabulate lanthanide ion and actinide ion crystal field parameters together as both series have partially filled $f$ shells and their trivalent ions frequently fit into the same sites. The main difference is that actinide ion crystal field parameters are typically of an order of magnitude larger than the crystal field parameters of the lanthanide ions in the same hosts.

One consequence of the actinide ions having larger crystal fields (and larger spin–orbit coupling) than the lanthanides, is that it can be difficult to label the energy levels in actinide spectra, making them far more difficult to analyse. Nevertheless, in the last decade, some workers have shown that it is possible to overcome these problems, and it is due to their dedication and perseverence that a considerable number of actinide crystal field parameters are available (e.g. see [CLWR91, Car92, KDM+97, IMEK97]).

The site symmetry at the $Pr^{3+}$ ion in crystalline $PrCl_3$ and $PrBr_3$ is $C_{3h}$, or effectively $D_{3h}$, as explained in Appendix 1, so there are four crystal field parameters (according to Table 2.3). The published neutron scattering results [SHF+87] are summarized in Table 2.4. The Stevens operator equivalent approach (see Chapter 3) was employed to obtain these results, but with the Stevens multiplicative factors $\theta_2 = \alpha$, $\theta_4 = \beta$, $\theta_6 = \gamma$, replaced by unity. The resulting parameter normalization is then characteristic of the $Pr^{3+}$ ion, and hence the parameters cannot be compared directly with the crystal field parameters for other ions in the same environment. This

Table 2.4. Crystal field parameters (in meV) for the ground multiplet of $Pr^{3+}$ in anhydrous halides, determined by neutron spectroscopy [SHF$^+$87].

| System | $A_{20}\langle r^2 \rangle$ | $A_{40}\langle r^4 \rangle$ | $A_{60}\langle r^6 \rangle$ | $A_{66}\langle r^6 \rangle$ |
|---|---|---|---|---|
| $PrCl_3$ | $-1.74(15) \times 10^{-1}$ | $4.7(3) \times 10^{-3}$ | $-2.73(6) \times 10^{-4}$ | $2.6(1) \times 10^{-3}$ |
| $PrBr_3$ | $-1.98(15) \times 10^{-1}$ | $5.9(1) \times 10^{-3}$ | $-2.33(1) \times 10^{-4}$ | $2.1(1) \times 10^{-3}$ |

Table 2.5. Inverse Stevens factors $(1/\theta_k)$, incorporating a change of units from meV to cm$^{-1}$, for the ground multiplets of $f^n$ configurations.

| Ion | Configuration | $k = 2$ | $k = 4$ | $k = 6$ |
|---|---|---|---|---|
| $Ce^{3+}$ | $f^1\ ^2F_{5/2}$ | $-1.412 \times 10^2$ | $1.270 \times 10^3$ | — |
| $Pr^{3+}$ | $f^2\ ^3H_4$ | $-3.839 \times 10^2$ | $-1.098 \times 10^4$ | $1.322 \times 10^5$ |
| $Nd^{3+}$ | $f^3\ ^4I_{9/2}$ | $-1.255 \times 10^3$ | $-2.771 \times 10^4$ | $-2.123 \times 10^5$ |
| $Sm^{3+}$ | $f^5\ ^6H_{5/2}$ | $1.954 \times 10^2$ | $3.225 \times 10^3$ | — |
| $Tb^{3+}$ | $f^6\ ^7F_6$ | $-7.985 \times 10^2$ | $6.588 \times 10^4$ | $-7.194 \times 10^6$ |
| $Dy^{3+}$ | $f^9\ ^6H_{15/2}$ | $-1.270 \times 10^2$ | $-1.362 \times 10^5$ | $7.793 \times 10^6$ |
| $Ho^{3+}$ | $f^{10}\ ^5I_8$ | $-3.630 \times 10^3$ | $-2.422 \times 10^5$ | $-6.235 \times 10^6$ |
| $Er^{3+}$ | $f^{11}\ ^4I_{15/2}$ | $3.176 \times 10^3$ | $1.817 \times 10^5$ | $3.897 \times 10^6$ |
| $Tm^{3+}$ | $f^{12}\ ^3H_6$ | $7.985 \times 10^2$ | $4.941 \times 10^4$ | $-1.439 \times 10^6$ |
| $Yb^{3+}$ | $f^{13}\ ^2F_{7/2}$ | $2.541 \times 10^2$ | $-4.658 \times 10^3$ | $5.450 \times 10^4$ |

normalization is often used in reporting neutron scattering results. It is important to be able to transform these parameters to the standard Stevens or Wybourne normalizations in order to make comparisons with parameters obtained from optical spectroscopy and with parameters for other ions in the same host crystal.

Neutron scattering results are normally recorded in meV, and the fitted parameters are commonly expressed in the same units, rather than wave numbers (i.e. cm$^{-1}$), which are the usual units used in optical spectroscopy. The multiplicative factors given in Table 2.5 incorporate this change of units. They are based on the assumption of $LS$ coupling, i.e. that the multiplets have the total orbital angular momentum $L$ and the total spin angular momentum $S$ as exact quantum numbers, as well as $J$. See Section 3.1.1 for further discussion of this point. The factors in Table 2.5 can be used

Table 2.6. Crystal field parameters (Stevens normalization, in $cm^{-1}$) for lanthanide and actinide ions in anhydrous chlorides and bromides.

| Ion | Crystal | $A_{20}\langle r^2 \rangle$ | $A_{40}\langle r^4 \rangle$ | $A_{60}\langle r^6 \rangle$ | $A_{66}\langle r^6 \rangle$ | Source |
|---|---|---|---|---|---|---|
| $Pr^{3+}$ | $PrBr_3$ | 78(6) | $-65(1)$ | $-36.0(2)$ | 323(14) | [SHF+87] |
| $Pr^{3+}$ | $PrCl_3$ | 69(6) | $-51(3)$ | $-42.2(9)$ | 407(15) | [SHF+87] |
| $Nd^{3+}$ | $LaCl_3$ | 97.6 | $-38.7$ | $-44.4$ | 443 | [Eis63b, Eis64] |
| $Eu^{3+}$ | $LaCl_3$ | 89 | $-38$ | $-51$ | 495 | [DD63] |
| $Dy^{3+}$ | $LaCl_3$ | 91.3 | $-39.0$ | $-23.2$ | 258 | [AD62] |
| $Ho^{3+}$ | $LaCl_3$ | 113.6 | $-33.9$ | $-27.8$ | 277 | [RK67, RK68] |
| $Er^{3+}$ | $LaCl_3$ | 93.9 | $-37.3$ | $-26.6$ | 265 | [Eis63a] |
| $Er^{3+}$ | $LaBr_3$ | 117 | $-39.6$ | $-19.2(2)$ | 212 | [KD66] |
| $Np^{3+}$ | $LaCl_3$ | 82(13) | $-69.9(55)$ | $-104.6(29)$ | 981(32) | [Car92] |
| $Pu^{3+}$ | $LaCl_3$ | 99(11) | $-73.3(48)$ | $-107.7(24)$ | 960(32) | [Car92] |
| $Cm^{3+}$ | $LaCl_3$ | 122 | $-108.9$ | $-77.0$ | 1031 | [IMEK97] |
| $Bk^{3+}$ | $LaCl_3$ | 140(20) | $-110.5(78)$ | $-80.8(43)$ | 940(38) | [Car92] |
| $Cf^{3+}$ | $LaCl_3$ | 153(15) | $-132.8(70)$ | $-90.1(30)$ | 894(34) | [Car92] |

to relate the neutron scattering results given in Table 2.4 to the optical spectroscopic results given in Table 2.6.

It is worth noting that intermediate coupling is usually incorporated in the analysis of optical spectra while *LS* coupling is often assumed in obtaining phenomenological crystal field parameters from neutron scattering results. The use of different coupling schemes in determining reduced matrix elements (see Appendix 1) introduces discrepancies of a few per cent between the crystal field parameters of the ground states determined from these two experimental methodologies (see Section 3.1.1). Table 2.5 can also be used to generate the Stevens multiplicative factors, if required, by using

   (i) $\theta_2 = \alpha = 8.066/(k = 2$ entry),
   (ii) $\theta_4 = \beta = 8.066/(k = 4$ entry),
   (iii) $\theta_6 = \gamma = 8.066/(k = 6$ entry).

For example, the value of $\alpha$ for $Pr^{3+}$ is $8.066/(-3.839 \times 10^2) = -2.1 \times 10^{-2}$.

### 2.4.1 Trivalent lanthanide ions in cubic sites

The spectra of magnetic ions in sites of cubic symmetry are the simplest from a theoretical point of view because there are only two non-zero crystal

Table 2.7. Crystal field parameters (Wybourne normalization, in $cm^{-1}$) for lanthanide ions substituted into cubic fluorides [Leś90].

| Ion | Parameter | $CaF_2$ | $SrF_2$ | $BaF_2$ | $CdF_2$ | $PbF_2$ |
|-----|-----------|---------|---------|---------|---------|---------|
| $Dy^{3+}$ | $B_0^4$ | $-2185$ | $-2029$ | $-1905$ | $-2245$ | |
| | $B_0^6$ | $733.6$ | $654.8$ | $590.4$ | $757.1$ | |
| $Er^{3+}$ | $B_0^4$ | $-1906$ | $-1755$ | $-1601$ | $-2007$ | $-1664$ |
| | $B_0^6$ | $650.5$ | $567.4$ | $504.6$ | $691.0$ | $540.2$ |

field parameters. Nevertheless, it is quite difficult to obtain good experimental spectra. The problem is that the physical inversion symmetry forbids electric dipole transitions in the first order, so that the optical spectra are necessarily weak. In the case of trivalent lanthanides there is the further problem that most cubic sites, the fluorites in particular, are occupied by divalent ions. Hence the substitution of a trivalent ion requires charge compensation somewhere in the crystal. The crystal grower then has to ensure that charge compensation is achieved sufficiently far away from the lanthanide ion for the cubic symmetry not to be affected.

In spite of these practical difficulties, the simplicity of cubic sites makes them of sufficient theoretical importance to be given detailed consideration in this book. In particular, systems with cubic symmetry have often been used in attempts to analyse local distortions around a substituted ion, sometimes in strained crystals. The main justification for our interest is that, although *exact* cubic symmetry is rare, many crystals, such as the garnets and high-$T_c$ materials, have sites with approximate cubic symmetry. Table 2.7 gives parameters for two trivalent lanthanides in cubic sites of the fluorite crystals (from Leśniak [Leś90]).

### 2.4.2 Trivalent ions in zircon structure crystals

Values of the crystal field parameters for several different lanthanide and actinide ions in phosphate and vanadate host crystals are given in Table 2.8. The last two sets of parameters in the table, for $Cm^{3+}$, are determined from fits to similar sets of experimental energy levels. The differences between them exemplify the magnitude of the uncertainties which are common in fitted actinide crystal field parameters.

Table 2.8. Crystal field parameters (Wybourne normalization, cm$^{-1}$) for trivalent lanthanide and actinide ions in zircon structure hosts. Division by the factors shown in the first row converts to Stevens normalization. Signs of $B_4^4$ and $B_4^6$ are correlated, although their overall sign cannot be determined by fitting. This indeterminacy is sometimes made explicit.

| System | $B_0^2$ | $B_0^4$ | $B_4^4$ | $B_0^6$ | $B_4^6$ | Source |
|--------|---------|---------|---------|---------|---------|--------|
|        | 2       | 8       | $4\sqrt{70}/35$ | 16 | $8\sqrt{14}/21$ | |
| Gd:LuPO$_4$ | 169 | 220 | $-1034$ | $-733$ | 961 | [SME$^+$95] |
| Er:YPO$_4$ | 279 | 155 | $-756$ | $-537$ | $-141$ | [HSE$^+$81] |
| ErPO$_4$ | 390 | 114 | $-711$ | $-697$ | $-62$ | [LSH$^+$93] |
| Er:LuPO$_4$ | 146 | 68 | $-760$ | $-643$ | $-89$ | [HSE$^+$81] |
| Ho:YPO$_4$ | 352 | 67 | $-673$ | $-757$ | $-4$ | [LSH$^+$93] |
| HoPO$_4$ | 374 | 60 | $-662$ | $-726$ | $-57$ | [LSH$^+$93] |
| Tm:LuPO$_4$ | 203 | 117 | $-673$ | $-705$ | $-16$ | [LSA$^+$93] |
| TmPO$_4$ | 315 | 220 | $-666$ | $-704$ | $-74$ | [LSA$^+$93] |
| Nd:YVO$_4$ | $-15$ | 220 | $-602$ | $-318$ | $-40$ | [ZY94] |
| Nd:YVO$_4$ | $-200$ | 628 | $-1136$ | $-1233$ | 149 | [GNKHV$^+$98] |
| Nd:YAsO$_4$ | $-164$ | 237 | $-1071$ | $-1043$ | $-10$ | [GNKHV$^+$98] |
| Nd:YPO$_4$ | 240 | 108 | $-1006$ | $-1190$ | $-90$ | [GNKHV$^+$98] |
| Er:YVO$_4$ | $-206$ | 364 | $\pm926$ | $-688$ | $\pm32$ | [LSA$^+$93] |
| Er:ScVO$_4$ | $-477$ | 423 | $\pm1003$ | $-942$ | $\pm28$ | [LSA$^+$93] |
| HoVO$_4$ | $-164$ | 302 | $-890$ | $-740$ | $-114$ | [Kus67] |
| Cm:LuPO$_4$ | 443 | 304 | $-1980$ | $-2880$ | 881 | [SME$^+$95] |
| Cm:LuPO$_4$ | 503 | 197 | $-1994$ | $-3007$ | 626 | [MEBA96] |

### *2.4.3 Lanthanide ions in garnet host crystals*

Table 2.9 provides a selection of the available crystal field parameter values for lanthanide ions in garnet host crystals. A more comprehensive listing may be found in [Hüf78]. The site symmetry is D$_2$, corresponding to the effective symmetry D$_{2h}$. This symmetry provides three equivalent twofold axes, any one of which may be taken as the principal axis in defining the crystal field parameters. Hence several quite distinct, but still perfectly valid, sets of parameters can be obtained by rotating the coordinate frame, in which the crystal field is defined, through 90$^0$ in the $x$–$z$ or $y$–$z$ planes. Appendix 4 describes a computer package for carrying out such rotations. A simple way to determine the relationship between the implicit coordinate frame of the crystal field and the crystal structure is to carry out a superposition model analysis, as described in Chapter 5.

Table 2.9. Crystal field parameters (Stevens normalization, $cm^{-1}$) of some trivalent lanthanide ions in aluminium garnet (AG) and gallium garnet (GG) hosts.

| | Nd:YAG | ErGG | Er:YGG | Er:YAG | DyGG | Dy:YGG | Dy:YAG |
|---|---|---|---|---|---|---|---|
| $A_{20}\langle r^2 \rangle$ | −208 | −52 | −14 | −156 | −34 | −22 | −178 |
| $A_{22}\langle r^2 \rangle$ | 305 | 72 | 89 | 267 | 192 | 141 | 298 |
| $A_{40}\langle r^4 \rangle$ | −339 | −222 | −238 | −281 | −274 | −282 | −287 |
| $A_{42}\langle r^4 \rangle$ | 409 | 246 | 255 | 221 | 195 | 181 | 262 |
| $A_{44}\langle r^4 \rangle$ | 1069 | 978 | 920 | 781 | 985 | 1040 | 1066 |
| $A_{60}\langle r^6 \rangle$ | 70 | 31 | 33 | 32 | 37 | 36 | 43 |
| $A_{62}\langle r^6 \rangle$ | −197 | −71 | −58 | −136 | −104 | −108 | −115 |
| $A_{64}\langle r^6 \rangle$ | 1113 | 657 | 645 | 649 | 681 | 725 | 731 |
| $A_{66}\langle r^6 \rangle$ | −41 | −55 | −70 | −178 | −93 | −57 | −80 |
| Source | [MWK76] | [OH69] | [CM67] | [MWK76] | [WL73] | [GHOS69] | [MWK76] |

Table 2.10. Crystal field parameters (meV), and Wybourne normalization ($cm^{-1}$) for $Ho^{3+}$ in $HoBa_2Cu_3O_{7-\delta}$.

| | $D_{2h}$, meV | $D_{4h}$, meV | $D_{2h}$, $cm^{-1}$ | $D_{2h}$, $cm^{-1}$ |
|---|---|---|---|---|
| $B_0^2$ | $-4.6(2) \times 10^{-2}$ | $-4.6(2) \times 10^{-2}$ | 334(14) | 435 |
| $B_2^2$ | $-2.0(4) \times 10^{-2}$ | | 59(12) | 77 |
| $B_0^4$ | $9.1(1) \times 10^{-4}$ | $9.5(1) \times 10^{-4}$ | −1763(16) | −1908 |
| $B_2^4$ | $-0.6(2) \times 10^{-4}$ | | 18(6) | −297 |
| $B_4^4$ | $-4.2(1) \times 10^{-3}$ | $-4.3(1) \times 10^{-3}$ | 973(27) | 1050 |
| $B_0^6$ | $-4.5(1) \times 10^{-6}$ | $-4.1(1) \times 10^{-3}$ | 449(10) | 472 |
| $B_2^6$ | $2.7(8) \times 10^{-6}$ | | −26(8) | −252 |
| $B_4^6$ | $-1.35(2) \times 10^{-4}$ | $-1.34(2) \times 10^{-4}$ | 1200(17) | 1305 |
| $B_6^6$ | $1.9(7) \times 10^{-6}$ | | −12(4) | −15 |
| Source | [FBU88b] | [AFB+89] | [SLGD91] [GLS91] | [SLGD91] |

### 2.4.4 Lanthanide ions in superconducting cuprates

Neutron spectroscopy has made it possible to study crystal field splitting in the ground multiplet of lanthanide ions substituted into yttrium or lanthanum sites in superconducting cuprates. Some of the superconducting cuprates have been found to occur in two structural forms, either with orthorhombic ($D_{2h}$ – see Appendix 1) lanthanide sites, or with tetragonal ($D_{4h}$) lanthanide sites, depending on their composition (see Allenspach *et al.* [AFB$^+$89]). In all cases the crystal field symmetry is close to cubic.

Many fits to crystal field parameters have been reported in the literature. A representative selection of results for trivalent holmium in HoBa$_2$Cu$_3$O$_{7-\delta}$ is given in Table 2.10. The transformation from parameters given in meV to the Wybourne normalization (in cm$^{-1}$) involves using the Stevens multiplicative factors given in Table 2.5, as well as the factors which change from Stevens to Wybourne normalization, given in Table 2.2.

The 'big' question is, of course, whether the observed crystal field splittings can throw any light on the mechanism of high-temperature superconductivity. This, and some other questions about the crystal field parameters for lanthanide ions in these materials, will be discussed in Chapters 5 and 9.

# 3

# Fitting crystal field parameters

## D. J. NEWMAN
*University of Southampton*

## BETTY NG
*Environment Agency*

This chapter is concerned with the most frequently encountered problem in crystal field theory, namely the calculation of a set of crystal field parameters from experimentally determined crystal field split energy levels. An iterative least-squares fitting procedure is generally required, as follows.

(i) Estimate initial values of the crystal field parameters for the system under consideration.

(ii) Construct the energy matrix using estimated or previously calculated crystal field parameter values.

(iii) Diagonalize the energy matrix to obtain its eigenvalues, which correspond to estimated positions of the energy levels.

(iv) Set up one-to-one correspondences between the experimental and calculated energy levels.

(v) Keeping the eigenvectors of the energy matrix fixed, determine the values of the parameters that minimize the sum of squares of the differences between the calculated and experimental energy levels.

(vi) Using the set of crystal field parameter values derived in step (v), return to step (ii) and continue iterating steps (ii)–(v) until the calculated and experimental energy levels are judged to be in good enough agreement.

Initial estimates of parameter values may be based on values already obtained for a similar system, such as those listed in Chapter 2. This and other ways of estimating parameter values required for step (i) are discussed next. Program ENGYFIT.BAS carries out steps (ii)–(vi) when all the energy levels are in a single multiplet.

It is sometimes necessary to calculate, from a given set of crystal field parameters, energy levels that have not been determined experimentally. This only involves steps (ii) and (iii). Program ENGYLVL.BAS carries out these two steps, and provides the additional facility of determining the eigenvector which corresponds to each eigenvalue. This is useful when irreducible representation labels for the energy levels are required (see Appendix 1).

The first calculation in both parameter fitting and energy level determination is step (ii), the construction of an energy matrix from the crystal field $V_{\rm CF}$. As was pointed out in Chapter 2, it is possible to restrict the crystal field parameters to those with $q \geq 0$, even for systems with site symmetry as low as $C_2$. This produces the following simplification of the crystal field expansion given in equation (2.7).

$$V_{\rm CF} = \sum_{k,q>0} B_q^k (C_q^{(k)} + (-1)^q C_{-q}^{(k)}) + \sum_k B_0^k C_0^{(k)}. \qquad (3.1)$$

The restriction to $q \geq 0$ provides the computational advantage that the energy matrix is real and symmetric and is incorporated into programs ENGYLVL.BAS and ENGYFIT.BAS. Furthermore, both of these programs assume that the energy matrix spans the $M_J$ values corresponding to a single $J$ multiplet.

The calculations described in this chapter are concerned only with the standard one-electron crystal field, the origins of which were discussed in Chapter 1. The use of various types of special and extended parameter schemes as aids in the estimation, fitting and interpretation of energy level splittings of magnetic ions in solids are treated in later chapters.

Descriptions of the QBASIC programs and data files discussed in this chapter are given in Appendix 2, which also gives web addresses from which they may be downloaded.

## 3.1 Determination of crystal field splittings from crystal field parameters

In lanthanides and actinides, the free-ion multiplets are normally well separated and can be labelled by the total orbital angular momentum $L$, the total spin $S$ and the total angular momentum $J$. While only $J$ is an *exact* quantum number, a particular $L$ and $S$ contribution is usually sufficiently dominant to provide an unambiguous labelling of the multiplet. In some cases (called $LS$ coupling) the $L$, $S$ labels are so good that admixtures of other $L$, $S$ states can be neglected entirely. This approximation is fairly good for the ground states of trivalent lanthanide ions, which provide all the examples discussed in this chapter.

Sometimes different $J$ multiplets are sufficiently close in energy for the crystal field to mix their states. More complete diagonalizations, involving

Hamiltonians that contain parametrized forms of the spin–orbit coupling and coulomb interaction, may then be necessary. Further discussion of this problem and a relevant computer package is given in Chapter 4.

It often happens that $L$, $S$ and $J$ do not, by themselves, provide a *unique* labelling for some of the many-electron states of a given $f^n$ configuration. In such cases the eigenstates of a free lanthanide or actinide ion are written in the form $|f^n \alpha L S J\rangle$, where the label $\alpha$ distinguishes states with the same $L$, $S$, $J$ labels. A systematic specification of the $\alpha$ labels, which simplifies the calculation of matrix elements, can be obtained using the Lie group approach pioneered by Racah (see, for example, the books by Judd [Jud63] and Condon and Odabaşi [CO80]). Discussion of the Lie group approach is beyond the scope of this book.

In the case of iron group ions, with $d^n$ configurations, the crystal field is generally larger than the spin–orbit coupling, so that the 'free-ion' states are defined in terms of the total orbital angular momentum $L$ and total spin $S$. Unless the spin–orbit coupling is small enough to be ignored, it is necessary to diagonalize $d^n$ energy matrices for both the crystal field and the spin–orbit coupling together, i.e. covering all the states with a given $L$ and $S$. In practice, the coulomb interactions may be included in these calculations as well, thereby extending the basis set to several $L$ and $S$ values. Hence it is often necessary to determine and diagonalize quite large energy matrices. A computer package which carries out such calculations is discussed in Appendix 3.

### 3.1.1 Reduced matrix elements

According to the Wigner–Eckart theorem (equation (A1.1)) the matrix elements of the tensor operators $C_q^{(k)}$, between many-electron open-shell states, can be factorized as

$$\langle f^n \alpha L S J M_J | C_q^{(k)} | f^n \alpha' L' S J' M_J' \rangle = (-1)^{J-M_J} \begin{pmatrix} J & k & J' \\ -M_J & q & M_J' \end{pmatrix}$$
$$\times (f^n \alpha L S J \| C^{(k)} \| f^n \alpha' L' S J').$$
(3.2)

The $3j$ symbols carry all the dependence on the labels $M_J$, $q$ and $M_J'$ while the so-called 'reduced' matrix element $(f^n \alpha L S J \| C^{(k)} \| f^n \alpha' L' S J')$ is independent of these labels, but depends on the normalization of the tensor operators.

In the $LS$ coupling limit, the $J$ multiplet is formed from a specific mixture of states with the same values of the total angular momentum $L$ and total spin $S$. In this case it is possible to carry out a second factorization of the matrix elements, viz.

$$(\alpha LSJ||C^{(k)}||\alpha'L'SJ') = (-1)^{S+k+J+L'}\sqrt{(2J+1)(2J'+1)}$$
$$\times \begin{Bmatrix} L & J & S \\ J' & L' & k \end{Bmatrix}(\alpha LS||C^{(k)}||\alpha'L'S). \quad (3.3)$$

The so-called $6j$ symbols $\begin{Bmatrix} L & J & S \\ J' & L' & k \end{Bmatrix}$ are discussed in Section A1.1.4. The 'doubly' reduced matrix elements $(f^n\alpha LS||C^{(k)}||f^n\alpha'L'S)$ are independent of $J$.

Numerical values of the 'singly' reduced matrix elements, in the $LS$ coupling limit, can be obtained using the program REDMAT.BAS. This inputs values of the doubly reduced matrix elements $(f^n\alpha LS||u^{(k)}||f^n\alpha'L'S)$ of the unit tensor operators $u^{(k)}$, which are normalized as shown in equation (A1.3). Unit tensor doubly reduced matrix elements have been tabulated for all $f^n$ (as well as $p^n$ and $d^n$) configurations by Nielson and Koster [NK63]. As entries in their tables are expressed in terms of powers of primes, there is a significant risk of making errors when transcribing them into decimal notation. For the convenience of readers, some of the commonly used doubly reduced matrix elements of the unit tensor operators for $L = L'$ are listed (using decimal notation) in Table 3.1. Some of the entries in this table have been incorporated into the program REDMAT.BAS.

Low-lying energy spectra, which span only the energy levels of the ground multiplet, cannot provide information about the extent of mixing between multiplets with the same $J$, but different $LS$ values. Hence it is not possible to determine the effects of this $J$ mixing, known as 'intermediate coupling', on the reduced matrix elements. One way of coping with this, especially in the neutron scattering literature, has been to set the reduced matrix elements to an arbitrary value. However, this precludes making direct comparisons between crystal field parameters for different magnetic ions in similar crystalline sites, and comparisons with relevant crystal field parameters obtained from optical spectra of the same ion (e.g. see [SHF+87]). An alternative approach is to suppose that there is no admixture of other $J$ states into the ground multiplet (i.e. that $LS$ coupling is a good approximation). It is better still to employ reduced matrix elements obtained from free-ion fits to optical data for the same magnetic ion, which incorporate the effects of intermediate coupling. This approach is fairly reliable because the free-

Table 3.1. Unit tensor $u^{(k)}$ doubly reduced matrix elements for some $f^n$ configurations. Tabulated values have been converted from power of primes notation used in [NK63].

| Config. | Term | $k = 2$ | $k = 4$ | $k = 6$ |
|---|---|---|---|---|
| $f^2$ | $^3H$ | −1.2367 | −0.7396 | 0.8951 |
| $f^3$ | $^4I$ | −0.4954 | −0.4904 | −1.1084 |
| $f^4$ | $^5I$ | 0.4540 | 0.4103 | 0.7679 |
| $f^5$ | $^6H$ | 0.8458 | 0.2979 | |
| $f^8$ | $^7F$ | −1.5159 | 0.7187 | −0.2277 |
| $f^9$ | $^6H$ | −1.6095 | −0.9135 | 1.0454 |
| $f^{10}$ | $^5I$ | 0.4438 | −0.6371 | 1.7827 |
| $f^{10}$ | $^5F$ | −0.5 | −0.5 | −0.5 |
| $f^{10}$ | $^5G$ | −0.0373 | 1.0536 | 0.8650 |
| $f^{10}$ | $^3K(1)$ | −0.7509 | −0.0171 | −0.6009 |
| $f^{11}$ | $^4I$ | 0.6438 | 0.6851 | 1.8000 |
| $f^{12}$ | $^3H$ | 1.5159 | 0.9583 | −1.3225 |

ion parameters, which determine the extent of intermediate coupling, are only weakly dependent on the host crystal.

Comparison of the intermediate coupled and $LS$ coupled reduced matrix elements given in Table 3.2 (labelled 'int') provides an indication of the uncertainties of the parameter values obtained when using $LS$ coupling. Intermediate coupling corrections were calculated using the factors given in table 9 of [Die68]. A set of intermediate coupling reduced matrix elements for 10 multiplets of $Er^{3+}$ is given in Table 8.5.

Energy levels correspond to the (necessarily) real eigenvalues of the energy matrices, and can be determined by their diagonalization. This process is sometimes simplified by using site symmetry properties to carry out a prior 'block diagonalization' (e.g. see [LN69]), but this procedure has not been incorporated into the programs which accompany this book. For some applications the eigenstates corresponding to each eigenvalue also need to be determined.

The energy levels for a single multiplet, and their corresponding eigenstates, can be obtained by diagonalizing the crystal field energy matrix defined by (3.1) and (3.2). Program ENGYLVL.BAS both constructs and diagonalizes such energy matrices, allowing the calculation of energy levels from given crystal field parameters. Before giving examples of the use of this program, however, it is worth comparing the method of constructing

Table 3.2. Reduced matrix elements for the ground multiplets of $f^n$ configurations. Lanthanide ion intermediate coupling values were obtained using factors given by [Die68].

| Ions | Config. | Coupling | $k = 2$ | $k = 4$ | $k = 6$ |
|------|---------|----------|---------|---------|---------|
| $Pr^{3+}$,$Pa^{3+}$ | $f^2\ ^3H_4$ | $LS$ | $-1.2367$ | $-0.7396$ | $0.8951$ |
| $Pr^{3+}$ | $f^2\ ^3H_4$ | int. | $-1.2057$ | $-0.7396$ | $0.7653$ |
| $Nd^{3+}$,$U^{3+}$ | $f^3\ ^4I_{9/2}$ | $LS$ | $-0.4954$ | $-0.4904$ | $-1.1084$ |
| $Nd^{3+}$,$Pu^{3+}$ | $f^3\ ^4I_{9/2}$ | int. | $-0.4731$ | $-0.4698$ | $-1.0618$ |
| $Sm^{3+}$ | $f^5\ ^6H_{5/2}$ | $LS$ | $0.8458$ | $0.2979$ | — |
| $Tb^{3+}$,$Bk^{3+}$ | $f^6\ ^7F_6$ | $LS$ | $-1.5159$ | $0.7187$ | $-0.2277$ |
| $Tb^{3+}$ | $f^6\ ^7F_6$ | int. | $-1.498$ | $0.705$ | $-0.2155$ |
| $Dy^{3+}$,$Cf^{3+}$ | $f^9\ ^6H_{15/2}$ | $LS$ | $-1.6095$ | $-0.9135$ | $1.0454$ |
| $Ho^{3+}$,$Es^{3+}$ | $f^{10}\ ^5I_8$ | $LS$ | $-0.6563$ | $-0.6797$ | $-1.7062$ |
| $Ho^{3+}$ | $f^{10}\ ^5I_8$ | int. | $-0.602$ | $-0.629$ | $-1.582$ |
| $Er^{3+}$,$Fm^{3+}$ | $f^{11}\ ^4I_{15/2}$ | $LS$ | $0.6438$ | $0.6851$ | $1.8000$ |
| $Er^{3+}$ | $f^{11}\ ^4I_{15/2}$ | int. | $0.6805$ | $0.6967$ | $1.7388$ |
| $Tm^{3+}$,$Md^{3+}$ | $f^{12}\ ^3H_6$ | $LS$ | $1.5159$ | $0.9583$ | $-1.3225$ |
| $Tm^{3+}$ | $f^{12}\ ^3H_6$ | int. | $1.5295$ | $0.9362$ | $-1.3264$ |

the energy matrix used in this book with the so-called 'operator equivalent' method.

### 3.1.2 Determination of the energy matrix

Traditionally, calculations of crystal field energy matrices for lanthanide ions have been carried out using the method of 'operator equivalents', pioneered by Stevens [Ste52]. Tables which facilitate the use of this method have been published in many books and papers (e.g. [Hut64, Die68, AB70, Hüf78]), and have been used by successive generations of research workers. The operator equivalent approach is well suited to hand calculations of the energy matrix, especially when only ground multiplet splittings are of interest. However, given that it is often necessary to carry out the diagonalization of the energy matrix on a computer, there is little point in constructing the matrix elements by hand, other than to provide an understanding of the manipulations involved. In practice, calculations of energy levels from known crystal field parameters (steps (i) and (ii) in the list at the beginning of this chapter) are most readily carried out using programs such as ENGYLVL.BAS, as illustrated in the next section.

Table 3.3. Values of the $3j$ symbols required to construct diagonal matrix elements for the ground multiplet of $Nd^{3+}$:$LaCl_3$.

| $m$ | $k = 2$ | $k = 4$ | $k = 6$ |
|-----|---------|---------|---------|
| 1/2 | −0.1557 | 0.1122 | −0.0864 |
| 3/2 | 0.1168 | −0.0187 | −0.0648 |
| 5/2 | −0.0389 | −0.1060 | 0.1080 |
| 7/2 | 0.0779 | 0.1371 | 0.1188 |
| 9/2 | 0.2335 | 0.1122 | 0.0324 |

In this book, the operator equivalent tables are replaced by the programs THREEJ.BAS and REDMAT.BAS (see Appendix 2), the outputs of which can be used to construct energy matrix elements if desired. As an example, THREEJ.BAS and REDMAT.BAS are used to construct the energy matrix for the $^4I_{9/2}$ ground multiplet of $Nd^{3+}$ in the $LaCl_3$ crystal field. The site symmetry in $LaCl_3$ is $C_{3h}$, allowing non-zero crystal field parameters $B_0^2$, $B_0^4$, $B_0^6$ and $B_6^6$ (see Table 2.3). Values of these parameters for $Nd^{3+}$:$LaCl_3$ are given in Table 2.6. Intermediate coupling reduced matrix elements for the ground multiplet are given in Table 3.2. Program THREEJ.BAS can be used to determine the $3j$ coefficients

$$\begin{pmatrix} 9/2 & k & 9/2 \\ -m & 0 & m \end{pmatrix}$$

needed in the calculation of the diagonal ($q = 0$) matrix elements. Their values are shown in Table 3.3. The off-diagonal matrix elements, arising from the parameter $B_6^6$, involve only two distinct $3j$ symbols with $k = 6$. They are calculated, using THREEJ.BAS, to be

$$\begin{pmatrix} 9/2 & 6 & 9/2 \\ -9/2 & 6 & -3/2 \end{pmatrix} = 0.1074,$$

and

$$\begin{pmatrix} 9/2 & 6 & 9/2 \\ -7/2 & 6 & -5/2 \end{pmatrix} = -0.1461.$$

The $LS$ coupled reduced matrix elements $(f^3LSJ||C^{(k)}||f^3LSJ)$ of the tensor operators $C^{(k)}$ for the ground multiplet $^4I_{9/2}$ (i.e. $L = 6$, $S = 3/2$ and $J = 9/2$) can be calculated using REDMAT.BAS. They are given, for $k = 2$, 4, 6, respectively, by −0.4954, −0.4804 and −1.1084. Combining the values of the $3j$ symbols given in Table 3.3 with these reduced matrix elements,

Table 3.4. Diagonal matrix elements of the $C_q^{(k)}$ for the $^4I_{9/2}$ ground multiplet of $Nd^{3+}$.

| $m$ | $k = 2$ | $k = 4$ | $k = 6$ |
|-----|---------|---------|---------|
| 1/2 | 0.07713 | −0.05502 | 0.09573 |
| 3/2 | 0.05786 | −0.00917 | −0.07180 |
| 5/2 | 0.01928 | 0.05196 | −0.11966 |
| 7/2 | −0.03857 | 0.06725 | 0.13162 |
| 9/2 | −0.11568 | −0.05502 | −0.03590 |

using equation (3.2), determines the diagonal matrix elements of the tensor operator components shown in Table 3.4. The off-diagonal matrix elements, calculated in the same way, are

$$\langle 7/2 | C_6^{(6)} | - 5/2 \rangle = -0.18186$$

and

$$\langle 9/2 | C_6^{(6)} | - 3/2 \rangle = -0.11906.$$

In order to construct the crystal field energy matrix, the tensor operator matrix elements must be combined with the crystal field parameters and summed over the $k$ values. The required crystal field parameters (in Stevens normalization) are given in Table 2.6. In Wybourne normalization they correspond to $B_0^2 = 195$, $B_0^4 = -310$, $B_0^6 = -710$ and $B_6^6 = 466$, all in units of $cm^{-1}$. Constructing the sums over $k$ allows us to determine the elements of the energy matrix for the ground multiplet of $Nd^{3+}:LaCl_3$, in $cm^{-1}$:

$$\langle \pm 1/2 | V_{CF} | \pm 1/2 \rangle = -35.87$$
$$\langle \pm 3/2 | V_{CF} | \pm 3/2 \rangle = 65.10$$
$$\langle \pm 5/2 | V_{CF} | \pm 5/2 \rangle = 72.61$$
$$\langle \pm 7/2 | V_{CF} | \pm 7/2 \rangle = -121.82$$
$$\langle \pm 9/2 | V_{CF} | \pm 9/2 \rangle = 19.99 \tag{3.4}$$
$$\langle \pm 3/2 | V_{CF} | \mp 9/2 \rangle = -55.48$$
$$\langle \pm 7/2 | V_{CF} | \mp 5/2 \rangle = -84.75$$
$$\langle \pm 5/2 | V_{CF} | \mp 7/2 \rangle = -84.75$$
$$\langle \pm 9/2 | V_{CF} | \mp 3/2 \rangle = -55.48.$$

All the other elements of the $10 \times 10$ matrix are zero. The symmetry of the remaining matrix elements under the interchange of $m$ with $-m$ shows that the $10 \times 10$ matrix can be block diagonalized into a single $1 \times 1$ matrix and two $2 \times 2$ matrices, as follows

$$-35.87, \begin{pmatrix} 65.10 & -55.48 \\ -55.48 & 19.99 \end{pmatrix}, \begin{pmatrix} 72.61 & -84.75 \\ -84.75 & -121.82 \end{pmatrix}.$$

Each of these matrices appears twice in the $10 \times 10$ matrix because of the double occurrence of all matrix elements (as shown above). This reflects the so-called 'Kramers' degeneracy of the crystal field splittings which occurs in half-integral $J$ multiplets. In order to determine the energy levels and eigenstates it is only necessary to solve two quadratic equations. However, rather than carrying out this evaluation, which is left as an exercise for the reader, we now turn to the program ENGYLVL.BAS, which both sets up and diagonalizes the energy matrix.

Readers who have had experience using the operator equivalent method may wish to check that they can obtain results in agreement with the above. Note that the factorization of the matrix element into $(q, m)$-dependent and $(q, m)$-independent parts is quite different in the two methods. In making numerical comparisons it is also necessary to remember the different normalizations of the Stevens and Wybourne parameters, given in Table 2.2.

### 3.1.3 Calculation of energy levels using ENGYLVL.BAS

As an example of the use of ENGYLVL.BAS we continue with the example introduced in the previous section, i.e. the calculation of the energy levels of the $^4I_{9/2}$ ground multiplet of $Nd^{3+}$ in the $LaCl_3$ crystal field. Following the 'RUN' command the program responds by asking for input data as follows:

```
THIS PROGRAM CALCULATES THE ENERGY LEVELS OF A J-MULTIPLET
FROM A SET OF CRYSTAL FIELD PARAMETERS
J VALUE (ANY REAL NO. <=8) FOR THE MULTIPLET =? 4.5
PROVIDE REDUCED MATRIX ELEMENTS FOR THIS MULTIPLET
REDUCED MATRIX ELEMENT FOR RANK 2 = ? -0.4954
REDUCED MATRIX ELEMENT FOR RANK 4 = ? -0.4904
REDUCED MATRIX ELEMENT FOR RANK 6 = ? -1.1084

FILENAME FOR CRYSTAL FIELD PARAMETERS = ? ND_LACL3.DAT

CONSTRUCTION OF THE ENERGY MATRIX BEGINS
PLEASE BE PATIENT.  IT MAY TAKE A WHILE .....
```

Appropriate input data, as shown, must be provided after each question
mark. The crystal field parameters are assumed to be in Wybourne normalization. Note that half integral $J$ values must be expressed in decimal
form in inputs, i.e. entered as 4.5 (rather than 9/2) in the above example.
After constructing the energy matrix, the program reports its progress and
asks for further inputs as follows:

```
FINDING THE EIGENVALUES
EIGENVALUES HAVE BEEN DETERMINED

OUTPUT WILL BE IN A SINGLE FILE CONTAINING
EIGENVALUES IN ARBITRARY ORDER.  DO YOU WISH
TO OUTPUT A SECOND FILE CONTAINING BOTH
EIGENVECTORS AND EIGENVALUES (y/Y FOR YES)? y
NAME OF EIGENVALUE FILE = ? ND_ENGY.DAT
NAME OF EIGENVECTOR AND EIGENVALUE FILE = ? ND_ENEI.DAT

OUTPUT FILE FOR EIGENVALUES = ND_ENGY.DAT
OUTPUT FILE FOR EIGENVECTORS AND EIGENVALUES = ND_ENEI.DAT
PROGRAM RUN IS COMPLETED SUCCESSFULLY
```

In the above example, the option of both eigenvalues and eigenvectors is
chosen. However, if only eigenvalues are required, the program will proceed
as follows:

```
OUTPUT WILL BE IN A SINGLE FILE CONTAINING
EIGENVALUES IN ARBITRARY ORDER.  DO YOU WISH
TO OUTPUT A SECOND FILE CONTAINING BOTH
EIGENVECTORS AND EIGENVALUES (y/Y FOR YES)? N
NAME OF EIGENVALUE FILE = ? ND_ENGY.DAT

OUTPUT FILE FOR EIGENVALUES = ND_ENGY.DAT
PROGRAM RUN IS COMPLETED SUCCESSFULLY
```

The file ND_ENGY.DAT should now contain the eigenvalues arranged in
a column, in which the first five entries are as follows: $-17.35$, $-153.58$,
$104.37$, $102.43$, $-35.87$. (The remaining five entries repeat these values in
reverse order – see Appendix 2.) Diagonalization of the two $2 \times 2$ matrices
given above gives the first four of these entries, while the last entry $(-35.87)$
agrees with the diagonal energy matrix entry for $\pm 1/2, \pm 1/2$.

The eigenvectors corresponding to the other eigenvalues can be read off
from the file ND_ENEI.DAT. Each eigenvector is written below its corre-

sponding eigenvalue as a column of 10 coefficients ordered as $-9/2$, $-7/2$, ... , $-1/2$, $1/2$, ... , $9/2$. For example, it shows that the two eigenvectors corresponding to the eigenvalue $-17.35$ are

$$0.8297\left|\mp\frac{9}{2}\right\rangle + 0.5583\left|\pm\frac{3}{2}\right\rangle.$$

### 3.1.4 Calculating energy levels of $Ho^{3+}$ from published crystal field parameters for a high-$T_c$ superconducting material

The crystal field splittings of the trivalent holmium ion are rich in examples of the special phenomena and problems of interpretation that form the subject matter of this book. An example of particular interest is the splitting of its $^5I_8$ ground multiplet in the superconducting cuprates, which has been observed, at least in part, by means of inelastic neutron scattering. This section is concerned with the analysis of the experimental energy levels and crystal field fits reported by Furrer and coworkers [FBU88b, AFB+89], and the modified crystal field fits reported by Soderholm *et al.* [SLGD91]. The relevant crystal field parameters are given in Table 2.10. These results refer to $Ho^{3+}$ ions at $D_{2h}$ and $D_{4h}$ sites in crystalline $HoBa_2Cu_3O_{7-\delta}$. The site symmetry is determined by the oxygen concentration, i.e. by the value of $\delta$.

With $\delta = 0.2$ the site symmetry is $D_{2h}$ and there are nine independent crystal field parameters (see Table 2.3). As the symmetry group $D_{2h}$ has only one-dimensional irreducible representations, there are no degenerate energy levels (see Appendix 1). Hence the $^5I_8$ ground multiplet splits into 17 distinct levels, which would be quite sufficient to determine the crystal field parameters if all the levels were observed. However, only 10 distinct transition energies have been observed [FBU88b].

Two sets of values of the nine crystal field parameters for $D_{2h}$ symmetry (in Wybourne normalization) are given in the fourth and fifth columns of Table 2.10. These are rounded off versions of the parameters given by Soderholm *et al.* [SLGD91]. The first set agrees with the parameters given by Furrer *et al.* [FBU88b] (column 2 of Table 2.10) if the factors in Table 2.5 are used in the conversion. The error estimates given by [FBU88b] justify the use of rounded off parameter values in the following analysis. The second set of parameters (column 4 of Table 2.10) was determined by Soderholm *et al.* [SLGD91], and is based on interchanging the irreducible representation labels for the two lowest-lying energy levels above the ground level. For convenience, the sets of parameter values given in columns 4 and 5 of Table 2.10 are contained in data files HO_BACO1.DAT and HO_BACO2.DAT,

Table 3.5. Experimental and predicted energy levels $(\text{cm}^{-1})$ for $\text{Ho}^{3+}$ in $\text{D}_{2h}$ and $\text{D}_{4h}$ sites in crystalline $\text{HoBa}_2\text{Cu}_3\text{O}_{7-\delta}$. The first column labels the energies in order of magnitude. Columns 3 and 7 give the irreducible representation labels for $\text{D}_{2h}$ and $\text{D}_{4h}$, respectively.

| Order | Set 1 | Γ | Set 2 | Exp. | Set 3 | Γ |
|------:|------:|--:|------:|-----:|------:|--:|
| 10 | 467.3 | 1 | 555.0 |      | 506.3 | 1 |
| 16 | 568.3 | 4 | 630.2 |      | 561.5 | 5 |
| 1  | 0.0   | 3 | 0     | 0.0  | 0.0   | 4 |
| 8  | 93.6  | 4 | 105.5 | 93.6 | 87.4  | 5 |
| 5  | 35.8  | 3 | 40.6  | 34.7 | 37.1  | 1 |
| 2  | 4.4   | 4 | 10.0  | 4.0  | 9.8   | 5 |
| 17 | 589.0 | 1 | 655.3 | 589  | 584.4 | 3 |
| 15 | 564.1 | 2 | 620.7 | 565  | 561.5 | 5 |
| 14 | 501.9 | 1 | 553.2 |      | 470.3 | 1 |
| 13 | 480.2 | 2 | 533.4 | 478  | 479.2 | 5 |
| 9  | 460.4 | 3 | 490.3 |      | 459.6 | 4 |
| 3  | 14.3  | 2 | 22.5  | 14.5 | 9.8   | 5 |
| 4  | 30.7  | 1 | 32.9  | 30.7 | 39.1  | 2 |
| 7  | 89.3  | 2 | 99.8  | 87.1 | 87.4  | 5 |
| 6  | 66.7  | 1 | 73.4  | 65.3 | 58.8  | 3 |
| 11 | 478.6 | 4 | 527.0 | 478  | 479.2 | 5 |
| 12 | 480.0 | 3 | 517.3 | 478  | 487.4 | 2 |

respectively. These files provide the starting point for the computational analysis described below.

The first step is to determine the energy levels corresponding to the original set of crystal field parameters determined by Furrer *et al.* [FBU88b] (listed in HO_BACO1.DAT). The energy level structure obtained by inputting this set of parameters into the program ENGYLVL.BAS, and using the *LS* reduced matrix elements (see Table 3.2), is given in the second column (labelled 'Set 1') of Table 3.5. Conversion of these to meV (i.e. dividing by 8.066), setting the lowest level to zero, and ordering in accord with their order of magnitude, shows them to generate the calculated energies given in [FBU88b]. Oddly enough, Soderholm *et al.* [SLGD91] calculated a quite different set of energies from the same parameters. This was presumably due to their use of different reduced matrix elements. Intermediate coupling matrix elements would give fitted crystal field parameters that were systematically greater than those obtained by Furrer *et al.* [FBU88b]. *This highlights the importance of providing explicit values of the reduced matrix elements used in calculating one-electron crystal field parameters from energy levels.*

The irreducible representation labels ascribed to the energy levels in Table 3.5 may be obtained by comparing the calculated eigenvectors with the standard forms for each label given in equation (4) of [FBU88b]. In brief, taking $m$ as an integer, the $J = 8$ basis states of the form

$$|J, M_J = 2m\rangle + |J, M_J = -2m\rangle \text{ transform as } \Gamma_1 \equiv A_1,$$
$$|J, M_J = 2m\rangle - |J, M_J = -2m\rangle \text{ transform as } \Gamma_3 \equiv B_1,$$
$$|J, M_J = 2m + 1\rangle + |J, M_J = -(2m + 1)\rangle \text{ transform as } \Gamma_2 \equiv B_2, \text{ and}$$
$$|J, M_J = 2m + 1\rangle - |J, M_J = -(2m + 1)\rangle \text{ transform as } \Gamma_4 \equiv B_3.$$

Labels obtained by these means (as listed in Table 3.5) agree with those given in [FBU88b].

Note that the energy levels given in Table 3.5 are listed in the order produced by the program ENGYLVL.BAS, with a common addition designed to bring the lowest-lying level (third in the table) to zero. Note the large gap between a cluster of eight lowest-lying levels (below 95 cm$^{-1}$) and a cluster of the nine remaining energy levels, between 460 cm$^{-1}$ and 600 cm$^{-1}$ in column 2 of Table 3.5. A simple explanation of this clustering phenomenon, in terms of the semiclassical model of crystal fields, is given in Chapter 9. Chapter 9 also demonstrates how the observed clustering can be used, in conjunction with the superposition model (described in Chapter 5), to make crude estimates of the crystal field parameters. Furrer *et al.* [FBU88b] followed an alternative procedure. They compared the splittings with those in the appropriate Lea, Leask and Wolf diagram [LLW62] to estimate $x = -0.4$ (see Chapter 2 for the definition of $x$), and hence obtained an estimate of the magnitudes of the dominant cubic contributions to the crystal field.

The fourth column of Table 3.5 (labelled 'Set 2') gives the energy levels generated from the crystal field parameters derived by Soderholm *et al.* [SLGD91]. These parameters were obtained by fitting to the same set of experimental energy levels, the only difference being an interchange of the irreducible representation labels on levels 2 and 3. *LS* coupling is assumed in our calculation, and it can be seen that the derived energy levels no longer provide good agreement with the experimental levels in column 5. This contrasts with the very good agreement shown in [SLGD91], again suggesting that these authors are not using *LS* coupled reduced matrix elements.

The third column of Table 2.10 gives the parameters obtained by Allenspach *et al.* [AFB$^+$89] by fitting to the experimental energy levels for the tetragonal (i.e. $D_{4h}$) $Ho^{3+}$ sites in $HoBa_2Cu_3O_{6.2}$. File HO_BACO3.DAT contains these parameter values, transformed to Wybourne normalization in the $LS$ approximation. This file and the program ENGYLVL.BAS have been used, in conjunction with $LS$ reduced matrix elements, to generate the energy levels shown in the sixth column of Table 3.5 (labelled 'Set 3'). With an appropriate change of units the energy levels obtained in this way agree with the calculated levels given in table I of [AFB⁺89]. The eigenvectors obtained from running the program ENGYLVL.BAS can also be used, in conjunction with the basis functions given in table 16.2 p. 531 of Butler's book [But81], to check the $D_{4h}$ irreducible representation labels given in column 7 of Table 3.5.

## 3.2 Fitting crystal field parameters to multiplet energy levels in lanthanides and actinides

This section is concerned with the practical problems that arise in using least-squares procedures to fit crystal field parameters to energy levels. Examples are based on the use of the program ENGYFIT.BAS, which is restricted to fitting single $J$ multiplets to crystal field parameters with $q \geq 0$.

The fitting of crystal field parameters is carried out by minimizing the sum of squares of differences between the given (normally experimental) set of energy levels and the set of energy levels generated by the parameter values. Before this can begin it is necessary to establish one-to-one correspondences between the given energies and those calculated using the estimated starting parameters. These correspondences are established by two methods.

(i) Identifying experimental and calculated energy levels that have the same symmetry label.

(ii) Identifying experimental and calculated energy levels that have a similar magnitude.

Symmetry labels may not be known and, in any case, often occur several times, so that similarities in magnitude must usually be taken into account.

Program ENGYFIT.BAS puts both the energy levels to be fitted and the energy levels generated from the estimated starting parameters in increasing numerical order. It then correlates levels with corresponding positions in the resulting sequences. If any information about the symmetry labels of the experimental levels is available (e.g. from transition intensities) it is necessary to check that this procedure generates the correct correspondences. This can be achieved by first generating the energy levels from the estimated starting crystal field parameters using program ENGYLVL.BAS, and determining the labelling. If this does not coincide with the labelling of the corresponding experimental levels a revised estimate of the starting parameters must be made.

### 3.2.1 Fitting crystal field parameters to the $^4I_{9/2}$ ground multiplet energy levels of $Nd^{3+}$ in the anhydrous chloride

In this section a parameter fit to the energy levels derived in the previous section is carried out to demonstrate the use of the program ENGYFIT.BAS. This should, of course, lead us back to the crystal field parameters which were previously used as input to the program ENGYLVL.BAS. Running ENGYFIT.BAS initially generates the following output on the screen. Every question mark requires an input from the user.

```
THIS PROGRAM FITS A SET OF CRYSTAL FIELD PARAMETERS
WITH Q>=0 TO THE ENERGY LEVELS OF A J-MULTIPLET

J VALUE (ANY REAL NO. <=8) FOR THE MULTIPLET =? 4.5
PROVIDE REDUCED MATRIX ELEMENTS FOR THIS MULTIPLET
REDUCED MATRIX ELEMENT FOR RANK 2 = ? -0.4954
REDUCED MATRIX ELEMENT FOR RANK 4 = ? -0.4904
REDUCED MATRIX ELEMENT FOR RANK 6 = ? -1.1084

FILENAME OF ENERGY LEVELS = ? ND_ENGY.DAT
```

Note that the energy level file must be complete, i.e. including degeneracies explicitly as multiple entries and guessed values for any missing levels. In other words, $2J + 1$ entries are required. Possible ways of coping with unknown levels are discussed at the end of this chapter. The crystal field parameters to be fitted are entered as follows:

```
NO. OF RANKS TO BE FITTED (2 OR 3)=? 3
INPUT VALUES OF K AND Q FOR FITTED PARAMETERS
RANK K = ? 2
NO. OF Q VALUES FOR THIS RANK = ? 1
Q= ? 0
RANK K = ? 4
NO. OF Q VALUES FOR THIS RANK = ? 1
Q= ? 0
RANK K = ? 6
NO. OF Q VALUES FOR THIS RANK = ? 2
```

```
Q= ?  0
Q= ?  6
```

It is clearly tedious to have to make all the above entries when carrying out many runs for the same system, or for systems with the same symmetry. In this case the user is recommended to "remark out" the input requests, and to assign values to the corresponding variables within the program. Before doing this, however, it is important to give your working program a different name, in case inadvertent errors are introduced.

For example, when carrying out fits to many systems with $C_{3h}$ symmetry it is unnecessary to designate which parameters are to be fitted in each run. In the present example, after renaming, lines 105–154 of the program can be remarked out and the following values of the variables specified:

```
NUMKF   =  3
LKF(1)=  2
LKF(2)=  4
LKF(3)=  6
NUMQF(1)=1
NUMQF(2)=1
NUMQF(3)=2
```

Having defined which phenomenological parameters are to be fitted, the program then allows the user to introduce constraints in the form of fixed parameter ratios. In the present example there are just sufficent energy levels (5) to determine the four crystal field parameters. (As the crystal field parameters are determined by energy *differences*, $N+1$ distinct energies are required to determine $N$ parameters.) However, the fitting program does not converge well when the number of fitted parameters equals the number of energy differences. To solve the problem, one either solves a set of simultaneous equations or introduces some constraints among the parameters in the fitting procedures. In this example, a constraint is introduced between the two rank 6 parameters. The ratio between these parameters for ions substituted into anhydrous chlorides is remarkably constant, as shown in Table 2.6. The reason for this is discussed in Chapter 5. For present purposes we assume that this ratio can be determined independently of the fitting procedure, and is $B_6^6/B_0^6 = -0.66$. Note that exactly the same result will be obtained with the ratio $+0.66$, which merely corresponds to a 30° rotation of the coordinate system used to describe the crystal field. There is no *correct* sign for $B_6^6$! The screen output continues

ANY CONSTRAINTS AMONGST THE PARAMETERS (Y/y FOR YES)? y

```
VALUE OF K AND Q FOR THE INDEPENDENT BKQ =? 6,0
VALUE OF K AND Q FOR THE DEPENDENT BKQ =? 6,6
RATIO OF THE DEPENDENT BKQ TO THE INDEPENDENT BKQ =? -0.66

THERE ARE 1 CONSTRAINTS
NO. OF INDEPENDENT FITTED PARAMETERS = 3
```

In order to begin the fitting procedure the program requires the user to estimate a set of starting parameter values. If these initial values are very far from the 'correct' values then the fitting process can lead to a false minimum of the sum of squared energy deviations. An advantage of using Wybourne normalization is that all the crystal field parameters are expected to have similar magnitudes, but the correct choice of signs may be important! For a first trial, values quite close to the expected result are used as inputs, based on a knowledge of the phenomenological parameters for other lanthanides in $LaCl_3$, shown in Table 2.6. The screen output continues

```
CHOOSE STARTING VALUES OF PARAMETERS BKQ
FOR K = 2 AND Q = 0
STARTING VALUE = ? 300
FOR K = 4 AND Q = 0
STARTING VALUE = ? -300
FOR K = 6 AND Q = 0
STARTING VALUE = ? -700

CONSTRUCTION OF THE ENERGY MATRIX BEGINS
PLEASE BE PATIENT.  IT MAY TAKE A WHILE .....
```

After constructing the energy matrix the program iterates through the fitting procedure up to 100 times. In each iteration the energy matrix is first diagonalized, and then a new set of parameters are calculated using the least-squares criterion. Parameter values are shown on the screen for each iteration, as indicated below.

```
FITTING, ITERATION 1
THE PROGRAM IS FINDING THE EIGENVALUES
K = 2  Q = 0  BKQ=  226
K = 4  Q = 0  BKQ= -296
K = 6  Q = 0  BKQ= -708
MEAN SQUARE DEVIATION = 7.2128
FITTING, ITERATION 2
THE PROGRAM IS FINDING THE EIGENVALUES
```

```
etc .......

FITTING, ITERATION 9
THE PROGRAM IS FINDING THE EIGENVALUES
K = 2   Q = 0   BKQ = 193
K = 4   Q = 0   BKQ =-310
K = 6   Q = 0   BKQ =-708
MEAN SQUARE DEVIATION = .00914

FITTING, ITERATION 10
THE PROGRAM IS FINDING THE EIGENVALUES
OUTPUT FILE NAME = BKQFIT.DAT
PROGRAM RUN IS COMPLETED SUCCESSFULLY
```

Comparison with the parameters that were used to generate the energy level file shows the error in the fitted parameters to be less than 2%. The accuracy of the final result is determined by the reduction of the mean square deviation of the differences between the fitted and experimental energy levels down to a certain limit, denoted CONLIM in the program. No further correction to the crystal field parameters occurs in the last iteration, where the mean square deviation is reduced below the value of CONLIM. This is assigned the value 0.005 in line 28 of the program, and this value will be used for all the examples given in this chapter. The user can increase or decrease CONLIM according to whether greater accuracy or quicker convergence is more important. It should be remembered, however, that the final accuracy of the parameters is limited by the accuracy of the input energy levels. Also, when there are fewer parameters than there are independent energy level differences, it will generally be impossible to obtain an exact fit. In the present example, the program runs to a full 100 iterations if CONLIM is set below 0.002, but no further improvement appears in the result after iteration 12. This is because, with the given constraint on the rank 6 parameters, the mean square deviation cannot be reduced below 0.002 195.

### 3.2.2 *Fitting crystal field parameters to the energy levels of* $Ho^{3+}$ *in a high-$T_c$ superconducting material*

Fitting a large number of parameters to a multiplet in which many of the energy level splittings are quite small presents a number of problems which will be explored in this section. In particular there is likely to be a large

Table 3.6. Fitted crystal field parameters (in $cm^{-1}$, Wybourne normalization) for $Ho^{3+}$ in $HoBa_2Cu_3O_{7-\delta}$. RMSD refers to the root mean square deviations in the energy levels.

|  | Original [FBU88b] | Set A start | Set A 100 its. | Set B start | Set B 9 its. |
|---|---|---|---|---|---|
| $B_0^2$ | 334 | 300 | 317 | 330 | 330 |
| $B_2^2$ | 59 | 50 | 60 | 60 | 59 |
| $B_0^4$ | −1763 | −1600 | −1687 | −1750 | −1761 |
| $B_2^4$ | 18 | 20 | 14 | 20 | 19 |
| $B_4^4$ | 973 | 1000 | 1026 | 970 | 974 |
| $B_0^6$ | 449 | 450 | 447 | 450 | 449 |
| $B_2^6$ | −26 | −20 | −21 | −25 | −25 |
| $B_4^6$ | 1200 | 1200 | 1204 | 1200 | 1200 |
| $B_6^6$ | −12 | −10 | −14 | −10 | −12 |
| RMSD |  |  | 0.77 |  | 0.005 |

number of closely spaced minima in the sum of squared energy deviations, not all of which will provide equivalent sets of crystal field parameters.

Table 3.6 compares the original set of crystal field parameters, obtained by Furrer *et al.* [FBU88b] for $Ho^{3+}$ in the orthorhombic (or $D_{2h}$) sites of $HoBa_2Cu_3O_{6.8}$, with those obtained by carrying out two fits to the energy levels (in file HO_ ENGY1.DAT) generated by using the Furrer *et al.* parameters and program ENGYLVL.BAS. The only difference in the two fits is the choice of starting parameters. Set A, shown in the third column of the table, does not produce convergence even after 100 iterations, as shown by the root mean square deviation (in column 4 of Table 3.6). Set B, on the other hand, converges well, producing parameter values close to the original set (with CONLIM ≤ 0.005) after only nine iterations.

The reason for the difficulty with the starting parameter values in Set A is that they do not generate the correct correspondences between the fitted and predicted energy levels. This can be demonstrated by generating the eigenvalues and eigenvectors of the starting Set A using program ENGYLVL.BAS. The resulting energy levels are given in column 4 of Table 3.7. The energy levels corresponding to the Set A starting parameters can be seen to interchange the order, and hence the labelling, of the 9th and 10th levels, and the 11th and 12th levels. The program ENGYFIT.BAS conserves the energy level order set up by the starting parameters, so that no amount

Table 3.7. Comparison of energy levels for two sets of $D_{2h}$ crystal field parameters.

| Order | Original | $\Gamma$ | set A |
|------:|---------:|:--------:|------:|
| 10 | 467.3 | 1 | 452.8 |
| 16 | 568.3 | 4 | 562.7 |
| 1  | 0.0   | 3 | 0.0   |
| 8  | 93.6  | 4 | 96.0  |
| 5  | 35.8  | 3 | 31.9  |
| 2  | 4.4   | 4 | 3.2   |
| 17 | 589.0 | 1 | 582.2 |
| 15 | 564.1 | 2 | 559.7 |
| 14 | 501.9 | 1 | 490.3 |
| 13 | 480.2 | 2 | 473.1 |
| 9  | 460.4 | 3 | 460.2 |
| 3  | 14.3  | 2 | 11.4  |
| 4  | 30.7  | 1 | 27.2  |
| 7  | 89.3  | 2 | 92.4  |
| 6  | 66.7  | 1 | 67.8  |
| 11 | 478.6 | 4 | 472.2 |
| 12 | 480.0 | 3 | 470.6 |

of iteration can produce the best fit when starting with the parameter values in Set A.

### 3.2.3 Choice of starting parameter values

The superconducting cuprate examples have made apparent the necessity of choosing appropriate starting parameter values. This problem is particularly difficult when the energy levels are closely spaced. While there is no foolproof way to determine starting values, information about the system, other than the energy levels to be fitted, may sometimes be used to advantage. Several techniques have been used in the literature.

(i) Use, or perhaps scale, the crystal field parameters obtained for another ion in the same series in an identical, or very similar, crystalline environment. The tables of parameter values for different ions in the same crystal, given in Chapter 2, suggest that this approach can be very effective.

(ii) Use the 'descent of symmetries' method, described below, when a sequence of approximate higher symmetries is known. When going from a higher to a lower symmetry, first estimates of the additional crystal field parameters can be obtained using second order perturbation theory.

(iii) Use the superposition model, described in Chapter 5, to estimate crystal field parameters from a detailed knowledge of the crystal structure. This approach generalizes the so-called 'point charge' ligand approximation which has frequently been used to make starting parameter estimates in the literature. As is demonstrated in Chapter 5, the superposition model can also provide a useful check that the final set of parameters is consistent with the crystal structure.

(iv) Combine the superposition model approach with the moment method (see Chapter 8), or the semiclassical model (see Chapter 9). These approaches have rarely been employed.

Here, we only comment on the descent of symmetries method. This method is only appropriate when the site structure is known to have an approximate higher symmetry. In such cases the energy levels that would be degenerate in the higher symmetry can be expected to be bunched together, or *clustered*, in the observed spectrum. A sequence of subgroups, starting with the group describing the approximate higher symmetry, and ending with the site symmetry group, is chosen. The advantage of this method is that fewer crystal field parameters are required in higher symmetry, reducing the dimensions of the fitting space. This also reduces the chances of converging on to a false minimum. For example, this method has been used to approximate the $S_4$ site symmetry of trivalent lanthanides in $LiYF_4$ as $D_{2d}$ [EBA$^+$79, GWB96, LCJ$^+$94]. For $LaF_3$, the site symmetry $C_2$ has also been approximated by $C_{2v}$ (see Chapter 8).

Program ENGYFIT.BAS is particularly suited to the descent of symmetry method because of the way it exploits the orthogonality of the tensor operators (see Appendix 1). This ensures that the gradual inclusion of additional parameters as the symmetry is lowered should make little difference to the values of the parameters that have already been determined.

The lanthanide sites are approximately cubic in the superconducting cuprates, suggesting that the sequence $O_h$, $D_{4h}$, $D_{2h}$ could be used. In general, however, it is not necessary that the parameter set used at each stage corresponds to a particular site symmetry. Any approach whereby one gradually increases the number of fitted parameters is feasible, so long as the largest parameters are included first.

### 3.2.4 Fitting incomplete sets of energy levels

The examples of crystal field fitting given previously do not deal with experimental sets of energy levels for which the labelling is incomplete and/or some of the levels have not have been determined. In principle, restricted sets of levels in a multiplet may be fitted if symmetry labels are known. The program ENGYFIT.BAS alone will not do this, however, because it assumes orthogonality properties of the operators that only hold for complete multiplets.

The trivalent holmium example dealt with above is a particularly difficult case because only 12 out of the 17 levels were observed by Furrer *et al.* [FBU88b] (see Table 3.5). This restricted information adds to the difficulty of obtaining good starting parameter values. However, if good starting values can be obtained, they can be used to fill in the missing experimental levels using program ENGYLVL.BAS. Programs ENGYFIT.BAS and ENGYLVL.BAS can then be used iteratively to calculate best fit parameters and fill in missing levels.

### 3.2.5 Fitting crystal field parameters to neutron scattering results

Inelastic neutron scattering experiments usually only provide information about energy levels in the lowest-lying multiplet of lanthanide and actinide ions. However, this is compensated to some extent by the information about line strengths that this experimental technique provides. This additional information places quantitative constraints on the form of the wavefunctions, and can therefore be used to supplement the energy levels in determining crystal field parameters. It also provides a useful check that the correct symmetry labels (see Apppendix 1) have been identified. While transition intensities provide a useful *post hoc* check on the wavefunctions determined using simple energy level fitting programs such as ENGYFIT.BAS, more reliable results can be obtained by using programs which use information on transition intensities in the fitting procedure.

Physicists who have access to the centralized facilities which are necessary to obtain neutron scattering results, will also have access to special program packages, such as the FOCUS package described in Section A3.3, which are necessary to analyse these results. It is therefore unnecessary to provide detailed descriptions of the use of such programs in this book.

# 4

# Lanthanide and actinide optical spectra

## G. K. LIU

*Argonne National Laboratory*

The basic theory used to interpret the electronic energy level structure observed experimentally in lanthanide ($4f^n$) and actinide ($5f^n$) ions in crystals has been considerably refined since the fundamental work of the 1950s [Jud63, Jud88, Wyb65a, Die68, Hüf78]. In addition to formal developments that fully explore the symmetry properties of the electronic structure of an $f$ element ion, theoretical modelling of the electronic structure utilizes Hartree–Fock methods to estimate the primary interactions within a free magnetic ion. For dealing with ion–ligand interactions, a 'crystal field' theory has been developed based on the one-electron approximation (see Chapter 1) together with the assumption that the crystal field interaction is weak in comparison with electronic interactions within the $f$ element ion. While the aspects of energy level splitting that are symmetry related are well understood, this is not true of the mechanisms that determine the magnitude of the splitting. Therefore, a semi-empirical approach has been employed which attempts to identify those effective interactions operating within the $f$ configurations that reproduce the observed energy level structure [Jud63, Die68] (see Chapter 2). Due to the systematic work of Carnall and coworkers since the 1970s [CBC+83, CGRR89, CLWR91, Car92, CC84, JC84] this phenomenological, or parametric modelling, approach has proven very powerful for analysing experimental spectra of both lanthanide and actinide ions.

Following the general pattern in this book, I attempt to summarize the practical procedures of parametric modelling and to provide general guidance for computational analysis of $f$ element crystal field spectra, specifically using the widely distributed computer program developed by Hannah Crosswhite. After a brief introduction to the effective operators of the free-ion and crystal field Hamiltonian in Section 4.1, matrix elements of the effective operator Hamiltonian that result from symmetry operations are summarized in Section 4.2. The practical procedures of establishing free-ion and crystal field parameters are discussed in Section 4.3. Finally, in Section 4.4, a comparison of spectroscopic properties between the $4f$ elements and the $5f$ elements is made.

## 4.1 The Hamiltonian

Historically, the development of a complete Hamiltonian for $f^n$ configurations was approached in two stages. The first dealt with the fundamental interactions in the gaseous free ion, and the second with the crystal field interactions that arise when the ion is in a condensed phase. Subsequently, additional effective operators dealing with higher order free-ion interactions were necessarily introduced to reproduce the energy level structures observed in experiments more accurately. The form of the free-ion Hamiltonian is assumed to be the same in both gaseous and condensed phases. The centres of gravity of groups of crystal field levels are interpreted on the same basis as the degenerate levels of the gaseous free ion. A commonly used effective operator Hamiltonian is

$$\mathbf{H} = \mathbf{H}_0 + \mathbf{H}_{ee} + \mathbf{H}_{so} + \sum_{i=1}^{4} \mathbf{H}_i(\text{corr}) + \mathbf{H}_{CF}, \qquad (4.1)$$

where the first three terms account for the primary interactions in a free ion. The first term represents the kinetic energy of the $f$ electrons and their coulomb interactions with the nucleus and electrons in filled shells. It contains only spin-independent spherically symmetric terms and does not remove any degeneracy within the $f^n$ configuration. Therefore, $\mathbf{H}_0$ is replaced by a constant in the effective operator Hamiltonian, and corresponds to the (arbitrary) mean energy of the spectrum. $E_0$ absorbs contributions from the spherical, or isotropic, part of the interactions between the $f$ electrons and their crystalline environment.

For ions with two or more $f$ electrons, the second term in (4.1) stands for electron–electron intra-shell coulomb interactions that split the $f^n$ configuration into $SL$ terms. The effective operator Hamiltonian is expressed as [Wyb65a]

$$\mathbf{H}_{ee} = \sum_{k=0,2,4,6} F^k(nf, nf) f_k, \qquad (4.2)$$

where $f_k$ are the angular parts of the interaction operators, and the $F^k$ are parameters corresponding to the Slater radial integrals. The $F^k$ are sometimes alternatively expressed as a set of four $F_k$ parameters (with different normalization), or as $E^k$ parameters (which are coefficients of Racah operators [Jud63]). They are related by

$$
\begin{aligned}
F^0 &= F_0, \\
F_0 &= (7E^0 + 9E^1)/7, \\
F^2 &= 225F_2, \\
F_2 &= (E^1 + 143E^2 + 11E^3)/42, \\
F^4 &= 1089F_4, \\
F_4 &= (E^1 - 130E^2 + 4E^3)/77, \\
F^6 &= (184,041/25)F_6, \\
F_6 &= (E^1 + 35E^2 - 7E^3)/462.
\end{aligned}
$$

The effective operator Hamiltonian for spin–orbit coupling is defined in terms of the magnetic dipole–dipole interactions between the spin and angular magnetic moments of $f$ electrons, i.e.

$$
\mathbf{H}_{\text{so}} = \zeta \sum_i \mathbf{s}_i \cdot \mathbf{l}_i, \tag{4.3}
$$

where $\mathbf{s}_i$ and $\mathbf{l}_i$ are the spin and angular momentum operators of the $f$ electron labelled $i$. The parameter $\zeta$ incorporates the physics of the spin–orbit interaction and is adjusted to fit the experimentally observed energies.

The electrostatic and spin–orbit interactions give the right order of magnitude for the energy level splittings of $f^n$ configurations. However, these primary terms do not accurately reproduce the experimental data. This is because the parameters $F^k$ and $\zeta$, which are associated with interactions within a $f^n$ configuration, cannot absorb all the effects of additional mechanisms such as relativistic effects and configuration interaction. Inclusion of new terms into the effective operator Hamiltonian is required to interpret the experimental data better. For example, Judd and Crosswhite [JC84] demonstrated that in fitting the experimental gaseous free-ion energy levels of $Pr^{2+}$ ($f^3$ configuration) the standard deviation could be reduced from 733 cm$^{-1}$ to 24 cm$^{-1}$ by adding nine more effective operators into the Hamiltonian.

As many as 15 corrective terms have been introduced which can be added to the effective free-ion Hamiltonian. Among these, a significant contribution to the $f^n$ energy level structure is from interactions between configurations of the same parity, which can be taken into account by a set of three two-electron operators recommended by Wybourne [Wyb65a], viz.

$$\mathbf{H}_1 = \alpha L(L+1) + \beta G(G_2) + \gamma G(R_7).\tag{4.4}$$

Here $\alpha$, $\beta$ and $\gamma$ are the so-called 'Trees' parameters, which are associated with the eigenvalues of Casimir operators for the Lie groups $R_3$, $G_2$ and $R_7$ [Jud63, LM86], viz. $G(R_3) = L(L+1)$, $G(G_2)$ and $G(R_7)$.

For $f^n$ configurations with $n \geq 3$, a three-body interaction term was introduced by Judd [Jud66, CCJ68], i.e.

$$\mathbf{H}_2 = \sum_i t_i T^i,\tag{4.5}$$

where $T^i$ are parameters associated with the three-particle operators $t_i$. This set of effective operators is needed in the Hamiltonian in order to represent the coupling of the ground configuration $f^n$ to excited configurations via the interelectronic coulomb interaction. It is usual to include only the six three-electron operators $t_i$ ($i = 2, 3, 4, 6, 7, 8$). When perturbation theory is carried beyond second order, eight additional three-electron operators $t_i$ (with $i = 11, 12, 14, 15, 16, 17, 18$ and $19$) are required [JL96]. A complete table of matrix elements of the 14 three-electron operators for the $f$ shell can be found in Hansen et al. [HJC96].

In addition to the magnetic spin–orbit interaction parametrized by $\zeta$, relativistic effects including spin–spin and spin–other–orbit interactions, both being represented by the Marvin parameters $M^0$, $M^2$ and $M^4$ [Mar47], are included in the third corrective term of the effective operator Hamiltonian in (4.1) [JCC68],

$$\mathbf{H}_3 = \sum_{i=0,2,4} m_i M^i.\tag{4.6}$$

As demonstrated by Judd et al. [JCC68, CBC$^+$83], the parametric fitting of $f$ element spectra can also be improved by introducing two-body effective operators to account for configuration interaction through electrostatically correlated magnetic interactions. This interaction can be characterized by introducing three more effective operators, giving the contribution

$$\mathbf{H}_4 = \sum_{i=2,4,6} p_i P^i,\tag{4.7}$$

where $p_i$ are operators and $P^i$ are additional parameters (see [JCC68]).

So far we have introduced 19 effective operators including those for two- and three-electron interactions. This set of effective operators is now used in most spectroscopic analyses of $f$ element ions in solids. Including the mean energy $E_0$, a total of 20 parameters associated with the free-ion operators are adjustable in non-linear least-squares fitting of experimental data.

Most crystal field analyses are carried out using a phenomenological one-electron crystal field theory in which the parameters are chosen appropriate to the given site symmetry. As discussed in Chapter 2 and Appendix 1, when a magnetic ion is in a crystal, the spherical symmetry is destroyed and each energy level splits under the influence of the crystalline environment. The degree to which the $(2J + 1)$-fold degeneracy of a free-ion $J$ multiplet is removed depends on the point symmetry at the ion. Based on the assumption that the crystal field can be treated as a perturbation and given that the unperturbed free-ion eigenfunctions have complete spherical symmetry, it is convenient to use Wybourne's formalism for the crystal field potential [Wyb65a] (see equation (2.7)).

## 4.2 Matrix element reduction and evaluation

In Chapter 3, crystal field fits to single multiplets were described, based on the presumption that either $LS$ coupling is a good approximation or that free-ion fits had been carried out to determine the reduced matrix elements in intermediate coupling. However, because of the advances in computational facilities, it is now feasible to diagonalize the complete free-ion and crystal field matrices simultaneously using the free-ion basis $|\nu LSJ\rangle$ and allowing for $J$-mixing. As described in the Introduction, these 'many-multiplet' fits follow the same general pattern as single-multiplet fits. In particular, before a diagonalization can be carried out, initial estimated values need to be assigned to the parameters and all matrix elements of the effective operator Hamiltonian must be evaluated.

Evaluation of matrix elements is apparently a tremendous effort particularly with the inclusion of effective two- and three-electron operators in the Hamiltonian. For a $f^n$ configuration with $3 < n < 11$, there are more than $10^4$ free-ion matrix elements and each of them may have as many as 20 terms to be evaluated on the basis of angular momentum operations [Jud63, Die68]. Fortunately, Hannah Crosswhite and coworkers have made this effort [CC84]. One may take advantage of the previous work by using the matrix elements evaluated and saved in a series of electronic files. These files, containing all values of matrix elements and other spectroscopic coefficients required for calculations of free-ion and crystal field energy level structure of the $4f^n$ and $5f^n$ configurations, are available from Argonne National Laboratory's web site: `http://chemistry.anl.gov` (downloadable from 'Internet access' in the left hand column of the home page).

### 4.2.1 Matrix elements of the free-ion Hamiltonian

The effective operators introduced in (4.2)–(4.7) have well-defined group theoretical properties [Jud63, Wyb65a]. Within the intermediate coupling scheme, all matrix elements can be reduced to new forms that are independent of the total angular momentum $J$, viz.

$$\langle \nu_1 SLJ | \mathbf{H}_i | \nu_2 S'L'J' \rangle = H_i \delta_{J,J'}\, c(SLS'L'J) \langle \nu_1 SL \| h_i \| \nu_2 S'L' \rangle, \qquad (4.8)$$

where $H_i$ is the parameter, $c(SLS'L'J)$ is a numerical coefficient, and $\langle \nu_1 SL \| h_i \| \nu_2 S'L' \rangle$ is the reduced matrix element independent of $J$. Indices $\nu_1$, $\nu_2$, etc. are used to distinguish between different Russell–Saunders states with the same quantum numbers $S$ and $L$. The tables of Nielson and Koster [NK63] can then be used to evaluate these reduced matrix elements. The matrix of the free-ion Hamiltonian is reduced to a maximum of 13 independent submatrices for $J = 0$, ..., 12 for even $n$ and $J = 1/2$, ..., 25/2 for odd $n$ of a $f^n$ configuration. Numbers of submatrices and their size can be determined from the values given in Table 4.1.

In the Crosswhite files named FnMP.dat for configurations of $f^n$ with $n = 2$–12, the free-ion matrix elements are given in terms of the 20 free-ion parameters. As an example, the coefficients of the parameters in the free-ion matrix for the $J = 1/2$ submatrix of the $f^3$ configuration are listed in Table 4.2. Column 5 of Table 4.2 lists the values of the product of $c(SLS'L'J)$ and $\langle \nu_1 SL \| h_i \| \nu_2 S'L' \rangle$, and column 6 lists the corresponding parameters. This is one of the simplest cases in which there are only two multiplets: $^4D_{1/2}$ and $^2P_{1/2}$. Except for $T^7$ and $T^8$, all the other 18 parameters appear in this $2 \times 2$ submatrix. Among the interaction terms that contribute to the energy level of these two multiplets, only $\zeta$, $M^0, M^2, M^4$ and $P^2, P^4$, $P^6$ mix the two $LS$ multiplets. As a result, $S$ and $L$ are no longer good quantum numbers. In addition, the $T^i$ terms induce a mixture of different multiplets with identical $S$ and $L$ but different $\nu$. Due to the $LS$–$L'S'$ mixing, the multiplet labelling $^{2S+1}L_J$ becomes nominal. Such a label only suggests that this multiplet may have a leading component from $^{2S+1}L_J$. Because spin–orbit coupling is stronger in the $5f$ shell, $LS$ mixing is more significant in $5f$ configurations than in $4f$ configurations. Diagonalization of each of the submatrices produces free-ion eigenfunctions of the form

$$\Psi(^{2S+1}L_J) = \sum_{\nu SL} a_{\nu SLJ} | \nu SL \rangle,$$

where the summation is over all $LS$ terms with the same $J$. As an example, we give the free-ion wavefunctions of the ground states $^4I_{9/2}$ for the $4f^3$ ion Nd$^{3+}$ and the $5f^3$ ion U$^{3+}$, which are generated using the free-ion

Table 4.1. Numbers of multiplets for each $J$-value in $f^n$ configurations.

| $J$ | $f^3, f^{11}$ | $f^5, f^9$ | $f^7$ |
|-----|-----|-----|-----|
| 25/2 |  |  | 1 |
| 23/2 |  | 1 | 3 |
| 21/2 |  | 3 | 5 |
| 19/2 |  | 5 | 11 |
| 17/2 | 1 | 9 | 18 |
| 15/2 | 3 | 16 | 26 |
| 13/2 | 3 | 20 | 35 |
| 11/2 | 5 | 26 | 42 |
| 9/2 | 7 | 29 | 46 |
| 7/2 | 7 | 30 | 50 |
| 5/2 | 7 | 28 | 42 |
| 3/2 | 6 | 21 | 31 |
| 1/2 | 2 | 10 | 17 |

| $J$ | $f^2, f^{12}$ | $f^4, f^{10}$ | $f^6, f^8$ |
|-----|-----|-----|-----|
| 12 |  |  | 2 |
| 11 |  |  | 2 |
| 10 |  | 2 | 8 |
| 9 |  | 2 | 11 |
| 8 |  | 7 | 20 |
| 7 |  | 7 | 24 |
| 6 | 2 | 13 | 38 |
| 5 | 1 | 14 | 37 |
| 4 | 3 | 19 | 46 |
| 3 | 1 | 13 | 37 |
| 2 | 3 | 17 | 37 |
| 1 | 1 | 7 | 19 |
| 0 | 2 | 6 | 14 |

parameters given in Tables 4.3 and 4.4:

$$\Psi(4f^3, {}^4I_{9/2}) = 0.984\,{}^4I_{9/2} - 0.174\,{}^2H_{9/2} - 0.017\,{}^2G_{9/2} + \text{etc.}$$

$$\Psi(5f^3, {}^4I_{9/2}) = 0.912\,{}^4I_{9/2} - 0.391\,{}^2H_{9/2} - 0.081\,{}^2G_{9/2}$$
$$+ 0.048\,{}^4G_{9/2} + 0.032\,{}^4F_{9/2} + \text{etc.}$$

In general, $LS$–$L'S'$ mixing becomes more significant in the excited multiplets of an $f^n$ configuration.

Table 4.2. Elements of the $J = 1/2$ submatrix of the free-ion Hamiltonian for $f^3$ configurations. In the first column, 1 stands for $J = 1/2$. In the second and third columns, 1 stands for $^4D$ and 2 for $^2P$. The coefficients of $T^2$ in column 5 are different from those of Hansen $et\ al.$ [HJC96] where an orthogonalized form of the operator $t_2$ was used.

| Submatrix | $\Psi$ | $\Psi'$ | Index | Coefficient | Parameter |
|---|---|---|---|---|---|
| 1 | 1 | 1 | 1 | 1.00 000 000 | $E_0$ |
| 1 | 2 | 2 | 1 | 1.00 000 000 | $E_0$ |
| 1 | 1 | 1 | 2 | 0.17 264 957 | $F^2$ |
| 1 | 2 | 2 | 2 | −0.04 957 265 | $F^2$ |
| 1 | 1 | 1 | 3 | 0.01 165 501 | $F^4$ |
| 1 | 2 | 2 | 3 | 0.00 155 400 | $F^4$ |
| 1 | 1 | 1 | 4 | −0.19 873 289 | $F^6$ |
| 1 | 2 | 2 | 4 | 0.07 321 738 | $F^6$ |
| 1 | 1 | 1 | 5 | −0.02 446 153 | $0.01\alpha$ |
| 1 | 2 | 2 | 5 | −0.02 846 153 | $0.01\alpha$ |
| 1 | 1 | 1 | 6 | −0.10 256 410 | $\beta$ |
| 1 | 2 | 2 | 6 | −0.26 923 077 | $\beta$ |
| 1 | 1 | 1 | 7 | −0.32 307 690 | $\gamma$ |
| 1 | 2 | 2 | 7 | 0.27 692 306 | $\gamma$ |
| 1 | 1 | 1 | 8 | 1.11 116 780 | $T^2$ |
| 1 | 2 | 2 | 8 | −0.25 253 800 | $T^2$ |
| 1 | 1 | 1 | 9 | 0.09 759 000 | $T^3$ |
| 1 | 2 | 2 | 9 | −0.58 554 000 | $T^3$ |
| 1 | 1 | 1 | 10 | −1.16 700 680 | $T^4$ |
| 1 | 2 | 2 | 11 | −1.52 362 350 | $T^6$ |
| 1 | 1 | 1 | 14 | −1.50 000 000 | $\zeta$ |
| 1 | 2 | 1 | 15 | 11.89 999 944 | $M^0$ |
| 1 | 1 | 1 | 16 | −18.09 999 640 | $M^2$ |
| 1 | 1 | 1 | 17 | −19.40 908 767 | $M^4$ |
| 1 | 1 | 1 | 18 | −0.12 777 776 | $P^2$ |
| 1 | 1 | 1 | 19 | −0.04 292 929 | $P^4$ |
| 1 | 1 | 1 | 20 | 0.03 399 380 | $P^6$ |
| 1 | 1 | 2 | 14 | −1.58 113 883 | $\zeta$ |
| 1 | 1 | 2 | 15 | −2.42 441 266 | $M^0$ |
| 1 | 1 | 2 | 16 | 2.31 900 337 | $M^2$ |
| 1 | 1 | 2 | 17 | −0.81 452 561 | $M^4$ |
| 1 | 1 | 2 | 18 | 0.06 441 676 | $P^2$ |
| 1 | 1 | 2 | 19 | 0.01 597 110 | $P^4$ |
| 1 | 1 | 2 | 20 | −0.03 583 260 | $P^6$ |
| 1 | 2 | 2 | 15 | 4.66 666 625 | $M^0$ |
| 1 | 2 | 2 | 16 | 3.00 000 097 | $M^2$ |
| 1 | 2 | 2 | 17 | −6.36 363 739 | $M^4$ |
| 1 | 2 | 2 | 18 | −0.09 629 627 | $P^2$ |
| 1 | 2 | 2 | 19 | 0.07 575 757 | $P^4$ |
| 1 | 2 | 2 | 20 | −0.04 532 506 | $P^6$ |

Table 4.3. Energy level parameters (in cm$^{-1}$) for trivalent lanthanide ions in LaF$_3$ [CGRR89]. Values in parentheses are errors in the indicated parameters. Values in square brackets were either not allowed to vary in the parameter fitting, or were fitted with constrained ratios. $M^0$ and $P^2$ were varied freely, with $M^2 = 0.56M^0$, $M^4 = 0.31M^0$, $P^4 = 0.5P^2$, $P^6 = 0.1P^2$. The standard deviation is denoted $\sigma$.

| | Pr$^{3+}$ | Nd$^{3+}$ | Tb$^{3+}$ | Dy$^{3+}$ | Ho$^{3+}$ | Er$^{3+}$ |
|---|---|---|---|---|---|---|
| $F^2$ | 68 878 | 73 018 | 88 995 | 91 903 | 94 564 | 97 483 |
| $F^4$ | 50 347 | 52 789 | [62 919] | 64 372 | 66 397 | 67 904 |
| $F^6$ | 32 901 | 35 757 | 47 252 | 49 386 | 52 022 | 54 010 |
| $\zeta$ | 751.7 | 885.3 | 1707 | 1913 | 2145 | 2376 |
| $\alpha$ | 16.23 | 21.34 | 18.40 | 18.02 | 17.15 | 17.79 |
| $\beta$ | −567 | −593 | −591 | −633 | −608 | −582 |
| $\gamma$ | 1371 | 1445 | [1650] | 1790 | [1800] | [1800] |
| $T^2$ | | 298 | [320] | 329 | [400] | [400] |
| $T^3$ | | 35 | [40] | 36 | 37 | 43 |
| $T^4$ | | 59 | [50] | 127 | 107 | 73 |
| $T^6$ | | −285 | −395 | −314 | −264 | −271 |
| $T^7$ | | 332 | 303 | 404 | 316 | 308 |
| $T^8$ | | 305 | 317 | 315 | 336 | 299 |
| $M^0$ | 2.08 | 2.11 | 2.39 | 3.39 | 2.54 | 3.86 |
| $P^2$ | −88.6 | 192(31) | 373(53) | 719(30) | 605(24) | 594(63) |
| $B_0^2$ | −218(16) | −256(16) | −231(24) | −244(18) | [−240] | −238(17) |
| $B_0^4$ | 738(40) | 496(73) | 604(49) | 506(43) | 560(27) | 453(90) |
| $B_0^6$ | 679(48) | 641(54) | 280(38) | 367(40) | 376(28) | 373(83) |
| $B_2^2$ | −120(13) | −48(12) | −99(16) | −65(12) | −107(10) | −91(14) |
| $B_2^4$ | 431(27) | 521(39) | 340(34) | 305(33) | 250(19) | 308(60) |
| $B_4^4$ | 616(27) | 563(41) | 452(31) | 523(25) | 466(19) | 417(56) |
| $B_2^6$ | −921(32) | −839(39) | −721(29) | −590(24) | −576(18) | −489(51) |
| $B_4^6$ | −348(41) | −408(35) | −204(29) | −236(27) | −227(20) | −240(51) |
| $B_6^6$ | −788(38) | −831(41) | −509(33) | −556(25) | −546(22) | −536(49) |
| $\sigma$ | 16 | 14 | 12 | 12 | 10 | 19 |

## 4.2.2 Matrix elements of the crystal field Hamiltonian

In the presence of crystal field interaction, which can be treated as a perturbation on the $2J+1$ degenerate free-ion states, each of the $^{2S+1}L_J$ multiplets splits into crystal field levels as described in Appendix 1. The selection rule of the $3j$ symbol in (3.2) indicates that crystal field operators induce mixing of states with different $J$ values, for which $M_J - M_{J'} = q$. Hence, the crystal field matrix is reduced into a number of independent submatrices each of

Table 4.4. Energy level parameters for trivalent actinide ions in LaCl$_3$ (in cm$^{-1}$) [Car92]. The values in parentheses are errors in the indicated parameters. The values in square brackets were either not allowed to vary in the parameter fitting, or were constrained to have a fixed ratio with another parameter. $P^2$ was varied freely, $P^4$ and $P^6$ were constrained by ratios $P^4 = 0.5P^2$, $P^6 = 0.1P^2$. The standard deviation is denoted $\sigma$.

|  | U$^{3+}$ | Np$^{3+}$ | Pu$^{3+}$ | Am$^{3+}$ | Cm$^{3+}$ | Cf$^{3+}$ |
|---|---|---|---|---|---|---|
| $F^2$ | 39 611 | 45 382 | 48 679 | [51 900] | [55 055] | [60 464] |
| $F^4$ | 32 960 | 37 242 | [39 333] | [41 600] | 43 938 | [48 026] |
| $F^6$ | 23 084 | 25 644 | 27 647 | [29 400] | 32 876 | [34 592] |
| $\zeta$ | 1626 | 1937 | 2242 | 2564 | 2889 | 3572 |
| $\alpha$ | 29.26 | 31.78 | 30.00 | 26.71 | 29.42 | 27.36 |
| $\beta$ | −824.6 | −728.0 | −678.3 | −426.6 | −362.9 | −587.5 |
| $\gamma$ | 1093 | 840.2 | 1022 | 977.9 | [500] | 753.5 |
| $T^2$ | 306 | [200] | 190 | 150 | [275] | 105 |
| $T^3$ | 42 | 45 | 54 | [45] | [45] | 48(11) |
| $T^4$ | 188 | 50 | [45] | [45] | [60] | 59(21) |
| $T^6$ | −242 | −361 | −368 | −487 | −289 | −529 |
| $T^7$ | 447 | 427 | 363 | 489 | 546 | 630 |
| $T^8$ | [300] | 340 | 322 | 228 | 528 | 270 |
| $M^0$ | [0.672] | [0.773] | [0.877] | [0.985] | [1.097] | [1.334] |
| $M^2$ | [0.372] | [0.428] | [0.486] | [0.546] | [0.608] | [0.738] |
| $M^4$ | [0.258] | [0.297] | [0.388] | [0.379] | [0.423] | [0.514] |
| $P^2$ | 1216 | 1009 | 949 | 613 | 1054 | 820(42) |
| $B_0^2$ | 287(32) | 164(26) | 197(22) | 242(34) | [280] | 306(29) |
| $B_0^4$ | −662(93) | −559(44) | −586(38) | −582(80) | [−884] | −1062(56) |
| $B_0^6$ | −1340(89) | −1673(49) | −1723(39) | −1887(83) | [−1293] | −1441(48) |
| $B_6^6$ | 1070(63) | 1033(34) | 1011(34) | 1122(49) | [990] | 941(36) |
| $\sigma$ | 29 | 22 | 18 | 21 | 23 | 19 |

which may be characterized by a *crystal quantum number* $\mu$, corresponding to an irreducible representation of the site symmetry group (see Appendix 1, and chapter 11 of Butler's book [But81]).

Inclusion of all $J$ multiplets produces extremely large matrices to be diagonalized, particularly for $f^n$ configurations with $4 \leq n \leq 10$. Diagonalization of the effective operator Hamiltonian on the entire $SLJM_J$ basis could be very time consuming and not actually necessary for analyses of experimental spectra. In practice, the crystal field energy level structure of an $f^n$ configuration is usually calculated for a selected energy region in which experimental data are available (typically for energies $< 50\,000$ cm$^{-1}$).

Theoretically, this is legitimate because crystal field coupling between two free-ion multiplets decreases as their energy difference increases. Given that the crystal field splitting of a free-ion multiplet is of the order of 100–1000 cm$^{-1}$, multiplets that are separated by $10^4$ cm$^{-1}$ should have no significant coupling. In the computer program developed by Hannah Crosswhite, an option is given to the user to truncate some of the free-ion states whose energy levels are far separated from the region of interest. This can readily be accomplished after diagonalization of the free-ion matrix to produce the free-ion energy level structure.

### 4.2.3 Structure of the Crosswhite computer program

Since the early 1980s, a FORTRAN program developed by Hannah Crosswhite at Argonne National Laboratory has been distributed internationally along with the supplementary data files containing matrix elements and other spectroscopic coefficients required as inputs to the main program. As mentioned above, the matrix elements themselves serve as a valuable database for various spectral analyses. The program itself, however, is not universally applicable without modification. The program was originally written in FORTRAN77. The FORTRAN codes downloadable from web site **http://chemistry.anl.gov** are revised for using with the VAX/VMS operating system, and can also be compiled and executed in Microsoft FOR-TRAN PowerStation. The generally useful information for using the Cross-white program is its main structure. Its structure facilitates inputs and outputs for independent calculations and non-linear least-squares fitting of free-ion and crystal field Hamiltonians.

As shown in Figure 4.1, the main program consists of two components, namely 'free-ion calculation' and 'least-squares fitting', that can be executed separately. The first part, including XTAL91.FOR and a set of supplementary subroutines in XTAL91SU.FOR, is designed to diagonalize the free-ion Hamiltonian and to construct the reduced crystal field matrix to be further diagonalized in the second part of the program. One must read the instructions given in the first page of XTAL91.FOR and know how to assign the Control Cards and the Input/Output channels.

Input I is a formatted data file that provides the values of the free-ion parameters and the control information, which includes the number of $f$ electrons, the number of parameters, crystal symmetry (expressed as the minimum value of $q$), the number of reduced crystal field matrices (or number of $\mu$ values), and selected $J$-values for truncation. One chooses the numbers of $J$ multiplets to be included in the crystal field matrices for each

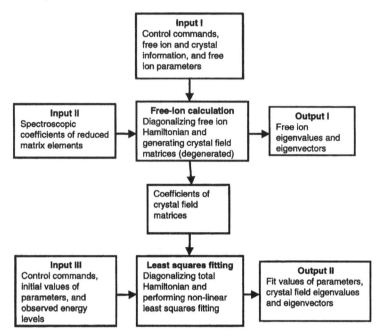

Fig. 4.1 Structure of the Crosswhite program for free-ion and crystal field calculation and non-linear least-squares fitting.

$J$ value from 0 to 12 for even $f^n$ configurations, and from $1/2$ to $25/2$ for odd $f^n$ configurations. Alternatively, one may construct two or three independent groups of matrices each covering a specified energy region. This is accomplished by skipping the lower energy multiplets that are included in the first truncated matrix, and choosing higher energy multiplets to form the second and third truncated matrices for each crystal field submatrix (labelled $\mu$).

For Input II, one needs to specify locations of two files of matrix elements, viz. coefficients FnMP.dat and reduced matrix elements of unit tensors FnNM.dat (where $n$ runs from 2 to 7). One of the two outputs from the execution of the free-ion program is used as an input to the second part: crystal field diagonalization and non-linear least-squares fitting. The other output contains free-ion eigenfunctions and eigenvalues.

The second part of the Crosswhite program also includes a main body, FIT91.FOR, and a subroutine file, FIT91SU.FOR. They are compiled and executed independently from the first part. The main function is to perform non-linear least-squares fitting of the experimentally observed crystal field energy levels and to generate the interaction parameters and crystal field eigenfunctions. However, it also gives the user an option for calculation only. An instruction is given in the first page of the FIT91.FOR codes, and the control commands are assigned through Input III. In addition to the control commands, Input III also contains initial values of free-ion and crystal field parameters and the experimentally observed transition energies that are accordingly assigned to the calculated energy levels. In Output II, the calculated energies, the observed energies, and the first two leading components of crystal field eigenfunctions are given, along with new and input values of parameters for each iteration of the least-squares fitting. By setting a control command in the input file Input III, the program can generate a separate output file for a complete set of crystal field eigenfunctions for each crystal field state.

## 4.3 Approaches to establishing physical parameter values

Because of the heavy mixing by spin–orbit interactions of $LS$-basis states for each $J$ value, the least-squares fitting method can converge to a false solution. A false solution can be recognized when there are supplementary data, such as Zeeman splitting factors or polarized spectra. However, this additional information alone does not provide the 'true' solution. The latter can only be found when initial parameter values, sufficiently close to the true solutions, are available for the least-squares fitting process.

Existing *ab initio* calculations of $f$ element spectra do not accurately reproduce the experimentally observed energy level structure of $f$ element ions in solids, and the calculated spectroscopic parameter values may be very different from the phenomenological ones that result from a fitting of the experimental data. Therefore, establishing realistic parameter values for the model Hamiltonian essentially relies on systematic analyses of the trends in parameter variation across the $f$ element series predicted by *ab initio* calculations.

### *4.3.1 Systematic trends in values of free-ion parameters*

A series of Hartree–Fock calculations has been carried out to predict free-ion parameter trends across lanthanide and actinide series [CC84, CBC+83]. The most important trends are those of the electrostatic interaction parameters $F^k$ and spin–orbit parameter $\zeta$. They both increase with the number of $f$ electrons, $n$. The Hartree-Fock values of the $F^k$s and $\zeta$ are always larger than those obtained by fitting experimental data [CC84].

Relativistic Hartree–Fock values of $\zeta$ agree remarkably well with empirical values, while the calculated $F^k$s are considerably larger than the empirical values. This is presumably because, in addition to relativistic effects, $f$-electron coupling with orbitals of higher-lying energies reduces the radial integrals assumed in the relativistic Hartree-Fock approximation. Moreover, experimental results are frequently obtained for an ion in a condensed phase, not for a gaseous free ion, which leads to an approximate 5% change [Cro77]. Because of the absence of mechanisms that absorb these effects in the relativistic Hartree-Fock model, calculated values of $F^k$s cannot be directly used as initial parameters for the least-squares fitting process. Figure 4.2 shows the comparison of relativistic Hartree-Fock values and experimental values for $F^2$ [CBC+83].

Although the relativistic Hartree-Fock values of $F^k$ are much larger than the experimental ones (see Figure 4.2), the differences between the relativistic Hartree-Fock and the experimental values of $F^k$ have been shown to be nearly constant. This is true for both lanthanides and actinides. With this characteristic, linear extrapolation of parameter values from one ion to another leads to values in good agreement with those obtained in the actual fitting process.

In addition to relativistic Hartree-Fock determinations of $F^k$ and $\zeta$, values of the $M^k$, $k$ =0, 2, 4, can also be computed [JCC68]. Values of these parameters do not vary dramatically across the $f$-series. In practice, experience has shown that they can be taken as given or varied as a single parameter by maintaining the relativistic Hartree-Fock ratios $M^2/M^0 = 0.56$ and $M^4/M^0 = 0.31$ [CGRR89]. For actinide ions, the ratio $M^4/M^0$ may be maintained at 0.38–0.4 (see Table 4.4).

For the remaining free-ion parameters, no direct Hartree-Fock values can be derived. Some values of $P^k$s for $Pr^{2+}$ and $Pr^{3+}$ have been calculated [CNT71]. In establishing systematic trends in the values of $P^k$ parameters for trivalent lanthanides substituted into $LaF_3$, Carnall et al. [CGRR88] constrained the $P^k$ parameters by the ratios $P^4 = 0.5P^2$ and $P^6 = 0.1P^2$, while $P^2$ was varied freely along with other parameters.

Once the systematic trends in the values of free-ion parameters are established, constraints can be imposed on other parameters that turn out to be relatively insensitive to the available experimental data. Some parameters such as $T^i$, $M^k$ and $P^k$ do not vary significantly across the series and, as a good approximation, can be fixed at the same values for neighbouring ions in the same series. In fact, most of the free-ion parameters are not host sensitive. The typical changes average about 1% between different lattice environments. Table 4.3 lists parameters for several trivalent lanthanides in $LaF_3$, and Table 4.4 for several trivalent actinides in $LaCl_3$. The free-ion parameters given in these two tables can be used as initial inputs for least-

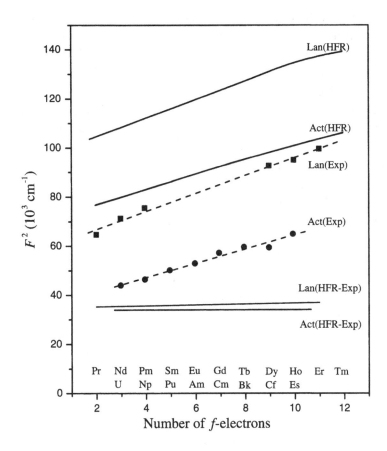

Fig. 4.2 Systematic trends in the values of free-ion parameters $F^2$ of trivalent lanthanide and actinide ions. The solid lines are relativistic Hartree-Fock calculations and the dashed lines are linear fit to the experimental data or differences between the relativistic Hartree-Fock values and experimental data.

squares fitting of energy level structure of a trivalent $f$ element ion in any crystalline lattice. If there are limited numbers of experimental data, one may only allow $F^k$ and $\zeta$ to vary freely along with the crystal field parameters, and keep other free-ion parameters fixed. For further improvement, $\alpha$, $\beta$, and $\gamma$ can be fitted. For final refinement, $M^0$ and $P^2$ may be varied freely, while keeping fixed ratios of $M^2$, $M^4$ to $M^0$, and fixed ratios of $P^4$, $P^6$ to $P^2$.

## 4.4 Empirical determination of crystal field parameters

For $f$ element ions in crystals providing a well-defined site symmetry, crystal field theory is widely used for predicting the number of energy levels and the selection rules for electronic transitions [Wyb65a, Hüf78]. Whereas the number of crystal field parameters can be determined from the site symmetry (see Appendix 1), their values are usually determined from the observed crystal field splitting. In general, supplementary spectroscopic information, such as polarized transitions allowed by electric or magnetic dipolar coupling, ensures the accuracy of the experimentally fitted crystal field parameters [LCJ+94]. In addition, temperature dependent spectra may be analysed to distinguish pure electronic lines from vibronic features. If multiple sites exist, site resolved spectra are required in order to distinguish between energy levels of ions at different sites [LCJ+94, LCJW94]. As discussed in Chapter 3, a correct assignment of observed energy levels to the calculated ones is crucial to avoid obtaining non-physical values for the fitted parameters. For spectra that lack other experimental information for unambiguous assignment, the procedure may involve several iterations of calculation and analyses that again requires understanding the basics of crystal field splitting of free-ion states [CGRR89, Car92]. Based upon the symmetry properties of the $3j$ symbols in equation (3.2), several criteria may be applied with the assumption that $JJ'$ mixing is small, which is valid for isolated multiplets.

(i) Splitting of $J = 1$ (or 3/2) states depends only on $B_q^2$ and that of $J = 2$ (or 5/2) depends also on $B_q^4$.

(ii) The $B_0^k$ parameters dominate splittings between the crystal field levels that have the same leading $M_J$ components.

(iii) The sign of crystal field parameters determines the ordering of crystal field levels.

As discussed in Chapter 3, one may simply use the values previously determined for different $f$ element ions in the same or similar host materials for setting the initial values in fitting crystal field parameters to observed energy levels. Comprehensive summaries of fitted crystal field parameter values have been given by Morrison and Leavitt [ML82] and Görller-Walrand and Binnemans [GWB96]. Values of crystal field parameters for lanthanides and actinides in a range of hosts can also be found in Chapter 2. Alternatively, the signs and magnitudes of crystal field parameters can be predicted using the superposition model (see Chapter 5).

## 4.5 Comparison between lanthanides and actinides

So far we have assumed that the approximations made for electrons in $4f^n$ configurations are also valid for electrons in $5f^n$ configurations. This is based on the results of spectroscopic analyses showing that, although significant differences exist between the two series, their spectra can be interpreted using the same theoretical framework. Historically, $f$ shell crystal field theory was primarily developed for lanthanides because early experimental advances in optical spectroscopy were made for trivalent lanthanides in crystals. Because the $5f$ electrons have more extended orbitals, they are expected to have smaller electrostatic interactions, stronger spin–orbit coupling, and stronger crystal field interactions [CC84, CLWR91, Kru87, LCJW94, LLZ$^+$98]. The outstanding question is whether the intermediate coupling scheme for the free-ion states and the perturbation approach to the ion–ligand interactions are still effective for actinides in condensed phases.

Experimental results are now available for comparing the spectroscopic properties of lanthanides and actinides. In Figure 4.3, the ratios of the experimentally determined parameter values are plotted as a function of the number of $f$ electrons for trivalent lanthanide ions and actinide ions in the same host material $LaCl_3$ [Cro77, Car92]. The ratio of spin–orbit coupling parameters $\zeta_{5f}/\zeta_{4f}$ is about 1.8, and the ratio of $F^2_{5f}/F^2_{4f}$ is about 0.6. These two ratios are almost constant across the series. The ratios of $F^4_{5f}/F^4_{4f}$ and $F^6_{5f}/F^6_{4f}$ have about the same values as $F^2_{5f}/F^2_{4f}$. Both the changes in spin–orbit coupling and electrostatic interaction increase the $LS$ mixing in the free-ion states of $5f^n$ configurations. This was discussed in Section 4.2.1 where free-ion wavefunctions for lanthanide and actinide ions in the same $LSJ$ multiplet were compared. The increase in $LS$–$L'S'$ mixing, however, has little effect on the validity of the intermediate coupling approximation so long as the complete sets of $LS$ terms are included in energy level calculations. In analyses of $5f$-electron energy level structure, one finds that labelling a free-ion state with $^{2S+1}L_J$ may no longer be meaningful because in most cases several $LS$ terms make similar contributions to a given $J$ multiplet.

The crystal field is sometimes characterized by a single parameter, known as the *crystal field strength* [AM83]. This is denoted $N_\nu$ or $S$ (see Section 8.2) and is defined as

$$N_\nu = \left[\sum_{k,q} \frac{(B^k_q)^2}{2k+1}\right]^{1/2}.$$

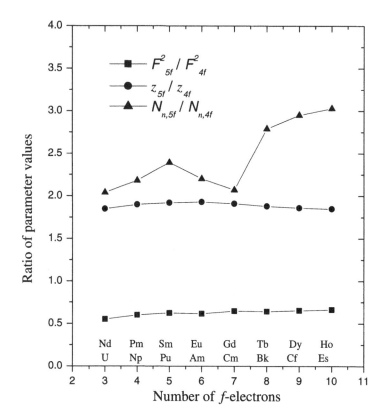

Fig. 4.3 A comparison of parameter ratios for trivalent lanthanide and actinide ions in LaCl₃.

As shown in Figure 4.3, for a given host crystal, the value of $N_\nu$ for an actinide ion is approximately twice its value for the lanthanide ion with the same number of $f$ electrons.

## Acknowledgement

This work was performed under the auspices of the Office of Basic Energy Sciences, Division of Chemical Sciences, US Department of Energy under Contract Number W-31-109-ENG-38. The author is grateful to Dr William T. Carnall for many valuable comments and for critically reading the manuscript.

# 5

# Superposition model

### D. J. NEWMAN
*University of Southampton*

### BETTY NG
*Environment Agency*

Calculations of crystal field parameters from first principles depend upon a knowledge of the crystal structure in the immediate neighbourhood of the magnetic ion. The calculated crystal field contributions for single ligands (see Chapter 1) can then be combined to estimate values of the crystal field parameters for real systems. Most calculations of this type [New71] have been based on the assumption that the combination process is purely additive, i.e. that the 'superposition principle' holds. At the present time, however, the superposition principle is mostly employed in the analysis of experimentally determined crystal field parameters. This phenomenological application is known as the 'superposition model'.

This chapter, taken together with some of the QBASIC programs described in Appendix 2, is designed to provide a comprehensive 'Do It Yourself' kit, supporting various applications of the superposition model in the calculation and interpretation of phenomenological crystal field parameters. The mathematical manipulations are relatively simple; some can even be carried out by hand using tables provided in this chapter.

Section 5.1 summarizes the physical assumptions that underlie the superposition model. It also gives the basic equations that describe the model and lists the various ways in which the model can be applied. A comprehensive survey of the empirical values of the *intrinsic*, or single-ligand, crystal field parameters is given in Section 5.2. Practical means of simplifying superposition model analyses, together with some examples of the use of the model to estimate empirical crystal field parameters from the intrinsic parameters, are described in Section 5.3. Section 5.4 describes how to determine the intrinsic parameters from phenomenological crystal field parameters. In Section 5.5, the use of the superposition model to study the effects of externally applied stress on crystal field parameters is described. Section 5.6 discusses the analysis and interpretation of the intrinsic parameters. Finally, in Section 5.7, the value and limitations of the superposition model are assessed.

## 5.1 Basic considerations

When geometrical information about the crystal structure at a magnetic ion is available, the superposition principle can be used to separate the two kinds of information that are contained in a set of phenomenological crystal field parameters. They are

(i) the site geometry; and

(ii) the physical interactions of the magnetic ion with its surrounding ligands.

The effectiveness of this separation depends on several general so-called 'superposition model' assumptions relating to the way in which the crystal field at a magnetic ion is built up of contributions from the other ions in the crystal. In addition to these general assumptions, it may be necessary to make assumptions which are specific to the system being analysed. For example, it is often necessary to assume that the site geometry at a substituted ion is the same as that in the pure crystal. All these assumptions must be borne in mind, as well as the accuracy of crystal field parameters and crystal structure data, when evaluating the results of superposition model analyses.

### *5.1.1 Physical assumptions*

The basic assumption of the superposition model is that the crystal field at a magnetic ion can be expressed as a simple *sum* of separate contributions from each of the other ions in the host crystal. Such an assumption would, of course, be correct if these were electrostatic contributions from point charges. However, as was demonstrated in the analysis of crystal field contributions in Section 1.5, the *superposition principle* remains valid for a far more realistic model of the crystal field, in which overlap and covalency dominate the contributions from neighbouring ions. In practice, the relative importance of these contributions makes it possible to make a second assumption, viz. that only contributions from the nearest neighbour ions, or ligands, need to be considered in practical applications of the superposition model.

These two assumptions can be expressed algebraically by writing the phenomenological crystal field $V_{\text{CF}}$ as

$$V_{\text{CF}} = \sum_L V_L, \qquad (5.1)$$

where each term $V_L$ represents the contribution from a neighbouring negative ion, or ligand, labelled $L$. Note that in the chemical literature the term 'ligand' is also used for charged complexes containing several atoms. As mentioned in Chapter 1, in this book 'ligand' will be used only to refer to single ions neighbouring a magnetic ion, and having significant overlap with the open-shell wavefunctions. These ions are also referred to as *coordinated* ions. In the case of ionic crystals, the coordinated ions generally form a well defined shell of (six to nine) negative ions surrounding the magnetic ion. Only in the case of very low-symmetry sites is there any practical difficulty in distinguishing coordinated from uncoordinated ions.

As demonstrated in Chapter 1, *all the major crystal field contributions from a single ligand are axially symmetric.* Using a coordinate system in which the $z$-axis points towards a particular ligand $L$, the contribution $V_L$ in equation (5.1) can be expressed as

$$V_L = \sum_k B_0^k C_0^{(k)}, \qquad (5.2)$$

i.e. in terms of $q = 0$ contributions alone. In order to distinguish the phenomenological parameters defined in Chapter 2 from the parameters used here to describe the axially symmetric crystal field contribution of a single ligand, the latter will be written as $\overline{B}_k(R_L)$, rather than $B_0^k$. In Stevens normalization, the single-ligand parameters will be written as $\overline{A}_k(R_L)$, instead of $A_{k0}\langle r^k \rangle$. In these expressions $R_L$ denotes the distance between the centres of the magnetic ion and its ligand $L$. It is useful to include $R_L$ explicitly as the single-ligand contributions depend on their distance from the magnetic ion (see Chapter 1). Using this notation, equation (5.2) becomes

$$V_L = V(R_L) = \sum_k \overline{B}_k(R_L) C_0^{(k)}. \qquad (5.3)$$

The single-ligand parameters $\overline{B}_k(R_L)$ and $\overline{A}_k(R_L)$ are usually referred to as *intrinsic* crystal field parameters, or just intrinsic parameters [New71, NN89b]. As discussed in Chapter 1, the major contributions to the intrinsic parameters from negatively charged ligands are expected to be *positive*, corresponding to a net repulsive interaction with the open-shell electrons.

Table 5.1. Spherical polar expressions for the coordination factors $g_{k,q}$ and $G_{k,q}$ defined in the text.

| $k$ | $q$ | $G_{k,q}/g_{k,q}$ | $G_{k,q}(\theta, \phi)$ |
|---|---|---|---|
| 2 | 0 | 1 | $(1/2)(3\cos^2\theta - 1)$ |
| 2 | 1 | $-2\sqrt{6}$ | $3\sin 2\theta \cos\phi$ |
| 2 | 2 | $\sqrt{6}$ | $(3/2)\sin^2\theta \cos 2\phi$ |
| 4 | 0 | 1 | $(1/8)(35\cos^4\theta - 30\cos^2\theta + 3)$ |
| 4 | 1 | $-4\sqrt{5}$ | $5(7\cos^3\theta - 3\cos\theta)\sin\theta\cos\phi$ |
| 4 | 2 | $2\sqrt{10}$ | $(5/2)(7\cos^2\theta - 1)\sin^2\theta\cos 2\phi$ |
| 4 | 3 | $-4\sqrt{35}$ | $35\cos\theta\sin^3\theta\cos 3\phi$ |
| 4 | 4 | $\sqrt{70}$ | $(35/8)\sin^4\theta\cos 4\phi$ |
| 6 | 0 | 1 | $(1/16)(231\cos^6\theta - 315\cos^4\theta + 105\cos^2\theta - 5)$ |
| 6 | 1 | $-2\sqrt{42}$ | $(21/4)(33\cos^5\theta - 30\cos^3\theta + 5\cos\theta)\sin\theta\cos\phi$ |
| 6 | 2 | $\sqrt{105}$ | $(105/32)(33\cos^4\theta - 18\cos^2\theta + 1)\sin^2\theta\cos 2\phi$ |
| 6 | 3 | $-2\sqrt{105}$ | $(105/8)(11\cos^3\theta - 3\cos\theta)\sin^3\theta\cos 3\phi$ |
| 6 | 4 | $3\sqrt{14}$ | $(63/16)(11\cos^2\theta - 1)\sin^4\theta\cos 4\phi$ |
| 6 | 5 | $-6\sqrt{77}$ | $(693/8)\cos\theta\sin^5\theta\cos 5\phi$ |
| 6 | 6 | $\sqrt{231}$ | $(231/32)\sin^6\theta\cos 6\phi$ |

### 5.1.2 Formulation

In combining the single-ligand contributions (5.3) to construct the phenomenological crystal field (expressed by equation (5.1)), it is necessary to use a common coordinate system for all the ligand contributions. The phenomenological crystal field normally has the $z$-axis coincident with the axis of highest rotational symmetry (as explained in Chapter 2 and Appendix 1). Hence, the $z$-axis of each of the local coordinate systems, in which the single-ligand contributions are axial symmetric, has to be rotated into alignment with this common $z$-axis. In cases where there are several equivalent axes of highest symmetry as, for example, in $D_2$ sites, it may not be known which of these axes is the axis implicit in a given set of fitted crystal field parameters. All possible alternatives may need to be considered.

Taking rotational factors into account, the Wybourne and Stevens crystal field parameters can be expressed as

$$B_q^k = \sum_L \overline{B}_k(R_L)g_{k,q}(\theta_L, \phi_L),$$

$$A_{kq}\langle r^k \rangle = \sum_L \overline{A}_k(R_L)G_{k,q}(\theta_L, \phi_L),$$

(5.4)

where $\theta_L$ and $\phi_L$ refer to the angular position of ligand $L$ with respect to the common coordinate system. The coefficients $g_{k,q}(\theta_L, \phi_L)$ and $G_{k,q}(\theta_L, \phi_L)$ are purely geometric, and will be referred to as *coordination factors*. Expressions for these factors are given in Table 5.1. Rudowicz [Rud87b] has also derived expressions for the $G_{k,q}(\theta_L, \phi_L)$ with $k = 1, 3, 5$. These are relevant in applications of the superposition model to transition intensities (see Chapter 10).

### 5.1.3 Distance dependence of intrinsic parameters

Well-defined values of intrinsic parameters have been determined from phenomenological crystal field parameters for a number of magnetic ions in various crystalline environments. The intrinsic parameters for *ionic crystals* are invariably positive, as mentioned previously. In many cases it has been possible to determine the empirical distance dependence of intrinsic parameters for a restricted range of ligand distances.

In most crystals, there are only a few distinct but similar distances between the magnetic ion and its ligands. Hence, reliable values of intrinsic parameters for a given magnetic ion and ligand can usually be obtained for only a small number of interionic distances $R_L$. This does not allow a functional form of $\overline{B}_k(R)$ to be determined empirically. Hence the functional forms of distance dependence that have been adopted in the literature are largely a matter of convention.

Because of the continuing interest in the point charge electrostatic model, in which the intrinsic parameters have specific power law dependences, a distance dependence of the form

$$\overline{B}_k(R) = \overline{B}_k(R_0)(R_0/R)^{t_k} \tag{5.5}$$

has been most widely assumed. This relates the distance dependence of each intrinsic parameter to a single power law *exponent* $t_k$. The expected reduction in magnitudes of intrinsic parameters with increasing ligand distance is then reflected by the inequality $t_k > 0$. Nevertheless, empirically determined values of $t_k$ are not generally in agreement with the electrostatic power law exponents $t_2 = 3, t_4 = 5, t_6 = 7$.

It is often the case that one can obtain intrinsic parameters for two ligand distances, $R_1$ and $R_2$, say. Taking the ratio of the equations obtained by substituting $R_1$ and $R_2$ for $R$ in (5.5), it becomes clear that the values of the power law exponents $t_k$ are independent of the choice of $R_0$. The most convenient equation to derive $t_k$ values can then be obtained by taking logarithms of (5.5):

$$t_k = \frac{\log \overline{B}_k(R_1) - \log \overline{B}_k(R_2)}{\log R_2 - \log R_1}. \tag{5.6}$$

An alternative description of the distance dependence, which we shall refer to as the 'two-power law' model, has been applied to rank 2 intrinsic parameters by Levin and coworkers [LC83a, LC83b, LE87]. This is of particular interest because of the special difficulties associated with the superposition model analysis of rank 2 crystal field parameters (discussed in Section 5.6), and the special role of rank 2 spin-Hamiltonian parameters (discussed in Chapter 7). Levin et al. used an argument, related to the theoretical analysis described in Chapter 1, to break the rank 2 intrinsic parameters into two parts (labelled $s$ and $p$), each with a specific distance dependence, viz.

$$\begin{aligned}
\overline{B}_2(R_L) &= \overline{B}_2^p(R_L) + \overline{B}_2^s(R_L) \\
&= \overline{B}_2^p(R_0)(R_0/R_L)^3 + \overline{B}_2^s(R_0)(R_0/R_L)^{10}. \tag{5.7}
\end{aligned}$$

Here $\overline{B}_2^p(R_L)$ is interpreted as the ligand point charge contribution and $\overline{B}_2^s(R_L)$ is the short range contribution, including mechanisms such as overlap and covalency.

Quite independent of such interpretations, however, the two-power law model can be treated simply as a phenomenological model. It can also be extended to other values of $k$. From a purely phenomenological point of view, using the two parameters $\overline{B}_k^p(R_L)$ and $\overline{B}_k^s(R_L)$ is no better or worse than using the two parameters $\overline{B}_k(R_L)$ and $t_k$. A possible disadvantage of the two-power law model is the arbitrary choice that has to made for the two exponents. In particular, equation (5.7) cannot be fitted to data for which the phenomenological power law exponent $t_k$ is greater than 10. Nevertheless, the linearity of equation (5.7) in the two phenomenological parameters can provide a distinct advantage in practical applications.

### 5.1.4 Types of application

Applications of the superposition model require a knowledge of the angles $\theta$ and $\phi$ for all the coordinated ions. These are usually derived from room temperature X-ray scattering results for the host crystal. However, they may not accurately represent the local environment of the magnetic ion in experiments used to obtain the crystal field. In particular, optical spectroscopic results are normally obtained at low (e.g. liquid helium) temperatures using dilute systems, in which the magnetic ions are substituted for a small proportion of positive non-magnetic ions in the host crystal. Both the bulk thermal changes in the crystal and the local distortion effects around substituted ions produce uncertainties in the coordination angles.

Crystal field parameters determined from neutron scattering are usually for 'concentrated' crystals (in which the magnetic ions occupy all the equivalent sites) at room temperatures. When good X-ray data are available, these allow the coordination angles to be determined with considerable reliability. However, neutron scattering determines only the low-lying energy levels, usually restricted to those within the ground $J$ multiplet. This means that the fitted crystal field parameters will generally include an indeterminate contribution from correlation effects (discussed in Chapter 6) in addition to the one-electron crystal field. Another problem in using concentrated crystals is that exchange interactions can modify the crystal field splittings. A clear example of this for layered high temperature superconductors has recently been investigated by Henggeler and Furrer [HF98].

When the coordination at the site of a magnetic ion is known, the superposition model may be used in several different ways. As remarked at the beginning of the chapter, the superposition principle was originally used in comparing experimental results with *ab initio* calculations. However, in accord with the aims of this book, this chapter focuses on the various ways in which the superposition model can be used to predict, analyse and interpret phenomenological parameters. These are as follows.

(i) The estimation of values of phenomenological crystal field parameters for a given host crystal from intrinsic parameters extrapolated from other hosts. Program INTRTOCF.BAS supports this application; an example is described in Section 5.3. Estimates produced in this way can provide starting values for least-square fits (see Chapter 3).

(ii) The analysis and interpretation of well-defined values of sets of phenomenological crystal field parameters. This application is supported by program CFTOINTR.BAS, described in Section 5.4.

(iii) Fitting energy levels directly to intrinsic parameters, allowing reliable crystal field parameters to be determined even when very few energy levels are known. Some applications of this type are discussed in Section 5.5 in relation to strained crystals.

(iv) Testing the validity of fitted crystal field parameters. This approach is used in the discussion of the multiplet dependence of crystal field parameters in Chapter 6.

(v) Estimation of the local distortions due to the substitution of magnetic ions. Most analyses of this type have been carried out for $S$-state ions (see Chapter 7).

## 5.2 Values of intrinsic parameters

In all applications of the superposition model, it is useful to have some idea of the magnitudes of the intrinsic parameters. The tables given in this section are intended to provide this information for a wide variety of magnetic ions and ligands. Details of the ways in which these parameters can be derived and used are given in later sections.

### 5.2.1 The spectrochemical series

Intrinsic parameters provide a means of ordering ligands, defining the so-called *spectrochemical series*. This series was first determined from the relative magnitudes of crystal field splittings in the spectra of divalent $3d$ transition metal ions, which are usually dominated by their (rank 4) cubic crystal field component $Dq$. Hence the spectrochemical series is directly related to the magnitude of the rank 4 intrinsic parameter $\overline{B}_4$ for $3d$ ions. The part of the resulting series relevant to magnetic ions in crystals is usually given as (e.g. see [GS73])

$$I^- < Br^- < S^{2-} < Cl^- < F^- < O^{2-}.$$

Gerloch and Slade [GS73] note the negative correlation between the spectrochemical series and ligand size. This is most readily explained by noting that smaller ligands produce greater local anisotropy. A similar ordering of ligands can be obtained from the lanthanide and actinide rank 4 intrinsic parameters although, as is clear from Table 5.2, this is complicated by considerable variations in the magnitudes of crystal field parameters for a given ligand. Given that, as was shown in Chapter 1, the rank 4 crystal field is dominated by the covalency and (closely related) overlap contributions, *the spectrochemical series is seen to provide a useful qualitative guide to the relative importance of covalency.*

The spectrochemical series also provides a rough and ready means for estimating magnitudes of intrinsic parameters. However, usually sufficient information is available to make the use of such a blunt tool unnecessary.

Table 5.2. Intrinsic parameters for $Er^{3+}$ with various ligands at $R(Å)$.
Bracketed values are uncertainties.

| Host | Ligand | $R(Å)$ | $\overline{B}_2(cm^{-1})$ | $\overline{B}_4(cm^{-1})$ | $\overline{B}_6(cm^{-1})$ | Ref. |
|------|--------|--------|---------------------------|---------------------------|---------------------------|------|
| $YVO_4$ | $O^{2-}$ | 2.24 | 725(25) | 441 | 378 | §5.4.2 |
| YGG | $O^{2-}$ | 2.34 | | 762 | 378 | §5.4.5 |
| YAG | $O^{2-}$ | 2.30 | | 885 | 403 | §5.4.5 |
| ZnS | $S^{2-}$ | 2.34 | 400(200) | 376(32) | 432(96) | [NN88] |
| $CaWO_4$ | $O^{2-}$ | 2.48 | 820(80) | 400(24) | 320(96) | [NN88] |
| $LaF_3$ | $F^-$ | 2.42 | 480(60) | 592(40) | 310(30) | [NN88] |
| $LaCl_3$ | $Cl^-$ | 2.42 | 438 | 264 | 162 | [NN88] |
| ErRh | Rh | | | 221 | −118 | [New83a] |
| ErCu | Cu | | | 152 | −94 | [New83a] |
| ErAg | Ag | | | 146 | −66 | [New83a] |
| ErZn | Zn | | | 64 | −112 | [New83a] |
| ErMg | Mg | | | 6 | −67 | [New83a] |
| $ErPd_3$ | Pd | | | 74 | −3.2 | [Div91] |
| $YPd_3$ | Pd | | | 100 | −3.2 | [Div91] |
| $ErNi_2$ | Ni | | | 296 | −365 | [Div91] |
| $ScAl_2$ | Al | | | 100 | −77 | [Div91] |
| $YAl_2$ | Al | | | 167 | −67 | [Div91] |
| $ErNi_5$ | Ni | | | 85(7) | | [Div91] |

### 5.2.2 Trivalent lanthanide and actinide ions

A representative sample of intrinsic parameters for trivalent erbium is collected in Table 5.2. As will be seen below, the signs and order of magnitudes of these intrinsic parameters are typical of trivalent lanthanides, although the parameters are somewhat larger for the lighter lanthanides. The values provided in Table 5.2 can therefore be used as a first approximation to the intrinsic parameters for all trivalent lanthanide ions, especially those with more than half-filled shells.

Table 5.2 can be used to illustrate some general properties of intrinsic parameters. Parameters of all ranks are positive for ionic crystals in which the ligands carry a negative charge. As was noted in Chapter 1, this can be interpreted as being due to a net repulsive interaction between the ligands and the open-shell electrons, which is dominated by overlap, covalency and electrostatic point charge contributions. The intrinsic parameters in non-ionic host crystals are, however, not necessarily positive. In particular, Table 5.2 shows a wide range of metallic ligands to have negative $\overline{B}_6$. From the discussion in Chapter 1, it seems possible that this is due to the

Table 5.3. Intrinsic parameters (in $cm^{-1}$) for lanthanides in $LnCu_2Si_2$
(derived from crystal field parameters of [GMO92, GO93, GOM94]).

| Ln | $\overline{B}_4(Cu)$ | $\overline{B}_6(Cu)$ | $\overline{B}_4(Si)$ | $\overline{B}_6(Si)$ |
|----|----|----|----|----|
| Ce | −27 | | 174 | |
| Pr | 30 | −0.6 | 84 | 43 |
| Nd | 92 | −148 | 32 | 136 |

dominance of charge penetration in the electrostatic contributions, combined with relatively small overlap and covalency contributions. Another source of negative contributions to the rank 6 parameters is $\pi$-bonding (see equation 5.29).

### 5.2.2.1 Silicon and copper ligands

Superposition model analyses for the compounds $LnCu_2Si_2$, where Ln denotes Ce, Pr or Nd, have been carried out by Goremychkin and coworkers [GMO92, GO93, GOM94]. The tetragonal lanthanide sites in these compounds have eight copper ligands and eight silicon ligands. The implied coordinate system used in the determination of the crystal field parameters is ambiguous, resulting in two possible choices of the coordinate system used in the superposition model calculation. Basing the choice on the expectation (see Table 5.2) that the copper rank 6 intrinsic parameters are negative, the copper and silicon intrinsic parameters shown in Table 5.3 are obtained.

### 5.2.2.2 Fluorine ligands

There are considerable experimental difficulties in substituting trivalent ions into the divalent metal ion sites in cubic fluorites without creating local charge compensation and consequent distortions from cubic symmetry. Nevertheless, several sets of cubic crystal field parameters have been determined. The superposition model can be very easily applied to these parameters (see Section 5.3.1). Table 5.4 summarizes the $F^-$ intrinsic parameters $\overline{A}_k$ ($k = 4, 6$) for several different lanthanide ions in cubic fluorite host crystals.

All the intrinsic parameters in Table 5.4 are positive, in keeping with the above discussion. The ratios $\rho = \overline{A}_6/\overline{A}_4$ lie in the very narrow range $0.28 \leq \rho \leq 0.32$ for all the trivalent ions and host crystals with the one exception

Table 5.4. Fluorine intrinsic parameters $\overline{A}_4$ and $\overline{A}_6$ (in $cm^{-1}$) obtained for cubic fluorite host crystals. Nominal ligand distances $R$ (in Å), corresponding to the ionic spacing in the host crystal, are also shown.

| Ion | Rank | $CaF_2$ | $SrF_2$ | $BaF_2$ | $CdF_2$ | $PbF_2$ | Source |
|-----|------|---------|---------|---------|---------|---------|--------|
| $Eu^{3+}$ | 4 | | 79.6 | | 90.6 | 79.9 | Table 2.7 |
| | 6 | | 23.9 | | 28.1 | 17.1 | |
| $Dy^{3+}$ | 4 | 87.8 | 81.5 | 76.5 | 90.3 | | Table 2.7 |
| | 6 | 25.8 | 23.0 | 20.7 | 26.6 | | |
| $Er^{3+}$ | 4 | 76.6 | 70.5 | 64.4 | 80.6 | 66.9 | Table 2.7 |
| | 6 | 22.9 | 19.9 | 17.8 | 24.3 | 19.0 | |
| $Gd^{3+}$ | 4 | 86.8 | 77.0 | 69.7 | | | [SN71b] |
| | 6 | 27.9 | 23.3 | 19.2 | | | |
| $Ho^{2+}$ | 4 | 83.0 | 70.2 | 59.1 | | | [SN71b] |
| | 6 | 18.7 | 15.5 | 13.3 | | | |
| $R =$ | | 2.366 | 2.512 | 2.685 | 2.333 | 2.567 | |

of ($Eu^{3+}$:$PbF_2$). This suggests that crystal field parameters determined for $Eu^{3+}$ in cubic sites in $PbF_2$ may be in error. Parameters for the divalent ion $Ho^{2+}$ have a smaller ratio $\rho$, which is almost independent of ligand distance.

Some determinations of intrinsic parameters for the very low-symmetry host crystal $LaF_3$ have also been made. For example, Yeung and Reid [YR89] obtain $\overline{A}_4(2.44\,\text{Å}) = 80\ cm^{-1}$, $\overline{A}_6(2.44\,\text{Å}) = 32\ cm^{-1}$ for $Pr^{3+}$:$LaF_3$, i.e. similar to the values for cubic hosts given in Table 5.4.

The stability of the parameter ratios $\rho$ is useful in estimating intrinsic parameters. The distance independence of parameter ratios shows that the distance dependences of the rank 4 and rank 6 intrinsic parameters must be very similar. The tabulated values show that both parameters decrease with increasing interionic distance. Nevertheless, it is difficult to determine the intrinsic parameters as *explicit* functions of ligand distance from data of this type, because the distortion of the crystal in the neighbourhood of the substituted magnetic ion is not known. This problem is discussed further in Section 5.5.

### 5.2.2.3 Ligands in group 5a of the periodic table

The cubic pnictides of praseodymium and neodymium provide values of rank 4 and rank 6 intrinsic parameters for some unusual ligands [New85]. These values, listed in Table 5.5, are invariably positive, but smaller than those for the same lanthanide ions with ionic ligands. The rank 4 parameters

Table 5.5. Intrinsic parameters $\overline{A}_k$ (in cm$^{-1}$) for praseodymium and
neodymium pnictides, from [New85].

| Ligand | $\overline{A}_4$(Pr) | $\overline{A}_4$(Nd) | $\overline{A}_6$(Pr) | $\overline{A}_6$(Nd) |
|--------|------|------|------|------|
| N  | 82   |      | 6.8  |     |
| P  | 30.6 | 25.2 | 3.8  | 7.1 |
| As | 27.3 | 23.5 | 3.7  | 6.1 |
| Sb | 18.4 | 16   | 1.8  | 4.0 |
| Bi | 15.3 | 14   | 2.5  | 5.5 |

are smaller for Nd$^{3+}$ than they are for Pr$^{3+}$. The opposite is the case for
the rank 6 parameters suggesting, as in the case of metals, the existence of
significant negative contributions.

### 5.2.2.4 Chlorine ligands

Much of the early optical spectroscopic work on lanthanide crystal fields
was carried out using anhydrous chloride host crystals. The relatively simple structure of the host crystal also made this system amenable to *ab-initio*
calculations of the type described in Chapter 1. As a result, a large amount
of data are available in regard to the intrinsic parameters for chlorine ligands. The following discussion is focussed on recent results.

Table 5.6 gives a selection of results for lanthanide and actinide ions
substituted into lanthanum chloride. The coordination angle $\theta$ is determined
from the ratio of the rank 6 crystal field parameters (see [NN89b]). The
results illustrate the general conclusion (see [Ede95]) that the rank 4 and
rank 6 intrinsic parameters for trivalent actinides are about twice those for
the corresponding lanthanides.

Table 5.7, which has been adapted from results collected by Reid and
Richardson [RR85], gives intrinsic parameters $\overline{A}_4$ and $\overline{A}_6$ for several trivalent
lanthanide ions with chlorine (Cl$^-$) ligands. These are derived for two very
different types of host crystal, so it is significant that reasonably consistent
results for the power law exponents are obtained. Note that $t_4 \geq t_6$, in
agreement with other phenomenological analyses (e.g. see the results for
the garnets obtained in Section 5.4.4).

Table 5.6. Trivalent lanthanide and actinide intrinsic parameters $\overline{B}_k$ (cm$^{-1}$) for chlorine ligands in anhydrous chloride host crystals. Results are based on optical spectra for ions substituted into LaCl$_3$. $\theta$ is the spherical polar coordinate angle of off-coplanar ligands. See [NN89b].

| Ion | $\theta$(deg) | $\overline{B}_2$ | $\overline{B}_4$ | $\overline{B}_6$ |
|-----|------|-----|-----|-----|
| Pr$^{3+}$ | 42.2 | 248 | 319 | 277 |
| Nd$^{3+}$ | 41.6 | 310 | 335 | 277 |
| Pm$^{3+}$ | 42.0 | 306 | 378 | 267 |
| Eu$^{3+}$ | 41.7 | 372 | 282 | 277 |
| Tb$^{3+}$ | 41.8 | 374 | 284 | 181 |
| Ho$^{3+}$ | 41.8 | 426 | 279 | 176 |
| Er$^{3+}$ | 41.8 | 438 | 264 | 162 |
| U$^{3+}$ | 42.6 | 688 | 480 | 605 |
| Np$^{3+}$ | 41.4 | 288 | 648 | 619 |
| Pu$^{3+}$ | 40.6 | 326 | 624 | 610 |
| Cm$^{3+}$ | 42.0 | 518 | 680 | 552 |

Table 5.7. Chlorine intrinsic parameters $\overline{A}_k$ (in cm$^{-1}$) for lanthanide ions. Set ($i$) are for cubic sites in Cs$_2$NaLnCl$_6$ and Cs$_2$NaYCl$_6$. Set ($ii$) are for C$_{3h}$ sites in anhydrous chlorides. Power law exponents ($t_4$ and $t_6$) are determined by taking into account the different ligand distances in the two hosts. Extracted from [RR85].

| Ion | $\overline{A}_4(i)$ | $\overline{A}_6(i)$ | $\overline{A}_4(ii)$ | $\overline{A}_6(ii)$ | $t_4$ | $t_6$ |
|-----|------|------|------|------|------|------|
| Pr$^{3+}$ | 82 | 22 | 40 | 17 | 16 | 6 |
| Nd$^{3+}$ | 70 | 24 | 40 | 18 | 13 | 7 |
| Sm$^{3+}$ | 58 | 23 | 32 | 17 | 14 | 8 |
| Tb$^{3+}$ | 58 | 14 | 34 | 11 | 12 | 5 |
| Ho$^{3+}$ | 60 | 15 | 33 | 11 | 13 | 6 |
| Er$^{3+}$ | 53 | 15 | 32 | 10 | 12 | 9 |

### 5.2.2.5 *Oxygen and sulphur ligands*

Table 5.8 collects some of the available rank 4 and rank 6 intrinsic parameters for trivalent europium with oxygen and sulphur ligands. For reasons explained later in this chapter, very few attempts have been made to deter-

Table 5.8. Rank 4 and rank 6 intrinsic parameters $(cm^{-1})$ for the $Eu^{3+}$ ion with oxygen or sulphur ligands at $R$ in various host crystals. In the first column YGG refers to yttrium gallium garnet.

| Host | Ion | $R(Å)$ | $\overline{B}_4$ | $\overline{B}_6$ | Ref. |
|------|-----|--------|--------|--------|------|
| $La_2O_2S$ | $O^{2-}$ | 2.42 | 706 | 302 | [MN73] |
| $La_2O_2S$ | $S^{2-}$ | 3.04 | 150 | 82 | [MN73] |
| $Gd_2O_2S$ | $O^{2-}$ | 2.27 | 713 | 315 | [NS71] |
| $Gd_2O_2S$ | $S^{2-}$ | 2.95 | 389 | 87 | [NS71] |
| $YVO_4$ | $O^{2-}$ | 2.24 | 400 | 432 | [NS71] |
| YGG | $O^{2-}$ | 2.34 | 475 | 493 | [NS69] |
| $LaAlO_3$ | $O^{2-}$ | 2.68 | 280 | 262 | [LL75] |

mine the rank 2 intrinsic parameters. Sulphur ligands have much smaller intrinsic parameters than those of oxygen reflecting their more diffuse electronic structure. The ratio of rank 4 and rank 6 parameters is also quite different.

Given that $O^{2-}$ has the same electronic structure as $F^-$ it is expected that the magnitudes of their intrinsic parameters will be very similar. While this is true for some crystalline hosts there is considerable variability in the oxygen ion intrinsic parameters, which is thought to reflect variations in the real ionicity of this nominally divalent ion.


### 5.2.3 Tetravalent actinides

The crystal fields for tetravalent actinides with oxygen and chlorine ligands in cubic and $D_{2d}$ sites have been analysed in detail by Newman and Ng [NN89a]. Table 5.9 provides a selection of the intrinsic parameters reported in that work. Bromine intrinsic parameters were found to be similar to those for chlorine and (very approximate) values of the power law exponents for both these ligands are $t_2 = 7$, $t_4 = 11$ and $t_6 = 8$.

More recently Carnall, Liu and coworkers [CLWR91, LCJW94] have determined the crystal field parameters for a wide range of actinide ions in tetrafluoride host crystals. Unfortunately, detailed superposition model calculations were not carried out, but the superposition model *prediction* of crystal field parameters in [CLWR91] shows that the rank 4 and rank 6 parameters quoted in Table 5.9 are in fair agreement with the empirical crystal field parameters for all the tetravalent actinides.

Table 5.9. Intrinsic parameters $\overline{A}_k$ for the tetravalent actinides $Pa^{4+}$, $U^{4+}$ and $Np^{4+}$ with various ligands [NN89a]. $R$ is the interionic distance.

| Ligand | $R$(Å) | $\overline{A}_2$(cm$^{-1}$) | $\overline{A}_4$(cm$^{-1}$) | $\overline{A}_6$(cm$^{-1}$) |
|---|---|---|---|---|
| chlorine | 2.7 | 1500(500) | 190(30) | 120(15) |
| oxygen | 2.45 | 1500(1000) | 190(30) | 160(30) |
| oxygen | 2.3 | 2500(500) | 320(40) | 250(50) |
| fluorine | — | 2200(500) | 300(30) | 150(30) |

### 5.2.4 Transition metals

Relatively few determinations have been made of the rank 2 and rank 4 intrinsic parameters for transition metal ions, although it is possible to derive values of the rank 4 intrinsic parameters from the many experimental determinations of the parameter $Dq$ quoted in the literature (e.g. see [GS73], p. 126). In the (most commonly occurring) sixfold cubic coordination

$$\overline{B}_4 = 6\,Dq = 8\overline{A}_4.$$

The corresponding expressions for other coordinations can be determined from the relationships given in Table 5.10. An additional complication in the case of transition metal ions is their tendency to form tightly bound "complexes" with ligands. The values of intrinsic parameters depend on the ionicity of these complexes and, for a given magnetic ion, can differ by as much as a factor of two. It is therefore not possible to provide typical values of the transition metal intrinsic parameters in the way we have done for lanthanides and actinides.

In most cases the cubic coordination at $3d$ transition metal ions is approximate. There are seldom sufficient experimental energy levels to determine all the additional phenomenological crystal field parameters produced by the broken symmetry. It is therefore appropriate to fit the intrinsic parameters directly to the energy levels. This approach was adopted in each of the three intrinsic parameter determinations quoted below.

Direct fitting of the intrinsic parameters to the energy level splittings of $Cr^{3+}$ ions in $LiNbO_3$ provided Chang *et al.* [CYYR93] with the estimates (in cm$^{-1}$), for Li sites,

$$\overline{A}_2 = 14\,184, \quad \overline{A}_4 = 1450,$$

and, for Nb sites,

$$\overline{A}_2 = 7063, \quad \overline{A}_4 = 989.$$

There is considerable uncertainty about the crystal structure in the neighbourhood of the $Cr^{3+}$ ion in these materials. Hence, as Chang *et al.* made clear, there is considerable uncertainty in the results quoted above. Also, consistent values of the power law exponents could not be determined.

A superposition model analysis of $Fe^{2+}$ in a naturally occurring garnet host has been carried out by Newman *et al.* [NPR78]. This work used the superposition model as an aid in the identification of observed energy levels. It determined $\overline{B}_4 = 8\overline{A}_4 = 4770$ cm$^{-1}$ and, assuming $t_2 = 3.5(5)$, $\overline{B}_2 = 2\overline{A}_2 = 12\,600(1\,600)$ cm$^{-1}$.

Wildner and Andrut [WA99] carried out a direct fit of energy levels of $Co^{2+}$ in $Li_2Co_3(SeO_3)_4$ to free-ion and superposition model parameters. Taking the reference metal–ligand (Co–O) distance as $R_0 = 2.1\,115$ Å, they obtain

$$\overline{B}_2(R_0) = 7000 \text{ cm}^{-1}$$
$$\overline{B}_4(R_0) = 4740 \text{ cm}^{-1}$$
$$t_2 = 5.5$$
$$t_4 = 3.1.$$

The value of the rank 4 intrinsic parameter obtained in their analysis is in good agreement with the values derived from the fitted $Dq$ parameters obtained by [Wil96] for the same magnetic ion and ligand. For example, $Dq = 826$ cm$^{-1}$ corresponds to $\overline{B}_4 = 4956$ cm$^{-1}$ for oxygen ligands at a mean distance of 2.09 Å.

## 5.3 Combined coordination factors

In most crystals the magnetic ions have more than one *set* of coordinated ions, or ligands, such that all the ligands in a given set are at the same distance. Each of these sets displays (at least) the full site symmetry. In applications of the superposition model, it is often convenient to sum the coordination factors for the ligands at a given distance, so as to form *combined* coordination factors for each set. Some examples of this procedure are given next.

Table 5.10. Combined $q = 0$ coordination factors for tetrahedral and cubic crystal fields in Stevens normalization. Values depend on the chosen direction of the $z$-axis.

| Coordination | $G^c_{4,0}$ | $G^c_{6,0}$ | $z$-axis |
|---|---|---|---|
| 12-fold | $-7/4$ | $-39/16$ | [100] |
| eightfold | $-28/9$ | $16/9$ | [100] |
| | $7/9$ | $-26/9$ | [110] |
| | $70/27$ | $256/81$ | [111] |
| sixfold | $7/2$ | $3/4$ | [100] |
| | $-7/8$ | $-39/32$ | [110] |
| | $-7/3$ | $4/3$ | [111] |
| fourfold | $-14/9$ | $8/9$ | [100] |
| | $28/7$ | $128/81$ | [111] |

### 5.3.1 Cubic coordination

In sites of cubic and tetrahedral symmetry, all the ligands are at the same distance. Hence a combined coordination factor can be determined for each rank, which takes account of *all* the ligands. Some combined coordination factors for $q = 0$ (in Stevens normalization) are given in Table 5.10.

In cubic symmetry there are other non-zero crystal field parameters for both ranks 4 and 6 (namely, those with $q \neq 0$), depending on the choice of $z$-axis. For example, evaluating the expressions in Table 5.1 shows that for 8-fold coordination, with the $z$-axis in the [100] direction (with 4-fold symmetry), there are two non-zero $q = 4$ parameters, given by

$$A_{44}\langle r^4 \rangle = -140\overline{A}_4/9 = 5A_{40}\langle r^4 \rangle$$

and

$$A_{64}\langle r^6 \rangle = -112\overline{A}_6/3 = -21A_{60}\langle r^6 \rangle.$$

In the above equations, the signs depend on choosing the $x$-axis to be aligned parallel to the edges of the cube. They change under a 45° rotation in the $x$–$y$ plane. The results in Table 5.10 are used to relate the intrinsic parameters to the $q = 0$ crystal field parameters.

If, on the other hand, the $z$-axis is chosen to be in a [111] direction (with threefold symmetry), the $q = 4$ parameters for a cubic eightfold coordinated system are zero, while the $q = 3$ and $q = 6$ parameters are non-zero. Clearly the crystal field itself must be invariant with respect to the choice of direction of the $z$-axis. This is most easily demonstrated by constructing the appropriate crystal field invariants (see Chapter 8).

The fluorites are eightfold coordinated systems, and the crystal field parameters listed in Table 2.7 correspond to the $z$-axis aligned in the [100] direction. Hence the fluorine $(F^-)$ intrinsic parameters can be obtained using the combined coordination factors for an eightfold system given in Table 5.10. The results of this calculation have already been included in Table 5.4.


### 5.3.2 Zircon structure crystals

Several of the examples discussed in this book relate to lanthanide ions in zircon structure crystals, e.g. vanadates, phosphates and arsenates. A selection of empirical crystal field parameters has been given in Table 2.8. A puzzling feature of these parameters is that although, as can be seen from Table 5.11, their site structures are very similar, the vanadates and phosphates have rank 2 crystal field parameters with different signs. It is interesting to see whether the superposition model throws any light on this.


#### 5.3.2.1 Calculation of combined coordination factors

The local geometry at the $D_{2d}$ trivalent ion sites in nine zircon structure crystals is shown, in terms of spherical polar coordinates, in Table 5.11. Each coordinate $(R_i, \theta_i)$ determines the position of four oxygen ligands at the corners of a tetrahedron. Taking $\phi_i = 0$ for each of the two oxygen sites specified in Table 5.11, the coordinates of the four oxygen ligands defining the vertices of each tetrahedron are $(\theta_i, 0)$, $(\theta_i, 180°)$, $(180° - \theta_i, 90°)$ and $(180° - \theta_i, 270°)$. Further details, including structural diagrams, can be found in papers by Löhmuller *et al.* [LSD+73], and by Newman and Urban [NU72].

Coordination angles for the eight nearest neighbour oxygen ions of a given trivalent ion in the zircon structure crystals can be found in files C_YVO4.DAT, etc. The format of these files (see Appendix 2) is appropriate for input into the programs CORFACS.BAS and CORFACW.BAS which calculate, respectively, the combined coordination factors in Stevens and Wybourne normalizations. Example coordination factor files generated using program CORFACW.BAS are called W_YVO4.DAT, etc. and those using program CORFACS.BAS are called S_YVO4.DAT, etc. These files are formatted appropriately to provide input files to the superposition model programs described below. Combined coordination factors listed in

Table 5.11. Ligand coordinates and combined coordination factors (Wybourne normalization) for nine zircon structure host crystals. The combined coordination factors correspond to oxygen tetrahedra, labelled $i = 1, 2$. Coordination factors are listed in files W_YVO4.DAT, etc.

| Compound | $R_i(\text{Å})$ | $\theta_i(\text{deg.})$ | $g_{20}^c$ | $g_{40}^c$ | $g_{44}^c$ | $g_{60}^c$ | $g_{64}^c$ |
|---|---|---|---|---|---|---|---|
| YVO$_4$ | 2.29 | 101.9 | $-1.745$ | 0.894 | 1.918 | $-0.272$ | $-0.685$ |
| | 2.43 | 32.8 | 2.239 | $-0.362$ | 0.180 | $-1.645$ | 0.818 |
| LuVO$_4$ | 2.24 | 101.8 | $-1.749$ | 0.903 | 1.920 | $-0.286$ | $-0.696$ |
| | 2.41 | 33.2 | 2.201 | $-0.423$ | 0.188 | $-1.654$ | 0.845 |
| ScVO$_4$ | 2.12 | 101.8 | $-1.749$ | 0.903 | 1.920 | $-0.286$ | $-0.696$ |
| | 2.37 | 33.8 | 2.143 | $-0.513$ | 0.200 | $-1.659$ | 0.886 |
| YPO$_4$ | 2.31 | 103.7 | $-1.663$ | 0.714 | 1.864 | $-0.015$ | $-0.479$ |
| | 2.37 | 30.2 | 2.482 | 0.060 | 0.134 | $-1.513$ | 0.648 |
| LuPO$_4$ | 2.26 | 103.5 | $-1.673$ | 0.735 | 1.870 | $-0.044$ | $-0.502$ |
| | 2.35 | 31.0 | 2.408 | $-0.074$ | 0.147 | $-1.570$ | 0.699 |
| ScPO$_4$ | 2.15 | 103.2 | $-1.687$ | 0.765 | 1.879 | $-0.087$ | $-0.538$ |
| | 2.28 | 31.6 | 2.353 | $-0.172$ | 0.158 | $-1.603$ | 0.738 |
| YAsO$_4$ | 2.30 | 102.2 | $-1.732$ | 0.865 | 1.909 | $-0.230$ | $-0.652$ |
| | 2.41 | 31.9 | 2.324 | $-0.220$ | 0.163 | $-1.617$ | 0.758 |
| LuAsO$_4$ | 2.25 | 102.0 | $-1.741$ | 0.884 | 1.915 | $-0.258$ | $-0.674$ |
| | 2.39 | 32.3 | 2.287 | $-0.284$ | 0.171 | $-1.632$ | 0.785 |
| ScAsO$_4$ | 2.13 | 101.7 | $-1.753$ | 0.913 | 1.923 | $-0.300$ | $-0.707$ |
| | 2.34 | 33.0 | 2.220 | $-0.393$ | 0.184 | $-1.651$ | 0.832 |

Table 5.11 were determined from the coordination angle files using the program CORFACW.BAS. For information about downloading program and data files, see Appendix 2.

Running program CORFACW.BAS produces screen outputs as follows:

```
THIS PROGRAM GENERATES COORDINATION
FACTORS FOR WYBOURNE PARAMETERS
FROM INPUT VALUES OF LIGAND DISTANCE LABELS AND
ANGULAR POSITIONS FOR UP TO 16 LIGANDS

NO. OF LIGANDS = ?8

DO YOU WANT TO INPUT LIGAND POSITIONS FROM A FILE?
TYPE IN Y/y (FOR YES), N/n (FOR NO)y

NAME OF INPUT FILE = ?
```

```
    12 CHARACTERS MAXIMUM (INCLUDING EXTENSION) c_yvo4.dat

NAME OF OUTPUT FILE = ?
    12 CHARACTERS MAXIMUM (INCLUDING EXTENSION) w_yvo4.dat

COORDINATION FACTORS FOR WYBOURNE PARAMETERS ARE
    OUTPUT TO FILE:                    w_yvo4.dat
PROGRAM RUN IS COMPLETED SUCCESSFULLY.
```

### 5.3.2.2 Algebraic expressions for combined coordination factors

Given that oxygen ligands in a tetrahedron are all at the same distance from the lanthanide ion (which is at its centre), and the $\phi$ angles are multiples of $90°$, it is possible to determine algebraic expressions for the combined coordination factors for each of the two oxygen tetrahedra as functions of the angle $\theta$ alone. The only non-vanishing combined coordination factors are for $q = 0$ and $q = 4$. In Stevens normalization, they may be expressed algebraically as

$$
\begin{aligned}
G_{20}^c(\theta) &= 2(3\cos^2\theta - 1) \\
G_{40}^c(\theta) &= (1/2)(35\cos^4\theta - 30\cos^2\theta + 3) \\
G_{44}^c(\theta) &= (35/2)\sin^4\theta \\
G_{60}^c(\theta) &= (1/4)(231\cos^6\theta - 315\cos^4\theta + 105\cos^2\theta - 5) \\
G_{64}^c(\theta) &= (63/4)(11\cos^2\theta - 1)\sin^4\theta.
\end{aligned}
\tag{5.8}
$$

It is of interest to note that, if the ligand coordinates are defined such that $\phi = 45°$ for the oxygen sites specified in Table 5.11, a factor of $\cos(4\phi) = -1$ will be introduced in the above expressions when $q = 4$. This shows that the choice of effective coordinate system is not uniquely determined by the site symmetry, resulting in there being two equivalent minima in the fitting space. Hence published parameters may have different signs (for all ranks) when $q = 4$, but the relative sign of these crystal field parameters for $k = 4$ and $k = 6$ should remain the same. This is reflected explicitly in the uncertain signs attached to these parameters by some authors (see Table 2.8). There is no 'correct' sign of the phenomenological parameters in such circumstances, when the orientation of the effective coordinate system is not uniquely determined by the site symmetry.

Given the above expressions for the combined coordination factors $G_{kq}^c(\theta)$, values of the five non-vanishing crystal field parameters can be determined by hand from known intrinsic parameters using the equations

$$A_{20}\langle r^2 \rangle = \overline{A}_2(R_1)G_{20}^c(\theta_1) + \overline{A}_2(R_2)G_{20}^c(\theta_2)$$
$$A_{40}\langle r^4 \rangle = \overline{A}_4(R_1)G_{40}^c(\theta_1) + \overline{A}_4(R_2)G_{40}^c(\theta_2)$$
$$A_{44}\langle r^4 \rangle = \overline{A}_4(R_1)G_{44}^c(\theta_1) + \overline{A}_4(R_2)G_{44}^c(\theta_2) \qquad (5.9)$$
$$A_{60}\langle r^6 \rangle = \overline{A}_6(R_1)G_{60}^c(\theta_1) + \overline{A}_6(R_2)G_{60}^c(\theta_2)$$
$$A_{64}\langle r^6 \rangle = \overline{A}_6(R_1)G_{64}^c(\theta_1) + \overline{A}_6(R_2)G_{64}^c(\theta_2).$$

Calculations of this type are particularly useful in obtaining a qualitative understanding of the signs and relative magnitudes of crystal field parameters.

### 5.3.2.3 Qualitative results

Several qualitative features of the crystal field parameters for trivalent lanthanide ions in zircon structure crystals can be understood from the combined coordination factors in Table 5.11. As remarked in Section 5.2.1, the intrinsic parameters are positive and both combined coordination factors for $B_0^6$ are negative for all structures. This is reflected by the negative empirical values of $B_0^6$ (see Table 2.8).

The positive empirical values of the $B_0^4$ parameters reflect the dominance of the nearest neighbour contribution with a positive intrinsic parameter $\overline{B}_4$. Both combined coordination factors for $k = 4$, $q = 4$ are positive for all lanthanides in all zircon structure host crystals. It can be seen from the values of $g_{44}^c$ in Table 5.11 that the nearest neighbour contributions to the parameter $B_4^4$ dominate. Hence the values of the intrinsic parameters can be bracketed by the two values obtained by: (i) assuming the nearest neighbour and next nearest neighbour intrinsic parameters are equal, or (ii) by assuming that the next nearest neighbour contributions can be neglected. For example, the crystal field parameters given in Table 2.8 and the coordination factors given in Table 5.11 for $Er^{3+}$:$YVO_4$ and $Er^{3+}$:$YPO_4$ allow us to determine upper and lower bounds of the nearest neighbour intrinsic parameters as 483 cm$^{-1}$ $\geq \overline{B}_4 \geq$ 441 cm$^{-1}$ and 406 cm$^{-1}$ $\geq \overline{B}_4 \geq$ 378 cm$^{-1}$, respectively.

For all crystal structures given in Table 5.11, $g_{20}^c(\theta_1)$ is smaller than $g_{20}^c(\theta_2)$, and has the opposite sign. Given that $R_1 < R_2$, the two contributions will tend to cancel. The phosphates have relatively greater positive values of $g_{20}^c(\theta_2)$ than the vanadates, which explains, at least qualitatively, the observed difference in the sign of $B_0^2$ for these two systems (see Table 2.8). As will be discussed below, this change of sign restricts the possible range of distance variation of intrinsic parameter $\overline{B}_2$.

## 5.4 Determination of intrinsic parameters from phenomenological crystal field parameters

It is often possible to determine intrinsic parameters from phenomenological crystal field parameters using (5.4). Such a calculation begins with the determination of combined coordination factors and setting up simultaneous equations for each rank $k$, like (5.9). If the number of unknown intrinsic parameters is equal to the number of crystal field parameters, as is the case for $k = 4$ and $k = 6$ in (5.9), the intrinsic parameters can be determined simply by solving the simultaneous equations. An example of this is given in the next section. Later sections deal with extensions of this procedure for cases in which the intrinsic parameters are either underdetermined or overdetermined for a single system.

The simultaneous equations can be written most conveniently in matrix form

$$\mathbf{y} = \mathbf{Gx} \tag{5.10}$$

where $\mathbf{y}$ is a column vector listing the given data, $\mathbf{G}$ is a matrix of known coefficients and $\mathbf{x}$ is the vector to be determined. In superposition model applications, $\mathbf{y}$ corresponds to the measured crystal field parameters for a given rank $k$, $\mathbf{G}$ is the matrix of combined coordination factors and $\mathbf{x}$ is a vector of intrinsic parameters for different ligand distances. When $\mathbf{G}$ is a square matrix, $\mathbf{x}$ can be determined from $\mathbf{y}$ by inverting $\mathbf{G}$. Formally

$$\mathbf{x} = \mathbf{G}^{-1}\mathbf{y}. \tag{5.11}$$

The success of this method depends on (5.10) being 'well conditioned'.

Numerical inversion of $\mathbf{G}$ can be carried out using standard programs, such as that provided in [PFTV86]. It is more convenient, however, to employ a high level program package such as Mathematica.

### 5.4.1 Calculation of rank 4 and rank 6 intrinsic parameters for the zircon structure crystals

Using the combined coordination factors given in (5.8) together with the values of $\theta$ in Table 5.11, equation (5.10) for the rank 4 and rank 6 parameters for lanthanide ions in $YVO_4$ takes the form,

$$\begin{bmatrix} A_{40}\langle r^4 \rangle \\ A_{44}\langle r^4 \rangle \end{bmatrix} = \begin{bmatrix} 0.89 & -0.37 \\ 16.0 & 1.5 \end{bmatrix} \begin{bmatrix} \overline{A}_4(R_1) \\ \overline{A}_4(R_2) \end{bmatrix} \tag{5.12}$$

and

$$\begin{bmatrix} A_{60}\langle r^6 \rangle \\ A_{64}\langle r^6 \rangle \end{bmatrix} = \begin{bmatrix} -0.27 & -1.65 \\ -7.69 & 9.21 \end{bmatrix} \begin{bmatrix} \bar{A}_6(R_1) \\ \bar{A}_6(R_2) \end{bmatrix}. \tag{5.13}$$

The corresponding equations for $YPO_4$ are

$$\begin{bmatrix} A_{40}\langle r^4 \rangle \\ A_{44}\langle r^4 \rangle \end{bmatrix} = \begin{bmatrix} 0.72 & 0.06 \\ 15.6 & 1.1 \end{bmatrix} \begin{bmatrix} \bar{A}_4(R_1) \\ \bar{A}_4(R_2) \end{bmatrix} \tag{5.14}$$

and

$$\begin{bmatrix} A_{60}\langle r^6 \rangle \\ A_{64}\langle r^6 \rangle \end{bmatrix} = \begin{bmatrix} -0.02 & -1.51 \\ -5.41 & 7.29 \end{bmatrix} \begin{bmatrix} \bar{A}_6(R_1) \\ \bar{A}_6(R_2) \end{bmatrix}. \tag{5.15}$$

The inverse matrices $\mathbf{G}^{-1}$ for $YVO_4$ are

$$\mathbf{G}_4^{-1}(YVO_4) = \begin{bmatrix} 0.21 & 0.05 \\ -2.21 & 0.12 \end{bmatrix} \tag{5.16}$$

and

$$\mathbf{G}_6^{-1}(YVO_4) = \begin{bmatrix} -0.61 & -0.11 \\ -0.51 & 0.02 \end{bmatrix}. \tag{5.17}$$

The inverse matrices for $YPO_4$ are

$$\mathbf{G}_4^{-1}(YPO_4) = \begin{bmatrix} -7.6 & 0.42 \\ 108.3 & -5.0 \end{bmatrix}, \tag{5.18}$$

and

$$\mathbf{G}_6^{-1}(YPO_4) = \begin{bmatrix} -0.88 & -0.18 \\ -0.65 & 0.00 \end{bmatrix}. \tag{5.19}$$

(Note that more significant figures are required in these matrices for accurate calculations.)

We now find ourselves in the common situation where the crystal field parameters (given in Table 2.8) are in Wybourne normalization, while the above numerical expressions are in Stevens normalization. The factors relating crystal field parameters in these two normalizations are given in Table 2.2. Using these conversion factors, the Stevens crystal field parameters for $Er^{3+}:YVO_4$ are

$$\begin{aligned} A_{20}\langle r^2 \rangle &= -103 \text{ cm}^{-1} \\ A_{40}\langle r^4 \rangle &= 45.5 \text{ cm}^{-1} \\ A_{44}\langle r^4 \rangle &= -968 \text{ cm}^{-1} \\ A_{60}\langle r^6 \rangle &= -43 \text{ cm}^{-1} \\ A_{64}\langle r^6 \rangle &= -23 \text{ cm}^{-1}. \end{aligned} \tag{5.20}$$

Similarly, the Stevens parameters for $Er^{3+}$:$YPO_4$ are

$$A_{20}\langle r^2 \rangle = \quad 140\,\text{cm}^{-1}$$
$$A_{40}\langle r^4 \rangle = \quad\ 19.4\,\text{cm}^{-1}$$
$$A_{44}\langle r^4 \rangle = -\,791\,\text{cm}^{-1} \qquad (5.21)$$
$$A_{60}\langle r^6 \rangle = \ -\,33.6\,\text{cm}^{-1}$$
$$A_{64}\langle r^6 \rangle = \ -\,99\,\text{cm}^{-1}.$$

Using the inverse matrices, and *changing the sign of the parameters with* $q = 4$, gives the following intrinsic parameters for $Er^{3+}$:$YVO_4$ (Stevens normalization)

$$\overline{A}_4(R_1) = 58.8\,\text{cm}^{-1}$$
$$\overline{A}_4(R_2) = 18.7\,\text{cm}^{-1}$$
$$\overline{A}_6(R_1) = 23.7\,\text{cm}^{-1} \qquad (5.22)$$
$$\overline{A}_6(R_2) = 22.3\,\text{cm}^{-1}$$

These parameters accord with general expectations, in that they are positive, and decrease with increasing ligand distance. Comparison with empirical values of intrinsic parameters (e.g. in Table 5.2) shows them to be of reasonable magnitude, and the value of $\overline{A}_4(R_1)$ is very close to the rough estimate made of $\overline{A}_4$ at the end of Section 5.3, which corresponds to $58 \pm 3$ cm$^{-1}$. However, one needs to be cautious about deducing the distance dependence of intrinsic parameters from a single set of phenomenological crystal field parameter values, because of the uncertainties in the crystal structure and possible errors in the observed parameters.

The effect of such uncertainties becomes very apparent when the inverse matrices given in equations (5.18) and (5.19) are used to evaluate the intrinsic parameters for $Er^{3+}$:$YPO_4$, giving

$$\overline{A}_4(R_1) = \quad\ 185\,\text{cm}^{-1}$$
$$\overline{A}_4(R_2) = -\,1854\,\text{cm}^{-1}$$
$$\overline{A}_6(R_1) = \quad\ 11.7\,\text{cm}^{-1} \qquad (5.23)$$
$$\overline{A}_6(R_2) = \quad\ 21.8\,\text{cm}^{-1}.$$

These results are very different to those obtained for the vanadate host and very far from expectations. This is a consequence of inverting poorly conditioned equations. In fact, the value $\overline{A}_4 = 51 \pm 2$ cm$^{-1}$, obtained at the end of Section 5.3, is far more accurate than the above results.

### 5.4.2 Calculation of rank 2 intrinsic parameters
### for the zircon structure crystals

For a given system, it is not possible to determine two intrinsic parameters (corresponding to two sets of oxygen ligands) from the single rank 2 crystal field parameter. However, if the further assumption, that the distance dependence of rank 2 intrinsic parameters is the same for all the vanadates, is made, it is possible to use crystal field parameters obtained for different host crystals to determine this distance dependence. An even stronger assumption would be that the same dependence is appropriate for all the zircon structure hosts.

The program CFTOINTR.BAS (see Appendix 2) can be used to carry out investigations of this type. For example, in order to determine a power law exponent and intrinsic parameter which is consistent with the rank 2 phenomenological parameters for $Er^{3+}$ substituted into $YVO_4$ and $YPO_4$, the program can be run twice as described below. In the case of $YVO_4$ the screen outputs and inputs read:

```
THIS PROGRAM CONVERTS A SET OF CRYSTAL FIELD
PARAMETERS TO A SET OF INTRINSIC PARAMETERS
IT ASSUMES THAT THE HIGHEST RANK IS 6

FILENAME OF CRYSTAL FIELD PARAMETERS = ? ER_YV1.WDT

IF THEY ARE WYBOURNE PARAMETERS, INPUT W/w? W

FILENAME OF COORDINATION FACTORS = ? W_YVO4.DAT

NO. OF DISTINCT DISTANCES = ? 2

OUTPUT FILE NAME OF INTRINSIC PARAMETERS = ? ER_011.DAT


FOR RANK        2
NO. OF INTRINSIC PARAMETERS > NO. OF CFPS
THERE ARE 2 DISTINCT LIGAND DISTANCES
PLEASE INPUT 1  INVERSE RATIOS OF THESE DISTANCES
RELATIVE TO THE NEAREST NEIGHBOUR DISTANCE
  I.E. R1/Ri WHERE R1 IS THE NEAREST NEIGHBOUR
  DISTANCE AND Ri IS THE iTH DISTANCE


R1/Ri = ? 1.061
```

```
HOW MANY tk VALUES (MAX.=6) DO YOU WANT TO TRY? 6
PLEASE PUT IN THE tk VALUES
tk = ? 3
tk = ? 6
tk = ? 7
tk = ? 7.1
tk = ? 7.3
tk = ? 8
INTRINSIC PARAMETERS ARE OUTPUT IN ORDER
OF INCREASING DISTANCE Ri


PROGRAM RUN IS COMPLETE
```

Table 5.12. Rank 2 intrinsic parameters calculated, as a function of $t_2$, for two zircon structure crystals. Values are for $\overline{B}_2$ in $cm^{-1}$.

|  | $R_1/R_2$ | $t_2 = 3$ | 6 | 7 | 7.1 | 7.3 | 8 |
|---|---|---|---|---|---|---|---|
| $YVO_4$ | 1.061 | −1590 | 1174 | 775 | 750 | 706 | 587 |
| $YPO_4$ | 1.026 | 446 | 609 | 689 | 698 | 717 | 790 |

Running this program twice, for $Er^{3+}:YVO_4$ and $Er^{3+}:YPO_4$, produces the results for rank 2 intrinsic parameters shown in Table 5.12. The most significant result is that the superposition model accounts for the difference in sign of the parameter $B_0^2$ in these two hosts, provided that the power law exponent is greater than 5. A more precise value of the power law exponent, $t_2 = 7.3$, can be estimated if it is assumed that the intrinsic parameters are the same in the two systems. However, the input data are not sufficiently robust for such precision. Analyses of other rank 2 parameters suggest that $t_2 = 7 \pm 1$ is more realistic. It is, nevertheless, possible to determine a well-defined value of the rank 2 intrinsic parameter for trivalent lanthanides in the zircon structure crystals, viz. $\overline{B}_2 = 725 \pm 25$ $cm^{-1}$.

### 5.4.3 Fitting intrinsic parameters

In fairly low symmetry sites, such as $D_{2h}$, there are more rank 4 and rank 6 crystal field parameters than the number of different intrinsic parameters (or different ligand distances). Hence, the vector **y** in (5.10) has more elements than the vector $\mathbf{x}$, and the matrix $\mathbf{G}$ is not square. It then becomes necessary to use a linear least-squares fitting procedure in order to determine the intrinsic parameters.

In a 'linear least-squares' fit, the weighted sum of squares of the deviations between the original and fitted crystal field parameters is minimized. It can be shown [MN87] that this is achieved by multiplying equation (5.10) on the left by $\mathbf{G}^T$ viz.

$$\mathbf{G}^T\mathbf{G}\mathbf{y} = \mathbf{G}^T\mathbf{x}, \tag{5.24}$$

which has the formal solution

$$\mathbf{y} = (\mathbf{G}^T\mathbf{G})^{-1}\mathbf{G}^T\mathbf{x}. \tag{5.25}$$

When the uncertainties (mean square deviations) of the crystal field parameters are known, they can be allowed for by including a 'weight' matrix $\mathbf{W}$ in the calculations. This has diagonal elements

$$W_{ii} = \frac{1}{\sigma_i^2}, \tag{5.26}$$

where $\sigma_i$ is the mean square deviation of the corresponding crystal field parameter. The fitted intrinsic parameters are then given by

$$\mathbf{y} = (\mathbf{G}^T\mathbf{W}\mathbf{G})^{-1}\mathbf{W}\mathbf{G}^T\mathbf{x}. \tag{5.27}$$

Fits to crystal field parameters seldom determine uncertainties in the individual crystal field parameters, so that $\mathbf{W}$ is normally chosen just to take account of parameter normalization. The Wybourne normalization is sufficiently uniform that $\mathbf{W}$ can be taken to be equal to the unit matrix. The Stevens normalization, on the other hand, is very irregular. In order to allow for this the $\sigma_i$ (used to determine the matrix $\mathbf{W}$) should be taken to be the ratios $G_{k,q}/g_{k,q}$ given in Table 5.1.

### 5.4.4 Determination of intrinsic parameters for lanthanide ions in yttrium gallium garnets (YGG) and yttrium aluminium garnets (YAG)

The site symmetry in garnet host crystals is $D_2$, but there is only a small distortion from cubic symmetry. The eight ligands are at the corners of the distorted cube and form two sets of four ligands. Precise determinations of the nine crystal field parameters have been made for several lanthanide ions in these crystals (see Table 2.9), and X-ray structural determinations have been made for several of the garnet hosts. As there are three rank 4 and

Table 5.13. Oxygen ligand positions in garnet host crystals. Positions of the other three ligands in each set are determined by symmetry operations. Angular coordinates are given in files C_YGG.DAT and C_YAG.DAT. Taken from [NS69], where primary references are given.

| Crystal | $R_1(\text{Å})$ | $\theta_1(\text{deg.})$ | $\phi_1(\text{deg.})$ | $R_2(\text{Å})$ | $\theta_2(\text{deg.})$ | $\phi_2(\text{deg.})$ |
|---------|-----------------|-------------------------|-----------------------|-----------------|-------------------------|-----------------------|
| YAG | 2.303 | 123.86 | −192.52 | 2.432 | 125.94 | 81.24 |
| YGG | 2.338 | 125.33 | −191.59 | 2.428 | 126.69 | 80.90 |

four rank 6 crystal field parameters, the garnets provide a good example to illustrate the use of linear least-squares fitting in the determination of intrinsic parameters.

The garnets, in common with other systems with orthorhombic site symmetry, have the complication that there are always six equivalent, but distinct, sets of phenomenological crystal field parameters, corresponding to the different possible choices of implicit coordinate system (e.g. see the discussion in [Rud91]). An explicit choice has to be made in applying the superposition model and it is necessary to ensure that the implicit and explicit choices coincide if any sense is to be made of the results. Once a correspondence has been found for one set of phenomenological crystal field parameters, it is usually apparent whether or not another set of crystal field parameters corresponds to the same choice of implicit coordinate system. Hence the superposition model provides a means of standardizing the choice of implicit coordinate system in relation to the local coordination. An alternative approach to standardization was taken in [Rud91]. This, and a means to transform the crystal field parameters between different implicit coordinate systems, are described in Appendix 4.

The positions of oxygen ligands in the yttrium gallium garnet (YGG) and yttrium aluminium garnet (YAG) are given in Table 5.13. Only angular coordinates are required in the determinations of intrinsic parameters and these are provided in files C_YGG.DAT and C_YAG.DAT, using a format appropriate for the determination of combined coordination factors using programs CORFACS.BAS (Stevens normalization) and CORFACW.BAS (Wybourne normalization). The combined coordination factor files obtained using these programs are called S_YGG.DAT, W_YGG.DAT, S_YAG.DAT and W_YAG.DAT. Details of these programs and files, and how they may be obtained, are given in Appendix 2.

Just as in the case of zircon structure crystals, program CFTOINTR.BAS can be used to obtain garnet intrinsic parameters from crystal field parameters and combined coordination factors. The only difference in this case is that program CFTOINTR.BAS uses a linear least-squares procedure for determining the rank 4 and rank 6 intrinsic parameters. For example, the intrinsic parameters for neodymium in yttrium aluminium garnet can be determined from the crystal field parameters by running CFTOINTR.BAS with the input files ND_YAG.DAT and S_YAG.DAT as shown below. Stevens normalization is used. Both program and data files can be downloaded from the web address given in Appendix 2.

Running CFTOINTR.BAS produces screen outputs as follows:

```
THIS PROGRAM CONVERTS A SET OF CRYSTAL FIELD
PARAMETERS TO A SET OF INTRINSIC PARAMETERS
IT ASSUMES THAT THE HIGHEST RANK IS 6

FILENAME OF CRYSTAL FIELD PARAMETERS = ? nd_yag.dat

IF THEY ARE WYBOURNE PARAMETERS, INPUT W/w?

FILENAME OF COORDINATION FACTORS = ? s_yag.dat

NO. OF DISTINCT DISTANCES = ? 2

OUTPUT FILE NAME OF INTRINSIC PARAMETERS = ? nd_o5.dat

INTRINSIC PARAMETERS ARE OUTPUT IN THE SAME
ORDER AS THE DISTANCES USED IN CONSTRUCTING
THE COORDINATION FACTORS

PROGRAM RUN IS COMPLETE
```

The intrinsic parameters are now in the file ND_O5.DAT. They should have the same order, and values, as the intrinsic parameters for $Nd^{3+}$:YAG shown in Table 5.14. The results are generally more reliable when there are sufficient parameters for a linear least-squares fitting procedure to be used.

Some of the results obtained in this analysis, and shown in Table 5.14 are unphysical. In particular, all the rank 2 parameters are either negative (as in the gallium garnets) or unreasonably large in magnitude (as in the

Table 5.14. Values of the intrinsic parameters $\overline{A}_k(\mathrm{cm}^{-1})$ for some trivalent lanthanide ions in garnet host crystals. Nearest neighbour ligands are labelled $i = 1$, and next nearest neighbour ligands are labelled $i = 2$. The format is the same as that in the data files generated by program CFTOINTR.BAS and used as input into program INTRTOCF.BAS.

| $k$ | $i$ | Nd:YAG | ErGG | Er:YGG | Er:YAG | Dy:YGG | Dy:YAG |
|---|---|---|---|---|---|---|---|
| 2 | 1 | 2901.7 | −350.8 | −75.5 | 2167.5 | −118.3 | 2474.7 |
| 2 | 2 | 2820.4 | −371.4 | −100.0 | 2096.3 | −157.1 | 2395.3 |
| 4 | 1 | 144.6 | 90.9 | 95.2 | 110.6 | 104.9 | 119.4 |
| 4 | 2 | 74.1 | 54.6 | 57.3 | 67.0 | 74.1 | 70.0 |
| 6 | 1 | 47.0 | 24.1 | 23.6 | 25.2 | 28.4 | 29.2 |
| 6 | 2 | 33.6 | 20.5 | 21.1 | 18.6 | 21.5 | 22.6 |

aluminium garnets). Such large discrepancies must be ascribed either to the use of inaccurate coordination factors or to the presence of sufficiently large long range contributions to invalidate the superposition model for $k = 2$. This question was investigated by Newman and Edgar [NE76], where it was shown that allowing for a small distortion in the neighbourhood of the substituted ion makes it possible to estimate rank 2 (nearest neighbour) intrinsic parameters as $\overline{A}_2 = 573$ cm$^{-1}$ for Nd$^{3+}$ and $\overline{A}_2 = 451$ cm$^{-1}$ for Er$^{3+}$, both in yttrium aluminium garnet.

All the rank 4 and rank 6 parameters are positive and satisfy the expected condition that the $i = 1$ parameters are larger than the $i = 2$ parameters. Power law exponents can be derived using equation (5.6). For example, in the case of Nd$^{3+}$:YAG, $t_4 = 12.2$ and $t_6 = 6.2$. These values exemplify a general empirical result, viz. $t_4 > t_6$, which is not in accord with the point charge electrostatic model.

The lanthanide ion sites in the superconducting cuprates are very similar to those in the garnets. In both cases the lanthanide ion is surrounded by eight oxygen ions, at two distances, situated at the corners of a slightly distorted cube. In the case of the superconducting cuprates, however, the site symmetry can be D$_{2h}$ or D$_{4h}$, as was discussed in Chapter 3. The lanthanide ions are also at the centre of a rectangular box, defined by eight equidistant copper ions at its corners. It is expected that the oxygen ions, as they are significantly closer to the lanthanide ion than are the copper ions, will provide the major contributions to the crystal field parameters. However, this has yet to be demonstrated by a superposition model analysis.

It is worth noting that the values of the oxygen intrinsic parameters in the superconducting cuprates are very similar to those in the garnets, suggesting that oxygen ligands have a similar electronic structure in both types of material.

## 5.5 Changes in crystal structure induced by stress

The superposition model analysis of the total crystal field into contributions of individual ligands, depending on their position in a coordinate system centred on the magnetic ion, provides a means of making predictions about the change in crystal field parameters when the crystalline lattice is distorted. Static distortions can be produced by externally applied stress and dynamic distortions are produced by lattice vibrations. The main difference, in practice, between these two types of distortion is that lattice vibrations couple to all possible distortion modes, while externally applied stresses only couple to a few modes. Examples of both types of application are given below to provide entry points into the literature.

### 5.5.1 Cubic sites distorted by lattice vibrations

Cubic sites provide the simplest applications of the superposition model to the coupling with lattice vibrations. In this case the number of modes of distortion is relatively small and, if only the ligands are involved, can be described (almost) uniquely in terms of the symmetry labels of the modes. A simplifying feature is that the amplitudes of distortions are small, so that the superposition model expressions for the 'dynamical' crystal field parameters can be obtained by differentiating the expressions for coordination factors of the static crystal field.

A formalism developed in this way has been presented in a series of papers to which reference should be made for further details. Basic results for cubic systems are given in [New80, CN81]. Analyses of experimental results for $Dy^{3+}$:$CaF_2$ and $Er^{3+}$:MgO can be found in [CN83] and [CN84b], respectively. A discussion of the relationship with the point charge model is given in [CN84a].

### 5.5.2 Effects of pressure on anhydrous chlorides

The sharp lines in lanthanide optical spectra show clear shifts of position when high pressures are applied to the crystal sample. These shifts can be shown to correspond to changes in both the free-ion and crystal field parameters. They can be used to provide tests of the superposition model if the pressure dependence of the crystal structure is known from X-ray determinations. A general formulation of the orbit–lattice interaction in these crystals has been given by Chen and Newman [CN82].

Gregorian, Holzapfel and coworkers [GdSH89, TGH93] studied changes of the crystal field of $Pr^{3+}$ and $Nd^{3+}$ in $LaCl_3$ when the crystal is subjected to pressures of up to 8 GPa. Corresponding changes in the crystal structure of the host crystal were obtained by X-ray diffraction, making it possible to estimate the corresponding changes in coordination factors. However, even with these X-ray results it is still not possible to provide precise determinations of the changes in coordination at the substituted $Pr^{3+}$ ions, because their ionic radii differ from those of the $La^{3+}$ ions in the host crystal.

In the experiments of Gregorian *et al.* [GdSH89] on $Pr^{3+}$:$LaCl_3$ under pressure, the rank 6 and rank 4 intrinsic parameters show a linear dependence on ligand distance, which can be expressed as

$$\overline{A}_k(R) = \overline{A}_k(R_0) + (R - R_0)d_k. \tag{5.28}$$

Allowing for local distortion effects, measured rank 6 crystal field parameters determine $d_6 = -47(9)\text{cm}^{-1}\text{Å}^{-1}$ and $\overline{A}_6(R_0) = 16(2)\text{cm}^{-1}$, where $R_0 = 2.95$ Å. This distance dependence can be shown [GdSH89] to correspond to a power law exponent $t_6 = 7.5(1.5)$. Because there is only one rank 4 crystal field parameter for sites with $C_{3h}$ symmetry, the distance dependence of rank 4 intrinsic parameters is not so well determined. Without correcting for local distortion, Gregorian *et al.* obtain $\overline{A}_4(R_0) = 38(2)$ cm$^{-1}$ and $t_4 = 7(2)$, corresponding to a value of $d_4 = -110(10)$ cm$^{-1}\text{Å}^{-1}$ in (5.28). A more recent determination by Tröster *et al.* [TGH93] of the intrinsic parameters and power laws for $Pr^{3+}$:$LaCl_3$ and $Nd^{3+}$:$LaCl_3$ gives, for both of these systems, $\overline{A}_4(R_0) = 30(4)$ cm$^{-1}$, $t_4 = 6(2)$, $\overline{A}_6(R_0) = 18(2)$ cm$^{-1}$ and $t_6 = 5.5(2)$.

Gregorian *et al.* [GdSH89] found a minimum in $A_{20}\langle r^2 \rangle$ of $Pr^{3+}$:$LaCl_3$ as the pressure is varied. This minimum is difficult to interpret in terms of the superposition model. They suggest that the breakdown of the superposition model in this case is due to the long-range nature of rank 2 contributions. An alternative explanation is suggested by comparison with the rank 2 spin-Hamiltonian parameters discussed in Chapter 7, where a minimum is explained in terms of competing correlation crystal field contributions (see also Chapter 6).

## 5.6 Analysis and interpretation of intrinsic parameters

Superposition model analyses of well-determined phenomenological crystal field parameters have several roles. In the first place, they provide useful tests of the accuracy of the superposition model itself. Taken together with other experimental evidence they allow the validity of neglecting long-range contributions to be assessed. Furthermore, the derived values of intrinsic parameters can be analysed to determine their distance and ligand dependence. They provide useful input data for testing the validity of other models, such as the angular overlap model (see Section 5.6.2).

In carrying out further analyses of intrinsic parameters it is often useful to relate them to the energies of single open-shell electrons $e_m$ (where $m$ is the magnetic quantum number). The required equations, for both $f$ and $d$ electrons, have already been given in Section 1.7.1. Expressions for the intrinsic parameters are obtained by replacing $\hat{B}_k$ with $\overline{B}_k$ throughout.

### 5.6.1 Separation of electrostatic contributions to the rank 2 crystal field parameters

In the case of $f$ electrons, the electrostatic contribution to the rank 4 and rank 6 (i.e. $k = 4, 6$) crystal field parameters in ionic crystals is small (see Chapter 1). In the approximation that these contributions can be entirely neglected, equation (1.29) can be used to estimate the electrostatic contribution to rank 2 crystal field parameters. The method is based on noting that covalency, overlap and exchange contributions to $e_2$ and $e_3$ vanish, as was pointed out in Chapter 1. Hence, neglecting the electrostatic contributions corresponds to writing $e_2 = e_3 = 0$ in (1.29).

With this simplification, and replacing capped by intrinsic parameters, (1.29) can be written

$$
\begin{aligned}
\overline{B}_0^c &= (1/7)(e_0 + 2e_1), \\
\overline{B}_2^c &= (5/14)(2e_0 + 3e_1), \\
\overline{B}_4 &= (3/7)(3e_0 + e_1), \\
\overline{B}_6 &= (13/70)(10e_0 - 15e_1).
\end{aligned}
\tag{5.29}
$$

Here the 'c' (for 'contact') superscripts indicate that the omitted electrostatic contributions are expected to be significant for the rank 0 and rank 2 parameters.

Eliminating $e_0$ and $e_1$ from the expressions for the intrinsic parameters given in (5.29) produces the following expression for the combined covalency, overlap and exchange contributions to the rank 2 intrinsic parameters:

$$\overline{B}_2^c = \frac{5}{11}\left(2\overline{B}_4 - \frac{7}{13}\overline{B}_6\right)$$
$$\overline{A}_2^c = \frac{40}{11}\left(\overline{A}_4 - \frac{7}{13}\overline{A}_6\right).$$

(5.30)

As discussed in Chapter 1, the net contributions from covalency, overlap and exchange to all intrinsic parameters are positive. Table 5.6 shows that, at least in anhydrous chlorides, both $\overline{B}_4$ and $\overline{B}_6$ are positive and have a similar order of magnitude. It follows, therefore, that $\overline{B}_2^c$ is expected to be rather less than three-quarters of $\overline{B}_4$. In practice most of the $\overline{B}_2$ values shown in Table 5.6 are greater than this. Hence the rank 2 crystal field parameters do contain significant electrostatic contributions. Such contributions will, in general, be of longer range than the contributions from covalency, overlap and exchange and hence will not satisfy the superposition model assumption that only ligands contribute to the crystal field.

Given reliable values of the rank 4 and rank 6 intrinsic parameters, the electrostatic contributions to phenomenological crystal field parameters can be estimated using

$$B_q^2(\text{electrostatic}) = B_q^2(\text{observed}) - \sum_L \overline{B}_2^c(R_L)g_{2q}(\theta_L, \phi_L),$$

(5.31)

and a similar equation for Stevens parameters. In (5.31) the intrinsic parameter $\overline{B}_2^c$ is calculated from (5.30) using values of $\overline{B}_4$ and $\overline{B}_6$ determined from the phenomenological crystal field parameters.

### 5.6.2 Angular overlap model

Chemists (e.g. Urland [Url78]) often express the analysis of the previous subsection in terms of $\sigma$ and $\pi$ bonding, where the respective bonding energies are identified as

$$e_\sigma = e_0, \ e_\pi = e_1.$$

Eliminating $\overline{B}_2^c$ and $\overline{B}_0^c$ from (5.29) produces, for $f$ electrons,

$$e_\sigma = (7/143)(13\overline{B}_4 + 2\overline{B}_6),$$
$$e_\pi = (14/429)(13\overline{B}_4 - 9\overline{B}_6).$$

(5.32)

This provides an alternative approach to the superposition model which only involves two parameters and entirely neglects the electrostatic contributions.

A particular version of this approach is referred to as the *angular overlap model*, because the distance dependence of the parameters $e_\sigma$ and $e_\pi$ is related to that of calculated overlap integrals between the open-shell states of the magnetic ion, and outer closed shell $s$ and $p$ orbitals on the ligands. This leads to the equations [NSC70]

$$e_\sigma = a_0 \langle f_0|p_\sigma \rangle^2 + a_s \langle f_0|s \rangle^2,$$
$$e_\pi = a_1 \langle f_1|p_\pi \rangle^2, \qquad (5.33)$$

where the suffices $\sigma$ and $\pi$ distinguish the orientations of the ligand outer shell $p$ orbitals. $a_0$, $a_s$ and $a_1$ are numerical coefficients.

The angular overlap model provides a more realistic method of treating ligand distance-dependence effects than the point charge electrostatic model (e.g. see [NSC70]). In practice, however, additional theoretical assumptions have to be made in order to estimate the values of the three coefficients $a_0$, $a_s$ and $a_1$. The introduction of calculated overlap integrals means that the advantages of working with purely phenomenological models are lost.

### 5.6.3 *Relationship between the nuclear quadrupole field and the rank 2 crystal field*

An alternative method of isolating the electrostatic contributions to the rank 2 crystal field parameters is based upon the observed values of nuclear quadrupole splitting parameters, usually denoted $Q_2^q$. The quadrupolar field at a nucleus is not subject to the effects of covalency and overlap, and can therefore be regarded as purely electrostatic in origin. Nevertheless, it is subject to screening effects which are quite different to those which modify the electrostatic field near the open-shell electrons. This difference is taken into account through the introduction of a screening ratio $\alpha$, defined as

$$\alpha = \frac{-\langle r^2 \rangle (1 - \sigma_2)}{2Q_N(1 - \gamma_\infty)}, \qquad (5.34)$$

where $\gamma_\infty$ is known as the Sternheimer nuclear antishielding factor and $\sigma_2$ is the electrostatic screening factor for open-shell electrons. $Q_N$ is the nuclear quadrupole moment and $\langle r^2 \rangle$ is the expectation value of $r^2$ for the open-shell electrons. $\sigma_2$ is in the range of $0 < \sigma_2 < 1$ and has the effect of reducing the electrostatic field seen by the open-shell electrons. On the other hand, the Sternheimer factor $\gamma_\infty$ is large and negative, and has the effect of magnifying the quadrupole component of electrostatic fields at the nucleus. Both of these factors and $\langle r^2 \rangle$ have been calculated from first principles for lanthanide ions (e.g. see [SBP68, AN78, AN80]), but the uncertainties in such calculations are significant, and they provide very poorly determined values of $\alpha$.

The observed second rank crystal field parameters can be written

$$B_q^2 = B_q^2(\text{covalency, overlap and exchange}) + B_q^2(\text{electrostatic}),$$
$$= \sum_L \overline{B}_2^c(R_L)g_{2q}(\theta_L, \phi_L) + \alpha Q_2^q. \tag{5.35}$$

Given the uncertainties in calculating $\langle r^2 \rangle$, $\gamma_\infty$ and $\sigma_2$, equations (5.30), (5.34) and (5.35) together can most appropriately be regarded as providing a means to test such calculations.

## 5.7 Assessment of the value and limitations of the superposition model

The stability of the values of both the intrinsic parameters and their distance dependences provide important checks on the validity of the superposition model. The neglect of contributions from distant ions to the crystal field parameters is partly based on the assumption that the magnitude of such contributions reduces rapidly with distance. This is justified by the empirically determined values of the power law exponents.

The results collected in this chapter provide overwhelming evidence for the effectiveness of the superposition model in the analysis of rank 4 and rank 6 crystal field parameters. Intrinsic parameters and power law exponents have been determined for many magnetic ions and ligands, and for some ligands these are found to be independent of the particular host crystal. A general result is that the intrinsic parameters are always positive for ionic ligands, such as $F^-$, $Cl^-$, $Br^-$, $O^{2-}$ and $S^{2-}$. When Wybourne normalization is used, a given ionic ligand has values of the intrinsic parameters which are similar in magnitude for all ranks. It is usually the case that the rank 4 power law exponent is greater than the rank 6 exponent for a given system.

While some values of the rank 2 intrinsic parameters have been obtained, superposition model analyses of the rank 2 crystal field parameters are generally difficult. One problem is that, except in the case of very low site symmetries, there is only one rank 2 parameter (viz. $B_0^2$). The best way to circumvent this problem is to use crystal field parameters determined for a given magnetic ion in more than one host, as described in Section 5.4.2.

Another problem is that near neighbour contributions to the rank 2 parameters often show strong cancellations, so that even small uncertainties in the angular coordination can materially affect the combined coordination factors. In addition, lattice sum calculations, such as that carried out by

Hutchings and Ray [HR63], suggest that long range electrostatic contributions are likely to be significant in the case of the rank 2 parameters. There is little hope, however, that such calculations will ever be made with sufficient accuracy to be of use in supplementing standard superposition model analyses.

When empirical values of the rank 2 intrinsic parameters cannot be obtained, the most practical approach in the case of lanthanides and actinides in ionic crystals is to use equation (5.30). This provides estimated lower bounds of rank 2 intrinsic parameters from the corresponding rank 4 and rank 6 intrinsic parameters, and provides a rough and ready means of estimating the electrostatic contribution to the rank 2 phenomenological parameters.

# 6

# Effects of electron correlation on crystal field splittings

M. F. REID

*University of Canterbury*

D. J. NEWMAN

*University of Southampton*

As was pointed out in Chapter 2, a considerable number of high-precision optical absorption and fluorescence spectra for lanthanide ions in ionic host crystals have been gathered since 1960. Crystal field fits, such as those described in Chapter 4, to the 100+ lowest-lying energy levels in lanthanide ions are generally good. However, fits to some particular 'problem' multiplets are invariably poor, e.g. the $^3K_8$ multiplet of $Ho^{3+}$, the $^1D_2$ multiplet of $Pr^{3+}$, and the $^2H_{11/2}$ multiplet of $Nd^{3+}$. In addition, some of the variations in the values of fitted parameters across the lanthanide series cannot be understood in terms of differences in ion size, or in terms of the ionic dependence of site distortions.

Problems with the one-electron crystal field parametrization also occur in fitting the energy levels of actinide and $3d$ transition metal ions. However, for these ions, the problems are not so easy to characterize as they are in the case of lanthanides. For this reason, while the main focus of this chapter is on the analysis of specific inadequacies of the one-electron model of lanthanide crystal fields, the conceptual aspects of this analysis should be understood as being relevant to all magnetic ion spectra.

Some of the difficulties found in fitting one-electron crystal field parameters can be associated with the use of inaccurate free-ion basis states resulting from the use of an inadequate parametrization of the free-ion Hamiltonian. Such inaccuracies produce errors in values of the intermediate coupled reduced matrix elements of the operators $C_q^{(k)}$ (defined in Section 3.1.1). The proper approach in these circumstances is to include more operators in the free-ion Hamiltonian, as described in Chapter 4.

It turns out, however, that many of the problems in fitting the one-electron crystal field parameters, including the specific examples mentioned at the beginning of this chapter, cannot be associated with inadequacies of the free-ion model. Hence, in spite of the success of the standard one-electron crystal field parametrization, there must still be something fundamentally wrong. It is thus appropriate to regard the one-electron model of the crystal field interaction as just a first approximation in the description of crystal field splittings of the open-shell energy levels of magnetic ions in crystals.

## 6.1 Generalizing the one-electron crystal field model

Failures of the one-electron approximation suggest that more information is contained in the observed energy levels of single magnetic ions than can be extracted by means of the techniques discussed in previous chapters. One way of going beyond the one-electron crystal field model is to carry out *ab initio* calculations using an extension of the formalism described in Chapter 1. However, the resulting calculations are extremely intricate and involve many approximations (see [NN86a, NN87b, NN87a]), which can only be validated by direct comparison with the results of phenomenological analyses. It is not practicable, therefore, *to replace* phenomenological analyses with *ab initio* calculations. Ways need to be found to obtain direct insights into the physical processes that cause the breakdown of the one-electron crystal field model, leading to the development of improved phenomenological models.

Two conceptually distinct approaches have been proposed.

(i) While retaining the one-electron crystal field concept, the basis set used in fitting is extended to include excited configurations which have electrons in empty shells and holes in filled shells of the ground configuration. A new set of crystal field parameters is introduced for each type of intra-shell and inter-shell one-electron matrix elements. In principle, there is no limit to the number of additional configurations (and parameters) that can be introduced in this way. However, because of the practical constraints of fitting to energy levels, attempts are usually made to severely limit the number of additional parameters through the use of *ab initio* calculations of relationships between the original and additional sets of crystal field parameters.

(ii) Keeping the same (open-shell) basis set, additional 'effective' operators, which are linearly independent of the one-electron operators, are introduced into the phenomenological crystal field. In this case, limitations on the possible ranks of operators ensure that only a finite, but possibly very large, number of additional operators are required.

### 6.1.1  Choosing the best approach

An example of approach (i) is provided by the work of Faucher and Moune [FM97] on the effect of $4f6p$ configurational admixtures in the ground $4f^2$ configuration of $LiYF_4:Pr^{3+}$. A Hartree–Fock calculation was used to determine the relative energy of the two configurations and to place constraints on the relative values of the coulomb integrals. Faucher and Moune showed that the root mean square deviation of their fit, which included both free-ion and crystal field parameters, could be reduced by more than a factor of two by including the three possible $4f/6p$ one-electron crystal field parameters and allowing the values of two coulomb integrals to vary freely. Similar improvements in fits to $U^{4+}$ and $Nd^{3+}$ energy levels were also cited.

While the results of Faucher and Moune are formally very good, they do not rule out the possibility of significant contributions from other configuration interaction processes. For example, admixtures of the configuration produced by exciting $5p$ electrons from the filled shell into the $4f$ shell would produce a similar weighting of contributions. This problem is further illustrated by another example of approach (i). The effect of allowing for $4f5d$ configurational admixtures into the $4f^2$ ground configuration of $Pr^{3+}$ has been investigated by Garcia and Faucher [GF89], who showed that the fit to the multiplet $^1D_2$, was considerably improved.

A drawback of the analyses discussed above is that only one of the possible excited configurations was considered in each case. In practice, many excited configurations can contribute to the breakdown of the one-electron crystal field parametrization. Attempts to make the analysis more realistic by including many additional excited configurations would require the introduction of a very large number of new parameters.

While such considerations undermine our confidence in using approach (i), there is a more general reason for rejecting it, on the basis that it involves a mixture of *ab initio* constraints and phenomenological parametrization. It has already been argued, in Chapter 1, that such hybrid theories can never provide clear tests of hypotheses about the nature of physical processes.

The remainder of this chapter, therefore, is only concerned with approach (ii), i.e. the search for possible ways of extending the parameter set used in the effective Hamiltonian acting on the states of the open-shell configurations. The problem is to find a generalization of the effective Hamiltonian which significantly improves the fit to experimental energy levels while, at the same time, introduces only a realistic number of additional parameters. If too many new parameters are introduced they will be so poorly determined as to make useful comparisons between values obtained for different systems impossible. On the other hand, too few parameters may not allow a proper representation of the physically significant effects.

When extending parameter sets it is important to choose, as far as possible, additional operators which are *orthogonal* to those in the original set (see Section A1.1.5) [New81, New82, JC84, JS84, Rei87a, JNN89]. This not only ensures linear independence of the parameters, but should also ensure that the original values of the one-electron crystal field parameters are unchanged in making the extension.

### *6.1.2 Multiplet dependence of the crystal field*

A simple way of dealing with the parametrization of poorly fitted multiplets is to use different, i.e. *multiplet-dependent*, crystal field parameters. This gives much the same results as the more sophisticated approach based on the formalism introduced in Section 6.2.4 when the multiplets are well spaced, and $LS$ coupling provides a fair approximation. It requires that the energy levels have been obtained for several multiplets of sufficient size to determine the crystal field parameters. There are several psooible pitfalls in applying the multiplet-dependent crystal field approach.

(i) The free-ion Hamiltonian may not be known with sufficient accuracy to determine the admixtures of $LS$ coupled states in the observed multiplets. In consequence, the intermediate coupled reduced matrix elements may be poorly determined, so that fitted crystal field parameter values are scaled to compensate.

(ii) When the crystal field parameters are not greatly overdetermined (i.e. the number of crystal field parameters is similar to the number of energy level differences), as is often the case when fitting to single multiplets, spurious minima in the fitting space may be found, leading to spurious parameter values.

(iii) In site symmetries where there is a multiplicity of true minima in the fitting space, parameter variations between multiplets may reflect different implicit choices of coordinate system (see Chapters 2 and 4), rather than different crystal field splitting mechanisms.

(iv) As discussed in Section 6.2.4, there can be additional 'effective' crystal field parameters with rank $k$ up to $4l$. Parameters with rank $2l < k \leq 4l$ have not hitherto been taken into account in multiplet-dependent crystal field fits.

The main advantage of single multiplet fitting is that a very simple program, such as ENGYFIT.BAS (see Chapter 3 and Appendix 2) can be employed. The superposition model (see Chapter 5) can be used to provide *post hoc* checks that physically realistic parameters have been obtained. In particular, it leads to the expectation that ratios of crystal field parameters of the same rank should be similar for all multiplets. The values of intrinsic parameters should also be similar for the same multiplet of a magnetic ion in different host crystals with the same ligands. However, while fits to separate multiplets may give some idea of the extent of the breakdown of the one-electron model for a specific magnetic ion, it does not provide a means to predict how the multiplets of other magnetic ions might behave.

The main practical application of the multiplet dependence approach is to determine the ranks of phenomenological parameters which are responsible for producing the difficulties in fitting particular multiplets. This would reduce the number of parameters that are required to extend the phenomenological parametrization as described in Section 6.2.4.

### 6.1.2.1 Trivalent holmium

The multiplet-dependent crystal field parameter approach has been applied to the crystal field splitting of trivalent holmium in five different host crystals by Pilawa [Pil91a, Pil91b]. This work provides a good illustration of the problems that are encountered, and the information that can be obtained, by fitting multiplet-dependent parameters. Particular attention is paid to the $^3K_8$ and $^5G_6$ multiplets.

It is instructive to compare Pilawa's results [Pil91b] for the multiplets $^5F_J$, $^5G_J$ and $^3K_J$ in the zircon structure hosts $YVO_4$ and $YAsO_4$. Pilawa noted that the fitted values of the $(k = 4, q = 4)$ and $(k = 6, q = 0)$ parameters were generally well defined, while the other crystal field parameters could vary significantly without making large differences to the fit.

The $(k = 4, q = 4)$ parameter is fairly constant for all the low-lying $^5F_J$ multiplets, but has significantly different values for all the $^5G_J$ and $^3K_J$ multiplets. The $(k = 6, q = 0)$ parameter varies only slightly across all the $^5F_J$ and $^5G_J$ multiplets, but has significantly larger values for the multiplets $^3K_8$ and $^3K_7$. Hence Pilawa's analysis provides evidence for significant 'anomalous' crystal field contributions of both the rank 4 and rank 6 crystal field parameters. Furthermore, the rank 6 contributions have a significant effect on the energy levels of the $^3K_J$ multiplets alone, suggesting that there is a rank 6 contribution which is sensitive to the total spin.

Pilawa [Pil91b] reported highly anomalous values of the $(k = 2, q = 0)$ parameter for the multiplet $^5G_6$. This could be due to the very small $LS$ coupled reduced matrix elements for the $^5G_J$ multiplets, making them very sensitive to errors in the free-ion wavefunction. Consequently, Pilawa's analysis does not provide conclusive evidence for a significant rank 2 contribution to the anomalous crystal field.

## 6.2 Generalizing the crystal field concept

In this section we discuss some plausible physical models for correlation effects, before introducing the general phenomenological approach which forms the basis of the mathematical treatment in subsequent sections. From a purely formal point of view, there are just two ways of generalizing a spin independent anisotropic one-electron operator: introduce spin dependence of the one-electron operator, or introduce anisotropic components of the two-electron interaction.

### 6.2.1 Relativistic crystal field

A possible origin for spin dependence of the crystal field is the influence of ligands on the isotropic free-ion spin–orbit coupling. This type of spin dependence can be expressed in terms of an anisotropic spin–orbit coupling. Like the spin–orbit coupling itself, this is a relativistic effect, so it is appropriate to call it the *relativistic crystal field* [Wyb65b, CNT73], rather than anisotropic spin–orbit coupling. However, apart from making a significant contribution to the splitting of ground states with zero orbital angular momentum (discussed in Chapter 7), its effects are generally found to be negligible.

### 6.2.2 Spin correlation

Another form of spin-dependent one-electron operator is analogous to the effective interaction between the electron spins, which occurs in the theory of the so called 'exchange interactions' between magnetic ions in solids. An interaction of the form $\mathbf{s}_i \cdot \mathbf{s}_j$ can be shown to be equivalent to allowing for the coulomb exchange interaction between unpaired electrons $i$ and $j$ on different magnetic ions. Similarly, an anisotropic exchange contribution to coulomb exchange between electrons in a given open shell can be shown to be equivalent to an effective operator of the form $\mathbf{s}_i \cdot \mathbf{s}_j(V_i + V_j)$, where $V_i$ and $V_j$ are one-electron potential functions.

The resulting 'spin-correlated crystal field' potential can be expressed in the form of a one-electron crystal field which depends on the relative direction of individual electron spins and the direction of the *total* spin on the magnetic ion (see Section 6.3.3 and [New70, New71, Jud77a]). The spin-correlated crystal field can also be shown to be equivalent to allowing for a specific type of the anisotropic two-electron interaction described in Section 6.2.4 (see equation (6.9) and [SN83]).

Apart from anisotropic exchange interactions, there are two other ways in which the ligand interactions with open-shell electrons could be affected by the alignment of the electron spin with the direction of the total spin of the magnetic ion.

(i) The wavefunction of the open-shell electrons with spins opposite to the total spin could be expanded relative to the wavefunctions for electrons with spins parallel to the total spin.

(ii) The energy denominators in (1.9) could depend on the relative spin alignments of the one-electron states.

It turns out (see [NSF82]) that possibility (i) would give the *wrong sign* for the abrupt change in the rank 6 crystal field parameters at half-filling (Table 2.6 shows the corresponding effect for the intrinsic parameters). Both the magnitude and sign of the observed change can, however, be explained in terms of changes in the energy denominators, confirming possibility (ii) as the main physical mechanism. Further discussion of the spin-correlated crystal field is given in Section 6.3.3.

## 6.2.3 Ligand polarization effects

It is well known that the coulomb interaction, which dominates the overall splitting of magnetic ion spectra, is reduced when ions are substituted into ionic crystals. Moreover, the magnitude of these reductions depends on the type of ligand. Whatever type of magnetic ion is involved, the reductions define the same so-called *nephelauxetic series*. In the case of lanthanides and actinides, this series is defined by the observed reductions in magnitudes of the rank 2 Slater parameter $F^{(2)}$ [Jor62, GS73, New77a, NNP84] relative to their free-ion values. These *reductions*, corresponding to a 'red shift' of the observed spectrum, satisfy the inequalities

$$0 < F^- < O^{2-} < Cl^- < Br^- < I^- < S^{2-}. \qquad (6.1)$$

The term 'nephelauxetic' (= cloud expanding) has been used to describe these reductions on the basis of the assumption that electron–electron (or coulomb) repulsions are reduced by a covalency-dependent expansion of the 'charge cloud' defined by the true open-shell wavefunctions. However, this assumption is in clear contradiction to the ordering of the magnitude of covalency, as expressed by the spectrochemical series (see Section 5.2.1).

In fact, it seems most likely that the observed reductions in the coulomb interaction are mainly due to screening by the ligand charge clouds (e.g. see [New73c, New74]). This is in accord with the empirical correlation between the magnitude of these reductions and *ligand polarizabilities* [New77a].

The observed contractions of lanthanide and actinide ion spectra due to the nephelauxetic effect are large compared with the magnitude of the crystal field splittings. This, in itself, does not mean that ligand polarization produces significant crystal field contributions, because the contractions are due to changes in the isotropic parameters. Nevertheless, the polarized ligand environment producing this effect is anisotropic, so that some anisotropic electron–electron contributions must be induced. Indirect evidence for significant anisotropic ligand polarization contributions is provided by the empirical correlation between ligand polarizability and values of the rank 4 spin-Hamiltonian parameters, discussed in Section 7.4.2.

Direct evidence of the importance of ligand polarization contributions to the correlation crystal field has been provided by the work of Denning *et al.* [DBM98]. In fits to one-electron crystal field parameters for terbium in elpasolite crystals with halide ligands, percentage deviations were found to increase in the order F < Cl < Br. This order corresponds to *increasing* ligand polarizability (as noted above), but *decreasing* overlap and covalency contributions as evidenced by the parameter values obtained by Denning *et al.* [DBM98] as well as the spectrochemical series (see Section 5.2.1). Hence the only reasonable interpretation of the results presented in [DBM98] is that ligand polarization makes significant contributions to the correlation crystal field. This is not reflected in existing *ab initio* calculations [NN87b] because they are based on a formalism [NN87a] which does not fully include ligand polarization effects.

A simplified phenomenological model of the contribution of ligand polarization to the anisotropic screening of coulomb interactions has been formulated by Newman [New77c]. While tests of this model have yet to be carried out, it is clear that any contributions must be regarded as additional to those of the spin-correlated crystal field.

### 6.2.4 The correlation crystal field

Several other extensions to the one-electron crystal field model have been proposed, including the orbitally correlated crystal field [YN87] and Judd's delta-function model [Jud78], which will be discussed in more detail in Section 6.3.4. It has become clear, however, that none of the models based on specific mechanisms is capable of explaining all the known cases where the one-electron spin-independent crystal field potential is inadequate.

As the dominant mechanisms which contribute to the anomalous crystal field are not known, it becomes necessary to introduce a full parametrization of the anisotropic two-electron interaction. This will be referred to as the correlation crystal field. It can be viewed either as a generalization of the one-electron crystal field, or as a generalization of the isotropic coulomb interaction between open-shell electrons.

Two-electron interactions are necessarily 'charge conjugation invariant', i.e. they have the same sign for two electrons or two holes. In contrast, one-electron interactions change sign when the electron is replaced by a hole. This difference in relative sign suggests that two-electron interactions would produce systematic differences between the crystal field splittings in, for example, magnetic ions with the $f^N$ and $f^{(14-N)}$ open-shell configurations.

The main problem with the general expression for the correlation crystal field is the very large number of parameters that are necessary to describe it: far more than can be determined by fitting to optical spectra. In order to give some idea of the problem, the relationships between various types of phenomenological crystal field in $f^N$ (lanthanide or actinide) configurations, and the numbers of parameters required for each type of parametrization, are shown in Figure 6.1.

In a site with no symmetry operations, there are 637 correlation crystal field parameters, i.e. more than 20 times the number of one-electron parameters. This ratio of the number of one-electron crystal field parameters to the number of correlation crystal field parameters is approximately the same for any site symmetry. At the present time there is no practical way in which full sets of phenomenological correlation crystal field parameters can be determined by fitting to optical spectra. Some approximations have to be made. The most obvious way to reduce the number of fitted parameters is to use the superposition model. Then, corresponding to $C_{\infty v}$ symmetry, 43 correlation crystal field intrinsic parameters (including the three one-electron intrinsic parameters) are required. It is impractical to fit even these 40 additional parameters to available experimental spectra. Moreover, given that each intrinsic parameter is associated with an indeterminate distance dependence, more than 40 parameters are required in practice. For this reason all attempts to determine correlation crystal field parameters hitherto have used physical arguments or just 'past experience' to reduce the number of fitted parameters.

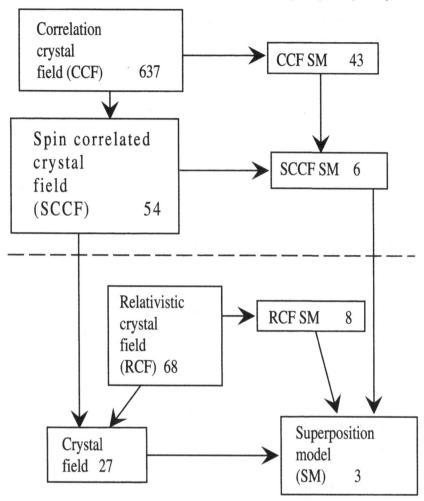

Fig. 6.1 Hierarchy of phenomenological crystal field parametrizations. The number of parameters given in each box corresponds to the case of no symmetry restrictions, and the number in square brackets corresponds to the number of parameters in cubic symmetry. In the case of the various forms of superposition model the symmetry is necessarily axial (power law exponents are not included). Note that the spin-correlated crystal field is just one example of possible models which reduce the number of phenomenological parameters. Ligand polarization, delta-function, or other models could be substituted.

## 6.3 A full parametrization

While it has not yet been possible to get a conceptual grasp of the physical mechanisms that produce the dominant contributions to the correlation crystal field, it is straightforward to formulate a complete phenomenological parametrization. It is only necessary to obtain a complete set of Hermitian, time-reversal-invariant, two-electron operators [New71]. The most obvious way to achieve this is to construct all products of two unit tensor operators $U_q^{(k)}$ (defined in Appendix 1), each of which acts on different electrons. The *correlation crystal field* Hamiltonian constructed in this way is:

$$H_{\text{CCF}} = \sum_{k_1 k_2 q_1 q_2} B_{q_1 q_2}^{k_1 k_2} \sum_{i>j} U_{q_1}^{(k_1)}(i) U_{q_2}^{(k_2)}(j). \qquad (6.2)$$

The unit tensors may be coupled (e.g. see [Jud63]) to give operators of well-defined total angular momentum as follows

$$H_{\text{CCF}} = \sum_{k_1 k_2 kq} B_q^k(k_1 k_2) \sum_{i>j} \left( \mathbf{U}^{(k_1)}(i) \mathbf{U}^{(k_2)}(j) \right)_q^{(k)}. \qquad (6.3)$$

Equation (6.3) has the advantage that the selection rules for the coupled operators are more obvious than those for the $U_{q_1}^{(k_1)} U_{q_2}^{(k_2)}$ operator products. The restrictions on $q$ are similar to the restrictions in the one-electron crystal field parametrization, which are discussed in Chapter 2 and Appendix 1. The relevant restrictions on $k_1$, $k_2$ and $k$ are discussed in detail by Newman [New71] and Wang and Stedman [WS94]. In brief, $k_1$ and $k_2$ must be less than or equal to $2l$ and $k$ less than or equal to $4l$. The sum $k_1 + k_2$ must be even and parameters with $k_1 = k_2$ must have $k$ even. The value of $k$ is not generally required to be even, but the other restrictions preclude any operators with $k = 1$. Note that, for $f^N$ configurations, values of $k$ up to 12 are possible.

The parametrizations given in (6.2) and (6.3) have the disadvantage that the operators are not orthogonal (see Section A1.1.5). This problem can, however, be overcome by changing the relative weights of the spin singlet and spin triplet two-electron matrix elements. In calculating matrix elements it is helpful to take advantage of the group theoretical properties of Racah's 'parentage' groups, $Sp_{14}, SO_7, G_2$ etc., described, for example, in [Jud63, CO80]. Judd [Jud77b] gave a prescription for the calculation of correlation crystal field operators, orthogonalized over complete $f^N$ basis sets, that makes full use of Racah's groups. Judd's $g_{iq}^{(k)}$ operators may be used to rewrite the correlation crystal field Hamiltonian as

$$H_{\text{CCF}} = \sum_{ikq} G_{iq}^k g_{iq}^{(k)}. \qquad (6.4)$$

Both (6.3) and (6.4) have the same number of parameters, and we may transform between the two parameter sets, $\left\{ B_q^k(k_1 k_2) \right\}$ and $\left\{ G_{iq}^k \right\}$. This transformation may be carried out using programs described in Section A3.2.

Note that the $\mathbf{g}_1^{(k)}$ are one-electron operators, proportional to the unit tensor operators $\mathbf{U}^{(k)}$. Reid [Rei87a] proposed a minor modification to the operators $\mathbf{g}_2^{(k)}$ to make them completely orthogonal to the conventional one-electron crystal field operators. These modified operators, denoted $\mathbf{g}'_2^{(k)}$, will be employed in this chapter.

### 6.3.1 Superposition model restrictions

The superposition model may be applied to the correlation crystal field parametrization. In analogy with the one-electron case (see Chapter 5) the $q = 0$ parameters for a single ligand at $R_0$ on the $z$ axis are called intrinsic parameters $(\bar{B}^k(k_1 k_2)$ or $\bar{G}_i^k)$. Then the parameters may be written as

$$B_q^k(k_1 k_2) = \sum_L \bar{B}^k(k_1 k_2)(-1)^q C_{-q}^{(k)}(L) \left(\frac{R_0}{R_L}\right)^{t_{k_1 k_2 k}}, \qquad (6.5)$$

or

$$G_{iq}^k = \sum_L \bar{G}_i^k(-1)^q C_{-q}^{(k)}(L) \left(\frac{R_0}{R_L}\right)^{t_{ik}}. \qquad (6.6)$$

In writing these expressions we have not assumed that the power law exponents, $t_{k_1 k_2 k}$ or $t_{ik}$, are the same as the one-electron exponents, $t_k$, or that they are fixed for a given $k$.

The intrinsic parameters are non-zero only if the identity irreducible representation of $C_{\infty v}$ is contained in the $O_3$ irreducible representation $k^+$ (since the correlation crystal field Hamiltonian has even parity). This is only the case if $k$ is even. Thus, in contrast to the case of the one-electron crystal field, the superposition model is more restrictive than the combined Hermiticity and time-reversal symmetries.

### 6.3.2 First principles calculations and models

Given that the number of correlation crystal field parameters is extremely large, steps must be taken to significantly reduce the number of parameters used in fitting observed spectra. In addition to the models described in Section 6.2, *ab initio* calculations can provide a useful guide in reducing the number of fitted parameters.

There have been various attempts to calculate correlation crystal field effects from first principles. The most comprehensive calculation is that carried out by Ng and Newman [NN87b]. These calculations assume the validity of the superposition model, and hence provide no information about the odd-rank parameters. Table XIV of [NN87b] shows that:

(i) parameters with $k > 6$ are predicted to be small;

(ii) some parameters are predicted to be much larger than others.

While these *ab initio* results are useful, they are not sufficient in themselves to reduce the number of phenomenological parameters to a manageable level. To this end we now consider the spin-correlated crystal fieldand delta-function models.

### 6.3.3 Spin-correlated crystal field

The origin of the energy difference between the crystal field parameters for electrons in which the spin is aligned and anti-aligned with the total spin has been discussed in Section 6.2.2. This can be expressed in terms of the phenomenological *spin-correlated crystal field* (SCCF) Hamiltonian

$$H_{\text{SCCF}} = \sum_i \mathbf{S} \cdot \mathbf{s}_i V_s(i) \qquad (6.7)$$

where $V_s(i)$, the potential acting on the $i$-th electron, can be expanded in terms of tensor operators in exactly the same way as the crystal field potential $V_{\text{CF}}$, viz.

$$V_s(i) = \sum_{k,q} a_{kq} C_q^{(k)}(i). \qquad (6.8)$$

A method of evaluating matrix elements of the operators $\sum_i \mathbf{S} \cdot \mathbf{s}_i C_q^{(k)}(i)$ is given in Section 8.3.3.

The spin-correlated crystal field operators appearing in (6.7) can also be written in terms of a small subset of the $\mathbf{g}_{iq}^{(k)}$ operators [Rei87a], viz.

$$\sum_i \mathbf{S} \cdot \mathbf{s}_i C_q^{(k)}(i) = \left(\frac{7-N}{8}\right) \mathbf{g}_1^{(k)} - \frac{\sqrt{30}}{8} \mathbf{g}'^{(k)}_2 + \frac{\sqrt{330}}{8} \mathbf{g}_3^{(k)}, \qquad (6.9)$$

where $k = 2, 4, 6$, and $N$ is the number of $f$ electrons. Standard techniques can then be used to evaluate the matrix elements of the $\mathbf{g}_i^{(k)}$ operators (see Section A3.2).

The same site symmetry considerations that determine the number of one-electron crystal field parameters can also be applied to the spin-correlated crystal field. Hence the introduction of spin correlation doubles the number of non-vanishing phenomenological parameters. Given that the same coordination factors are relevant, it is expected that the ratios of parameters of the same rank should be effectively the same for the spin-correlated crystal field as they are for the one-electron crystal field. The accuracy of this expectation does, of course, depend on the distance dependences of the spin-correlated crystal field intrinsic parameters being very similar to those of the crystal field intrinsic parameters. If this approximation is correct it allows the spin-correlated crystal field to be expressed in terms of the parameter ratios

$$c_k = a_{kq}/B_q^k, \tag{6.10}$$

where $k = 2, 4, 6$ for lanthanides and actinides, and $k = 2, 4$ for $3d$ transition metal ions. Such a small increase in the number of fitted parameters makes it practicable to include the $c_k$ in most crystal field fits to optical spectra. Judd [Jud77a] identified several cases where taking account of the $c_k$ would produce improved fits, and suggested that these parameters should be routinely included in fits to energy levels determined by optical spectroscopy. Crosswhite and Newman [CN84c] showed that a value of $c_6$ of about 0.15 could explain both the anomalous crystal field splitting in the $^3K_8$ multiplet of $Ho^{3+}$:$LaCl_3$ and the abrupt change in values of the trivalent lanthanide rank 6 parameters at half-filling (see Table 2.6). This is a significant advance over the multiplet-dependent parametrization of Section 6.1.2, with the addition of a single parameter removing most of the anomalies in the crystal field fit.

### 6.3.4 Delta-function model for $f$ electrons

Judd [Jud78] introduced a model that incorporates a slightly larger number of $g_i^{(k)}$ operators than the spin-correlated crystal field. This model, known as the delta-function model, is based on the assumption that two $f$ electrons interact with a given ligand at a single point, denoted $\mathbf{R}'_L$, located between the magnetic ion and the ligand. This interaction can be written

$$H_\delta = -A\delta(\mathbf{r}_i - \mathbf{R}'_L)\delta(\mathbf{r}_j - \mathbf{R}'_L). \tag{6.11}$$

Here $A$ is a parameter to be fitted and $\mathbf{r}_i$ and $\mathbf{r}_j$ are the positions of the $f$ electrons. A possible justification for taking this model seriously is that shielding effects tend to reduce the range of coulomb interactions. The operator in (6.11) may be written in terms of a subset of the $g_i^{(k)}$ operators, with $i = 1, 2, 3$, and 10. This has been discussed in detail by Lo and Reid [LR93], and McAven *et al.* [MRB96]. *Ab initio* calculations ([NN87b], table XIV) predict that parameters associated with these operators are quite large.

If all radial parts are treated as parameters, the model Hamiltonian can be written as

$$H_\delta = \sum_{kq} D_q^k \delta_q^{(k)}, \tag{6.12}$$

where, $k$ runs over the even integers from 2 to 12. The operators $\delta_q^{(k)}$ may be taken from the tables of McAven *et al.* [MRB96]. For example, the $k = 4$ operators are given by

$$\delta_q^{(4)} = -\frac{21\sqrt{105}}{2\sqrt{11}} g'_{2q}^{(4)} + \frac{63\sqrt{105}}{22} g_{3q}^{(4)} + \frac{84\sqrt{42}}{\sqrt{715}} g_{10Aq}^{(4)} + \frac{8232\sqrt{3}}{11\sqrt{1105}} g_{10Bq}^{(4)}. \tag{6.13}$$

Here we have ignored the one-particle part $\left(g_{1q}^{(4)}\right)$ since that is absorbed by the one-electron crystal field.

### 6.3.5 The strong crystal field parametrization for d electrons

The treatment of correlation crystal field effects in magnetic ions with partially filled $3d$ shells has been studied from a quite different point of view. Only systems with cubic site symmetry have been considered. Griffith [Gri61] presents a formalism in which the 10 coulomb integrals formed from the $e_g$ and $t_{2g}$ $d$ orbitals in cubic symmetry are distinguished. Because it is expressed explicitly in terms of anisotropic coulomb integrals this is known as the *strong* crystal field parametrization.

The strong crystal field parametrization is not dependent on any assumptions about the physical mechanisms involved and can therefore be related to the correlation crystal field parametrization developed above. Ng and Newman [NN84] tabulate the transformation coefficients relating the two parametrizations, and give the parameter constraints appropriate to the spin-correlated crystal field and ligand dipole polarization models. They also show that a two-electron contribution to the cubic (one-electron) crystal field parameter ($\Delta = 10Dq$) can be expressed as a linear combination of six of the anisotropic coulomb integrals.

In practice, it is not possible to fit the 10 additional parameters to the rather few energy levels that can be determined for $3d$ transition metal ions. Hence, Ng and Newman [NN86b] attempted to fit the single spin-correlated crystal field parameter $c_4$ (defined in Section 6.3.3) to the optical spectra of $Mn^{2+}$ in a number of fluorides, $MnCl_2$, $MnBr_2$ and $MnI_2$. The results obtained were, however, not sufficiently consistent to determine whether or not spin correlation effects are significant in $d$ electron systems.

## 6.4 Parameter fits

To test correlation crystal field models properly it is desirable to perform fits using large data sets (with preferably over 100 observed levels) and a sophisticated treatment of the free-ion Hamiltonian so that the fits are not distorted by a poor representation of the atomic interactions (see Chapter 4). Fits to smaller data sets may have value in some special cases, but in general they are rather pointless. If only a small number of levels are used in the fit then the $\mathbf{g}_i^{(k)}$ operators are not orthogonal, and may not even be linearly independent, over the data set. This results in poorly convergent fits and the fitted parameters have little value, since many different parameter sets can be found that fit the data equally well.

Fits to $Nd^{3+}$ spectra are of particular interest, since it is quite common to be able to observe energy levels up to $40\,000$ cm$^{-1}$, missing only four multiplets containing 16 Kramer's doublets. Thus some studies, such as that of $Nd^{3+}$:$LaF_3$ by Carnall *et al.* [CGRR89] and $Nd^{3+}$:YAG by Burdick *et al.* [BJRR94] fit over 140 of the possible 160 energy levels of the $4f^3$ configuration.

Most parametrization work has concentrated on trying to improve fits to certain "problem" multiplets in the spectra of $Ho^{3+}$, $Gd^{3+}$, $Nd^{3+}$, $Pr^{3+}$ and $Er^{3+}$, such as those mentioned at the beginning of this chapter. This has provided several clues as to which operators are important.

The spin-correlated crystal field model [CN84c] provides marked improvements in fits for certain multiplets of $Ho^{3+}$ and $Gd^{3+}$ in $LaCl_3$. Reid [Rei87a] identified which individual operators in the spin-correlated crystal field parametrization, defined by (6.7)–(6.9), were responsible for the improvements. Out of the possible operators the $g_{3q}^{(6)}$ were found to be the most important.

Many studies have been performed on $Nd^{3+}$. Faucher and co-workers [FGC$^+$89, FGP89] pointed out that the poor fits to the $^2H_{11/2}$ multiplet at about $16\,000$ cm$^{-1}$ could be improved by (arbitrarily) modifying the reduced matrix elements of the one-electron crystal field operator $\mathbf{C}^{(4)}$, in the spirit of the multiplet-dependent crystal field approach of Section 6.1.2. The reduced matrix elements of this operator are unusually small for the $^2H_{11/2}$ multiplet of $Nd^{3+}$, and hence very sensitive to the quality of the free-ion fit.

Li and Reid [LR90] pointed out that the operators $\mathbf{g}_{10A}^{(4)}$ and $\mathbf{g}_{10B}^{(4)}$ have large matrix elements for the $^2H_{11/2}$ multiplet of $Nd^{3+}$ and performed fits using those operators. They also found that the $\mathbf{g}_2^{(4)}$ operator was effective in improving the fits to some other multiplets. In performing the fits Li and Reid fixed the ratios of correlation crystal field parameters with different values of $q$ to have the same ratio as the one-electron crystal field parameters. Thus

$$H_{CCF} = \sum_{ik} G_i^k \left( g_{i0}^{(k)} + \sum_{q \neq 0} g_{iq}^{(k)} \frac{B_q^k}{B_0^k} \right), \qquad (6.14)$$

with the ratio $B_q^k/B_0^k$ taken from previous fits. This was done, in the spirit of the superposition model, to reduce the number of freely varied parameters and should be considered as an approximation, to be relaxed if possible.

Table 6.1 compares the parameter fits of Burdick *et al.* [BJRR94] to the $Nd^{3+}$ :YAG spectrum, with and without the extra correlation crystal field parameters $G_2^4$, $G_{10A}^4$ and $G_{10B}^4$. Table 6.2 compares the observed and fitted energy levels to selected multiplets. The fit to the $^2H_{11/2}$ multiplet is greatly improved by the inclusion of the correlation crystal field parameters, whereas the fit to the $^4G_{5/2}$ changes very little. In general, the extra operators improve the fits to 'problem' multiplets, without disturbing the good fit to the rest of the spectrum.

Note that the ratios of correlation crystal field parameters to one-electron crystal field parameters look large. However, correlation crystal field operators are normalized differently, and the $B_q^k$ should be multiplied by approximately $14 \times (12/77)^{1/2}$ ($\approx 5$) for a fair comparison, indicating that the correlation crystal field is significantly smaller than the one-electron crystal field.

Table 6.3 summarizes the ratios of the $k = 4$ one- and two-electron crystal field parameters. All ratios are quite close to the ratios predicted by the *ab initio* calculations of Ng and Newman [NN86a]. They are also similar to the ratios predicted by Judd's delta-function model (using equation (6.13)). Other fits to $Nd^{3+}$ spectra in various compounds give similar ratios [LR90].

Burdick and Richardson [BR98b, BR98c] used the delta-function model to fix the ratios between correlation crystal field parameters of the same rank (see equations (6.12)–(6.14)) and thus carried out fits with only one extra parameter for each value of $k$. Fits including only two ($k = 2$ and 4) delta-function correlation crystal field parameters have been quite successful for both $Pr^{3+}$ [BR98b, BR98c] and $Nd^{3+}$ [QBGFR95] systems. It would be worthwhile to test this model on other ions.

Jayasankar and coworkers [JRTH93] studied the effect of pressure on the one-electron and correlation crystal field parameters for $Nd^{3+}$:$LaCl_3$. The ratio of correlation crystal field to one-electron parameters changed dramatically when pressure was applied. A possible explanation in the context of

Table 6.1. Parameters (in cm$^{-1}$) obtained from crystal field (CF) and correlation crystal field (CCF) analyses of Nd:YAG energy level data [BJRR94].

| Parameter[a] | CF | | | CF + CCF | | |
|---|---|---|---|---|---|---|
| $E_0$ | 24 097 | ± | 11 | 24 095 | ± | 6 |
| $F^2$ | 70 845 | ± | 156 | 70 809 | ± | 78 |
| $F^4$ | 51 235 | ± | 338 | 51 132 | ± | 175 |
| $F^6$ | 34 717 | ± | 145 | 34 819 | ± | 71 |
| $\alpha$ | 21.1 | ± | 0.4 | 20.8 | ± | 0.2 |
| $\beta$ | −645 | ± | 19 | −629 | ± | 10 |
| $\gamma$ | 1660 | ± | 43 | 1656 | ± | 22 |
| $T^2$ | 345 | ± | 57 | 366 | ± | 29 |
| $T^3$ | 46 | ± | 7 | 46 | ± | 3 |
| $T^4$ | 61 | ± | 9 | 66 | ± | 5 |
| $T^6$ | −272 | ± | 17 | −270 | ± | 8 |
| $T^7$ | 318 | ± | 30 | 324 | ± | 15 |
| $T^8$ | 271 | ± | 38 | 307 | ± | 18 |
| $\zeta$ | 876 | ± | 4 | 873 | ± | 2 |
| $M^0$ | 1.62 | ± | 0.41 | 1.76 | ± | 0.22 |
| $M^2$ | 0.558 $M_0$ | | | 0.558 $M_0$ | | |
| $M^4$ | 0.377 $M_0$ | | | 0.377 $M_0$ | | |
| $P^2$ | 107 | ± | 85 | 209 | ± | 44 |
| $P^4$ | 0.75 $P_2$ | | | 0.75 $P_2$ | | |
| $P^6$ | 0.50 $P_2$ | | | 0.50 $P_2$ | | |
| $B_0^2$ | −405 | ± | 29 | −387 | ± | 15 |
| $B_2^2$ | 179 | ± | 25 | 172 | ± | 12 |
| $B_0^4$ | −2823 | ± | 84 | −2766 | ± | 45 |
| $B_2^4$ | 540 | ± | 93 | 529 | ± | 45 |
| $B_4^4$ | 1239 | ± | 67 | 1275 | ± | 36 |
| $B_0^6$ | 955 | ± | 101 | 972 | ± | 51 |
| $B_2^6$ | −390 | ± | 87 | −333 | ± | 45 |
| $B_4^6$ | 1610 | ± | 56 | 1611 | ± | 27 |
| $B_6^6$ | −281 | ± | 78 | −229 | ± | 39 |
| $G_2^4$ [b] | — | | | −804 | ± | 135 |
| $G_{10A}^4$ [b] | — | | | 1290 | ± | 80 |
| $G_{10B}^4$ [b] | — | | | 609 | ± | 108 |
| $N$ [c] | 144 | | | 144 | | |
| $n$ [d] | 25 | | | 28 | | |
| $\sigma$ [e] | 31.1 | | | 15.3 | | |

[a]Parameter notation follows that of Chapter 4.
[b]Parameter ratios for the correlation crystal field parameters were constrained as in (6.14).
[c]Total number of energy levels used in the parametric data analyses.
[d]Total number of freely varied parameters.
[e]Standard deviations (in cm$^{-1}$) calculated for least-squares energy level fits.

Table 6.2. Comparison of experimental energy levels for Nd $^{3+}$:YAG with those calculated using the fitted parameters given in Table 6.1 [BJRR94]. Energy deviations are denoted by $\Delta$ and root mean square deviations for the multiplets are denoted by $\sigma$. All quantities are in cm$^{-1}$.

| Multiplet | | Expt. | CF only | $\Delta$ | CF+CCF | $\Delta$ |
|-----------|---|-------|---------|----------|---------|----------|
| $^2H_{11/2}$ | | 15 741 | 15 862 | −121 | 15 757 | −16 |
| | | 15 831 | 15 882 | −51 | 15 842 | −11 |
| | | 15 865 | 15 909 | −44 | 15 864 | 1 |
| | | 15 950 | 15 920 | 30 | 15 945 | 5 |
| | | 16 088 | 16 005 | 83 | 16 087 | 1 |
| | | 16 104 | 16 022 | 82 | 16 119 | −15 |
| | $\sigma$ | | | 75 | | 10 |
| $^4G_{5/2}$ | | 16 842 | 16 864 | −22 | 16 848 | −6 |
| | | 16 982 | 16 982 | 0 | 16 984 | −2 |
| | | 17 038 | 17 057 | −19 | 17 071 | −33 |
| | $\sigma$ | | | 17 | | 19 |

Table 6.3. Ratios of selected parameters for Nd $^{3+}$, calculated with respect to $-G^4_{10A}$.

| | $B^4_0$ | $G^4_2$ | $G^4_{10A}$ | $G^4_{10B}$ |
|-----------|---------|---------|-------------|-------------|
| Experiment (Table 6.1) | 2.14 | 0.62 | −1 | −0.47 |
| Calculation [NN87b] | 2.00 | 0.58 | −1 | −0.30 |
| Delta model (6.13) | — | 1.59 | −1 | −1.90 |

the superposition model was that the power law dependence for the CCF parameters is very different from that for the one-electron parameters. If this is the case it is clear that other fits described in this chapter, which assume that the one-electron and two-electron parameters share the same ratios between different $q$ components, may be oversimplified.

## 6.5  Future directions

It has been demonstrated that parameter fits to experimental data can be considerably improved by the addition of a relatively small number of parameters. Workers in this area should always consider the possibility of adding these parameters, and trying other parameters. Programs described in Chapter 4 and Appendix 3 may be used for this purpose.

However, although considerable progress has been made over the last two decades the correlation crystal field is still poorly characterized. The enormous number of correlation crystal field parameters makes it extremely difficult to satisfactorily test the theory by fitting experimental data. Until much larger data sets can be obtained it will be difficult to make further progress. This will become possible only by using experimental techniques such as excited state absorption and UV synchrotron studies [WDM+97].

More thought should also be given to identifying the most important mechanisms that contribute to the correlation crystal field, and carrying out thorough tests of the available models. For example, little has been done to test the ligand polarization model, although there are clear indications (see Sections 6.2.3 and 7.4.2) that this mechanism is important. Improved *ab initio* calculations, along the lines of the work by Ng and Newman [NN86a], could also contribute to the identification of important mechanisms, as well as providing more accurate estimates of the parameter values.

# 7

# Ground state splittings in $S$-state ions

## D. J. NEWMAN

*University of Southampton*

## BETTY NG

*Environment Agency*

Magnetic ions with half-filled open shells have ground states with zero total orbital angular momentum $L$. Hence they are often referred to as '$S$-state ions'. The most studied $S$-state ions with half-filled $f$ shells are the lanthanide ions $Gd^{3+}$, $Eu^{2+}$, $Tb^{4+}$, and the actinide ions $Cm^{3+}$, $Bk^{4+}$; all of which have the spin octet ground state ${}^8S_{\frac{7}{2}}$ (i.e. total spin $S = 7/2$). Important $S$-state ions in the transition metal series are $Fe^{3+}$ and $Mn^{2+}$. These have a half-filled $d$ shell, with the spin sextet ground state ${}^6S_{\frac{5}{2}}$ (i.e. total spin $S = 5/2$).

Given that *pure* $L = 0$ states have no spatial anisotropy, it might appear that the crystal field cannot split the ground state. In practice, however, energy level splittings are observed, but these are generally very much smaller than the crystal field splittings in the higher lying multiplets. These ground state splittings are often referred to as 'zero-field splittings', because they occur even in the absence of external magnetic fields. Zero-field splittings are produced by small admixtures of states with other $L$ and $S$ values into the $L = 0$ ground state, so that $J$ provides an exact description of the ground state rather than $L$ or $S$.

While the contribution of the one-electron crystal field to the ground state splittings is usually significant, it has been found that the observed splittings can only be explained when two additional types of crystal field contribution are taken into account. These are the relativistic crystal field and the correlation crystal field, which have already been introduced in Chapter 6. The mechanisms which give rise to the relativistic crystal field are outlined in Section 7.5.1, and the relative importance of the various contributions to zero-field splittings is discussed in Section 7.5.2.

## 7.1 The spin-Hamiltonian

Given the appropriate (intermediate coupling) reduced matrix elements, the ground multiplets of $S$-state ions can be fitted to the same one-electron crystal field Hamiltonian as the excited multiplets. However, the significant correlation and relativistic crystal field contributions to the ground state splitting make it likely that the crystal field parameters which provide a good fit to the energy levels of the excited multiplets will not, at the same time, provide a good fit to the ground multiplet energy levels. In addition, experimental techniques capable of determining the ground multiplet energy levels are often not appropriate for determining the splitting of the higher-lying multiplets, and vice versa. Hence it is customary to parametrize the ground multiplet splittings separately.

The energy matrix corresponding to the zero-field splittings can be constructed from the matrix elements $\langle S, M_S | \mathcal{H}_S | S, M'_S \rangle$ of a so-called 'spin-Hamiltonian' $\mathcal{H}_S$, which is expressed in terms of the total spin operator $\mathbf{S}$. In the presence of a magnetic field $\mathbf{H}$, and ignoring any coupling to the nuclear spin of the magnetic ion in question, or to spins on other ions, this Hamiltonian can be written in the form

$$\mathcal{H}_S = \beta \mathbf{S} \cdot \mathbf{g} \cdot \mathbf{H} + \sum_{k,q} B_k^q O_k^q(\mathbf{S}), \tag{7.1}$$

where $\beta = e/mc$ is the Bohr magneton. Note that the spin-Hamiltonian parameters $B_k^q$ must be distinguished from the symbols $(B_q^k)$ used to denote crystal field parameters in Wybourne normalization. The first term in equation (7.1) expresses the energy contribution from the coupling to an externally applied magnetic field and the second term represents the energies arising from the coupling between the spin and the crystalline environment. The $3 \times 3$ matrix, or tensor, $\mathbf{g}$ allows for the possibility of anisotropic coupling between the total spin and the magnetic field. This anisotropy is small, and seldom observable (see [AB70, New77b]).

The operators $O_k^q(\mathbf{S})$ in (7.1) are conventionally expressed in terms of spin operators. Explicit expressions can be found in Abragam and Bleaney [AB70], Newman and Urban [NU75] and Newman and Ng [NN89b]. These expressions have the same form as the operators used in the Stevens operator equivalent method for the one-electron crystal field, with the substitution of $\mathbf{S}$ for $\mathbf{J}$. It follows that the parameters $B_k^q$ are normalized in the same way as the Stevens parameters, with the corresponding operator equivalent factors taken to be unity. In $LS$ coupling these operator equivalent factors would be identically zero.

It is conventional to introduce scaled parameters $b_k^q$ as follows

$$b_2^q = 3B_2^q, \; b_4^q = 60B_4^q, \; b_6^q = 1260B_6^q.$$

These parameters, which are the most commonly used in the literature, will be adopted as standard in this book. The use of lower case '$b$' parameters also has the advantage of avoiding any possibility of confusion between spin-Hamiltonian parameters ($B_k^q$) and the crystal field parameters ($B_q^k$) in Wybourne normalization. A discussion of the various notations and normalizations of spin-Hamiltonian and crystal field parameters has been given by Rudowicz (see Appendix 4 and [Rud87a]).

In any site symmetry, the non-vanishing spin-Hamiltonian parameters $b_k^q$ have exactly the same $k, q$ labels as the non-vanishing crystal field parameters $B_q^k$. However, the energy matrix corresponding to the zero-field splittings has dimensions $(2S + 1) \times (2S + 1)$ (where $S = (2l + 1)/2$), while the one-electron crystal field energy matrix has dimensions $(2l+1) \times (2l+1)$.

In the case of transition metal ions (with a $d^5$ half-filled shell) the rank 6 (i.e. $k = 6$) spin-Hamiltonian parameters vanish identically. In cubic symmetry the only surviving parameter is

$$a = \frac{2}{5} b_4^4. \tag{7.2}$$

Most host crystals for $3d$ ions are nearly cubic, so that this parameter often dominates the zero-field splitting. Deviations from cubic symmetry which retain a twofold axis are conventionally parametrized using $a$, as defined above, together with

$$F = 3(b_4^0 - b_4^4/5)$$
$$D = b_2^0 \tag{7.3}$$
$$E = b_2^2/3.$$

When a threefold axis is retained, $E = 0$ and the above definitions of $a$ and $F$ are replaced by

$$a = -3b_4^3/(20\sqrt{2})$$
$$F = 3(b_4^0 - b_4^3/(20\sqrt{2})). \tag{7.4}$$

Note that spin-Hamiltonians in the form (7.1) can also be used in a formal analysis of the splitting of any isolated group of energy levels (e.g. see [AB70]). Nevertheless, in this book they will only be used as operators acting on the complete ground multiplets of $S$-state ions. For this special case the spin operators transform like vectors under arbitrary rotations. This is a necessary condition for applying the superposition model (see Chapter 5 and Section 7.4).

## 7.2 Experimental results

In the case of lanthanides and $3d$ transition metals it is not usually possible to determine the ground state splitting of $S$-state ions by optical spectroscopy. The standard procedure is to use one of the magnetic resonance techniques. The procedures for determining the energy levels and spin-Hamiltonian parameters are then quite different to the procedures (for fitting crystal field parameters) described in Chapter 3. It is not necessary to go into the details here because these procedures are fully described in the literature. For example, see chapter 3 of Abragam and Bleaney [AB70] for the methods of determining spin-Hamiltonian parameters from conventional (fixed frequency) paramagnetic resonance experiments, and Newman and Urban [NU75] for the determination of spin-Hamiltonian parameters by using the variable frequency method.

### 7.2.1 Ions with the $4f^7$ configuration

Lanthanide spin-Hamiltonian parameters are generally very much smaller than the corresponding crystal field parameters. Their magnitudes also decrease rapidly with increasing $k$. Hence, except where symmetry forces the rank 2 spin-Hamiltonian parameters to be zero, they generally dominate lanthanide zero-field splittings, as is illustrated in Table 7.1. It is often practicable, therefore, to neglect some, or all, of the higher rank parameters. As a consequence, far more empirical values are available for the rank 2 spin-Hamiltonian parameters of $Gd^{3+}$ and $Eu^{2+}$ than for the higher rank parameters. Very few rank 6 parameters have been determined at all.

### 7.2.2 Ions with the $5f^7$ configuration

Trivalent curium has attracted considerable attention in recent years. Its overall ground state splittings are in the range 2–50 cm$^{-1}$ [IMEK97], i.e. an order of magnitude larger than the ground state splittings in $Gd^{3+}$ and $Eu^{2+}$. Several recent papers [LBH93, KEAB93, THE94, MEBA96, IMEK97, LLZ$^+$98] show that $Cm^{3+}$ ground state splittings are most easily determined by optical spectroscopy.

Parameter fittings, which include only the one-electron crystal field contributions and are carried out over complete sets of experimental energy

Table 7.1. Spin Hamiltonian parameters for $Gd^{3+}$ in some scheelite host crystals (from [VP74]). All entries are given in $10^{-4}$ cm$^{-1}$.

| System | $b_2^0$ | $b_4^0$ | $b_4^4$ | $b_6^0$ | $b_6^4$ |
|--------|---------|---------|---------|---------|---------|
| $CaMoO_4$ | $-855$ | $-16.9$ | $-92$ | $0.0$ | $3.2$ |
| $SrMoO_4$ | $-807$ | $-13$ | $-67(15)$ | $0$ | $0$ |
| $CaWO_4$ | $-894$ | $-22.8$ | $-140$ | $0.25(3)$ | $0.2$ |
| $SrWO_4$ | $-868$ | $-17$ | $-119$ | $0$ | $0$ |

Table 7.2. Comparison of some fitted and experimental ground state energy levels for $Cm^{3+}$ in cubic sites in $ThO_2$ [THE94], $D_{2d}$ sites in $LuPO_4$ [MEBA96] and $C_{3h}$ sites in $LaCl_3$ [LBH93]. All energies are given in cm$^{-1}$. Symmetry labels are defined in Appendix 1.

| System | No. of parameters | Fitted energies | Experimental energies | Symmetry label |
|--------|-------------------|-----------------|-----------------------|----------------|
| $ThO_2$ | 2 | $-1.7$ | $0$ | $\Gamma_6$ |
|  |  | $10.8$ | $15$ | $\Gamma_8$ |
|  |  | $40.8$ | $36$ | $\Gamma_7$ |
| $LaCl_3$ | 4 | $-2.25$ | $0.00$ | $5/2$ |
|  |  | $0.49$ | $0.50$ | $1/2$ |
|  |  | $0.93$ | $1.55$ | $5/2$ |
|  |  | $2.99$ | $1.97$ | $3/2$ |
| $LuPO_4$ | 5 | $-8.2$ | $0.00$ | $\Gamma_7$ |
|  |  | $-4.1$ | $3.49$ | $\Gamma_6$ |
|  |  | $7.8$ | $9.52$ | $\Gamma_7$ |
|  |  | $8.7$ | $8.13$ | $\Gamma_6$ |

levels, show the ground multiplet levels to be generally poorly fitted. However, the comparisons given in Table 7.2 show the cubic host, $ThO_2$, to be an exception to this rule. There is about a factor of 2 discrepancy between the fitted and experimental values of the overall splitting in non-cubic hosts.

It is, of course, always possible to fit the ground state splittings to spin-Hamiltonian parameters. The problem in this approach is that, as shown in Table 7.2, there are at most four doubly degenerate energy levels, i.e. sufficient to determine only three parameters. In the cubic case ($ThO_2$) there

Table 7.3. Comparison of crystal field parameters (in $cm^{-1}$) fitted to the ground multiplet energy levels alone (see Table 7.2) and to the complete spectrum for $Cm^{3+}$:$LaCl_3$ ([LBH93]). $B_6^6$ is constrained to be $-0.42B_0^6$ in the ground multiplet fit.

|         | All multiplets | Ground multiplet |
|---------|:--------------:|:----------------:|
| $B_0^2$ | 153            | 36               |
| $B_0^4$ | $-721$         | $-385$           |
| $B_0^6$ | $-1488$        | $-770$           |

are only three energy levels, but these are sufficient to determine the two cubic parameters. In most other systems, such as $LaCl_3$ and $LuPO_4$, fits to spin-Hamiltonian parameters can only be carried out if constraints on the values of some of the parameters are imposed. In the case of lanthanides this problem is less important because of the dominance of the rank 2 spin-Hamiltonian parameters. However, for reasons explained in Section 7.3.2, it is not possible to neglect even the rank 6 parameters in the case of curium. Fitting to curium spin-Hamiltonian parameters may only be possible if constraints, such as those provided by the superposition model (Section 7.4), can be applied to the ratios of parameters of the same rank.

Liu *et al.* [LBH93] have compared crystal field parameters fitted to the ground multiplet and to the whole spectrum for curium in $LaCl_3$. Their results, summarized in Table 7.3, show that while the effective (ground multiplet) rank 2 parameter is reduced by a factor of about 4, the rank 4 and rank 6 parameters are also reduced by a factor of almost 2. The discrepancies in the fitted crystal field parameters shown in Table 7.3 suggest that curium ground state splittings provide a useful means of exploring the relative importance of relativistic crystal field and correlation crystal field contributions in actinide ions.

Analyses of experimental results for $Bk^{4+}$ (e.g. [LCJW94]), show that the fit to the $^8S_{7/2}$ energy levels is every bit as good as that to the levels in the higher-lying multiplets. This is presumably because the admixture of excited state free-ion wavefunctions into the ground state of $Bk^{4+}$ is significantly greater than that in $Cm^{3+}$. In these circumstances the relativistic and correlation crystal field contributions are not expected to play a particularly important role in determining the zero-field splitting. Hence the ground state splitting of $Bk^{4+}$ has no special interest.

Table 7.4. Spin Hamiltonian parameters (in $10^{-4}$ cm$^{-1}$) for Mn$^{2+}$ in some scheelite host crystals (from [BHSB80]).

| System | $b_2^0$ | $b_4^0$ | $b_4^4$ |
|--------|---------|---------|---------|
| CaMoO$_4$ | 33.0 | −0.5 | −4.9 |
| SrMoO$_4$ | 23.4 | −0.4 | −4.8 |
| CaWO$_4$ | −134.8 | −1.5 | −11.5 |
| SrWO$_4$ | −108.9 | −1.2 | −5.3 |

### 7.2.3 Ions with the 3d$^5$ configuration

The divalent manganese, or manganous, ion (ionic radius 0.80 Å) and the ferric ion (ionic radius 0.64 Å) are much smaller than many of the ions that they substitute in crystals. For example, divalent calcium and divalent strontium have radii 0.99 Å and 1.12 Å, respectively. Hence substituted manganese ions fit very loosely into their host crystals when substituted for these ions as, for example, in the case of the scheelites. Loose fits can result in the substituted ion being situated in a site (possibly of lower symmetry) displaced from the position of the ion which it substitutes. However, the only cases where this has been demonstrated clearly are systems where a local vacancy provides charge compensation, e.g. see [SM79b].

Biederbick *et al.* ([BHS78, BHSB80]) have provided a comprehensive survey of the temperature and structure dependence of the spin-Hamiltonian parameters of Mn$^{2+}$ in scheelite (i.e. tungstate and molybdate) host crystals. Some typical parameter values are given in Table 7.4. Significant differences between the tungstate and molybdate parameters are apparent. In particular the rank 2 parameters have different signs. This sign change does not occur when Gd$^{3+}$ is substituted into the same host crystals (see Table 7.1).

## 7.3 Relationship between crystal field and spin-Hamiltonian parameters

Considerable effort has been expended in trying to understand the physical processes which, in addition to the one-electron crystal field, generate contributions to the ground state splitting of *S*-state ions. As most of this work has been concerned with the rank 2 spin-Hamiltonian parameters for Gd$^{3+}$, it is appropriate to focus attention on this case.

Table 7.5. A selection of rank 2 spin-Hamiltonian (units $10^{-4}$cm$^{-1}$) and crystal field parameters (units cm$^{-1}$) for Gd$^{3+}$ from table 1 of [LG92].

| Host crystal | $b_2^0$ | $A_2^0\langle r^2\rangle$ |
|---|---|---|
| $Y_2O_3$ | 1604 | $-850$ |
| $LiNbO_3$ | 1260 | $-417$ |
| $LaAlO_3$ | 479 | $-110$ |
| $YVO_4$ | $-479$ | $-54$ |
| $YPO_4$ | $-728$ | 181 |
| $CaMoO_4$ | $-855$ | 247 |
| $CaWO_4$ | $-920$ | 233 |

### *7.3.1 Rank 2 contributions to the Gd$^{3+}$ spin-Hamiltonian*

One approach to obtaining an understanding of the spin-Hamiltonian parameters has been to seek a phenomenological relationship between their values and the values of the corresponding crystal field parameters. In this spirit Malhotra and Buckmaster [MB82] compared ratios of the rank 2 crystal field and spin-Hamiltonian parameters for Gd$^{3+}$ with oxygen ligands in various host crystals. They could not explain the range of positive and negative ratios obtained. Recent work by Levin and Gorlov [LG92] used a more thorough approach in which the experimental results for each type of ligand were investigated independently and more general linear correlations were examined. Their work also studied the correlations between both sets of parameters and the nuclear quadrupole interaction (Section 5.6.3).

The results obtained by Levin and Gorlov [LG92] can best be illustrated by using a small sample of the large number of experimentally determined parameters tabulated in their work. An extract of their table for oxygen ligands, comparing the spin-Hamiltonian parameters $b_2^0$ for Gd$^{3+}$ with the corresponding rank 2 crystal field parameters for a variety of trivalent lanthanide ions of similar radii, is given in Table 7.5. The entries in this table are ordered in terms of decreasing values of $b_2^0$, making it clear that this order correlates well with increasing values of $A_2^0\langle r^2\rangle$, although their *ratios* vary widely. In fact, a fairly good straight line can be drawn close to all the points on a graph of $A_2^0\langle r^2\rangle$ versus $b_2^0$. Levin and Gorlov [LG92] analyse these results using the 'two-power law' model described in Section 5.1.3.

### 7.3.2 Form of the ground states

Crystal field contributions to $L = 0$ ground state splitting arise because of the breakdown of pure $LS$ coupling, leading to a (largely spin–orbit induced) admixture of multiplets with $L \neq 0$. These admixtures tend to be small, particularly in the case of lanthanides, because the lowest-lying excited states (with $L \neq 0$) are at energies which are large compared with the magnitude of the spin–orbit coupling. As will become apparent in Section 7.5.2, the dominant $^6P_{\frac{7}{2}}$ admixture contributes to the ground state splitting by the relativistic and correlation crystal field, but not the one-electron crystal field.

Following Wybourne [Wyb66], the ground state of ions with $f^7$ configurations can be constructed out of states with exact $L$, $S$ and $J = 7/2$ quantum numbers as follows

$$|\mathcal{S}_{\frac{7}{2}}\rangle = s|^8S_{\frac{7}{2}}\rangle + p|^6P_{\frac{7}{2}}\rangle + d|^6D_{\frac{7}{2}}\rangle + f|^6F_{\frac{7}{2}}\rangle + g|^6G_{\frac{7}{2}}\rangle + \cdots . \qquad (7.5)$$

The coefficients in this equation can be determined from fits to the energy levels obtained from optical spectroscopy (as detailed in Chapter 4) and are fairly independent of the host crystal because the energy of the first excited multiplet is very much greater than the crystal field energies. In the case of $Gd^{3+}$, it is a good approximation to use the values [Wyb66, NU75]

$$s = 0.9866, p = 0.162, d = -0.0123, f = 0.0010, g = -0.00014. \qquad (7.6)$$

The $Eu^{2+}$ ground state has not been determined with the same precision. A reasonable estimate of the dominant contributions (see [NU75]) is

$$s = 0.991, p = 0.134, d = -0.0102. \qquad (7.7)$$

Actinide ions show very much larger admixtures of excited multiplets into the ground multiplet, accounting for the fact that the observed ground state splittings are larger. Liu, Beitz and Huang [LBH93] have reported the extent of this mixing in the ground state of $Cm^{3+}$ substituted into $LaCl_3$. Their results for the coefficients defined above are

$$s = 0.8866, p = 0.4222, d = -0.0922, f = 0.0226, g = -0.0084. \qquad (7.8)$$

These are very similar to the free-ion values given by Edelstein [Ede95], viz.

$$s = 0.8859, p = 0.4232, d = -0.0926, f = 0.0227, \qquad (7.9)$$

where some corrections in the relative signs have been made. The close agreement between these two sets of coefficients leads us to expect that the coefficients given by Liu *et al.* [LBH93] provide a fairly accurate description

of the trivalent curium ground state in *any weak crystal field*. Nevertheless, they cannot be expected to be accurate for strong crystal fields, as in
the case of $Cm^{3+}$ in $ThO_4$. Liu *et al.* [LBH93] also give the values of 11
further coefficients which might need to be taken into account in accurate
calculations.

The coefficients in (7.5) reduce with increasing $L$ for both lanthanides
and actinides. This reduction is much more pronounced in the case of
lanthanides, and explains the very small rank 4 and rank 6 contributions to
their zero-field splittings.

## 7.4 Superposition model

The physical mechanisms which produce $S$-state ion ground state splittings
are far more complicated than those which produce the one-electron crystal
field. Nevertheless, because these additional mechanisms can all be associated with single ligands interacting with the magnetic ion, the theoretical
justification for applying the superposition model to spin-Hamiltonian parameters is much the same as in the case of the crystal field parameters.
Moreover, empirical evidence is accumulating which shows the superposition
model to work quite well for spin-Hamiltonian parameters, and provides a
useful tool in determining the relative importance of the various contributions to the ground state splitting.

### 7.4.1 Formulation and interpretation

The only fundamental requirement for using the superposition model is that
the Hamiltonians to which it is applied act on sets of states which form
irreducible representations of the full rotation group $SO_3$ (see Appendix 1).
This ensures that the set of states is closed with respect to the arbitrary
rotations that are necessary to express the individual ligand contributions
in a common coordinate system. Both the $2J + 1$ dimensional basis used in
crystal field theory, and the $2l + 2 = 2S + 1$ dimensional basis on which the
corresponding spin-Hamiltonian acts, satisfy this criterion.

Given that the transformation properties of spin tensor operators are
identical to those of orbital tensor operators, it follows that the superposition model for spin-Hamiltonians is formally the same as the model for
crystal fields, described in Chapter 5. Hence it is not necessary to reiterate
the discussion in that chapter, but to merely rewrite equation (5.4) in terms

of spin-Hamiltonian parameters

$$b_k^q = \sum_L \bar{b}_k(R_L) G_{k,q}(\theta_L, \phi_L). \qquad (7.10)$$

Here the coefficients $G_{k,q}$ are the so-called 'coordination factors' (in Stevens normalization) tabulated in Table 5.1. The *intrinsic spin-Hamiltonian parameters* $\bar{b}_k(R_L)$ represent the axially symmetric contributions (for each $k$) of ligand $L$.

It is possible to follow the analogy with crystal fields one step further, and express the distance dependence of the intrinsic parameters $\bar{b}_k(R_L)$ in terms of power laws

$$\bar{b}_k(R) = \bar{b}_k(R_0) \Big(\frac{R_0}{R}\Big)^{t'_k}. \qquad (7.11)$$

The prime is used to distinguish spin-Hamiltonian from crystal field power law exponents. It turns out, however, that the distance variations of the rank 2 and rank 4 intrinsic spin-Hamiltonian parameters often do not have a simple power law dependence. There is also evidence that the intrinsic parameters $\bar{b}_k$ can have different signs, even for a given ligand. As we shall see, these variations in the sign, and the related breakdown of the power law dependence of the parameters on ligand distance, do not signal a breakdown of the superposition model. These peculiar properties are a straightforward consequence of strong cancellations between the several different mechanisms which contribute to the ground state splitting.

In the case of lanthanides, the most precise information is available for the rank 2 spin-Hamiltonian parameters. As was pointed out in Chapter 5, however, several difficulties arise in applying the superposition model to the rank 2 crystal field parameters. This was partly because of the strong cancellations that generally occur between the contributions from different ligands, and partly because of the possible significance of non-local electrostatic contributions. Similar problems might also be expected in the case for the rank 2 spin-Hamiltonian parameters. Certainly, given that the same coordination factors are used in both types of analysis, it is clear that strong cancellation effects between single-ion contributions may also occur in this case.

It has been suggested, especially in the case of $Mn^{2+}$, that there are significant contributions to the spin-Hamiltonian which are quadratic in the crystal field parameters. This might certainly be expected to occur when there is a crystal field induced admixture of (free-ion) excited states into the ground state. In these circumstances the superposition model would not provide a valid procedure for analysing the empirical results. However, given

Table 7.6. A selection of derived intrinsic spin-Hamiltonian parameters $\bar{b}_k$ $(cm^{-1})$ and power law exponents $t_2'$ for $Gd^{3+}$ in various host crystals.

| Host | Ligand | $\bar{b}_2$ | $t_2'$ | $\bar{b}_4 \times 10^4$ | Source |
|------|--------|-------------|--------|-------------------------|--------|
| $CaF_2$ | $F^-$ | −0.065 | | 15.0 | [NU75] |
| $LiYF_4$ | $F^-$ | −0.13 | −0.6 | | [VBG83] |
| YAG | $O^{2-}$ | −0.26 | 0 | 17.5 | [New75, NE76] |
| $CaWO_4$ | $O^{2-}$ | −0.18 | | 11(2) | [New75] |
| $CaMoO_4$ | $O^{2-}$ | −0.16 | | 8(3) | [New75] |
| CaO | $O^{2-}$ | −0.20 | 0.4 | −3.5 | [SN74, NU75] |
| SrO | $O^{2-}$ | −0.20 | 0.8 | −1.4 | [SN74, NU75] |
| $YVO_4$ | $O^{2-}$ | −0.10 | 1.0 | −2.5 | [New75] |
| $LaCl_3$ | $Cl^-$ | 0.0 | | −1.6 | [New75] |
| $LaBr_3$ | $Br^-$ | | | −8.5 | [New75] |
| CdS | $S^{2-}$ | −0.13 | | | [USH74] |

the inherent uncertainties in making quantitative estimates of the various contributions to the spin-Hamiltonian parameters, it seems best to regard the superposition model, as applied to the spin-Hamiltonian parameters, as an empirically testable model in its own right. If it can be shown to be valid then, just as in its application to crystal fields, it provides an effective means of separating information about crystal structure from information about electronic structure.

### 7.4.2 Applications to the zero-field splitting of $Gd^{3+}$

The superposition model accounts, at least semi-quantitatively, for the experimentally determined values of the $Gd^{3+}$ rank 2 spin-Hamiltonian parameters in a wide range of host crystals [NU75, NN89b]. This suggests that the net non-local contributions to these parameters are not very large. As there is usually only one rank 2 spin-Hamiltonian parameter for a given system, the main test of the applicability of the superposition model to rank 2 parameters is whether different hosts produce similar values of the intrinsic parameter $\bar{b}_2$. The results collected in Table 7.6 show this expectation to be fulfilled in the case of $Gd^{3+}$: its rank 2 intrinsic spin-Hamiltonian parameters are invariably negative and of the order of $10^{-1}$ $cm^{-1}$. More precise tests of the superposition model can be made for the rank 4 spin-Hamiltonian parameters, as will be demonstrated next.

The discussion in the previous section showed that, while some regularity appeared in the relationship between the $B_0^2$ and $b_2^0$ parameters, the physical basis for this relationship was far from obvious. Is the corresponding relationship between the values of $\bar{b}_2$ and $\overline{B}_2$ capable of being understood in terms of the underlying physical mechanisms? In the case of $Gd^{3+}$, the ratio between these intrinsic parameters for the nearest ligand has been found [NU75] to be of the order

$$\bar{b}_2/\overline{B}_2 = \bar{b}_2/2\overline{A}_2 = -1.4 \times 10^{-4} \qquad (7.12)$$

for several host crystals with oxygen ligands [NU75]. However, the very different power laws for $\bar{b}_2$ and $\overline{B}_2$ mean that this ratio must be regarded as approximate.

The proposed explanation for the low, and sometimes negative, values of $t_2'$ shown in Table 7.6 is that there is a very close cancellation of contributions from the crystal field and relativistic crystal field, both of which are negative, with a positive contribution from the correlation crystal field (see Section 7.5.2). This results in a minimum in the function $\bar{b}_2(R)$ close to the nearest ligand distance (see figure 9 of [NU75]).

Determinations of the rank 4 intrinsic parameters and power law exponents have produced even more surprises than the rank 2 spin-Hamiltonian parameters. The main puzzle is the very large range of values that is obtained for these parameters, as shown in Table 7.6. Overall, a systematic variation occurs, in which the more polarizable ions give the more negative parameters. The oxygen intrinsic parameter $\bar{b}_4$ can take either positive or negative values, depending on the host crystal. This is presumably due to the well-known variability of the polarizability of $O^{2-}$ in different crystalline environments. It suggests that $\bar{b}_4$ could be used as a measure of ligand polarizability, although a quantitative framework for this technique has yet to be drawn up.

Given that highly polarizable ligands produce negative values of $\bar{b}_4$, it is clear that there is a large negative contribution due to ligand polarizability. This is most likely to be a correlation crystal field contribution, arising from the dependence of some coulomb integrals on screening (see Section 6.2.3). The sensitivity of the rank 4 spin-Hamiltonian parameters to this contribution suggests that the rank 4 correlation crystal field may be largely due to ligand polarization.

Table 7.7. Selected rank 2 $Gd^{3+}$ spin-Hamiltonian and $Er^{3+}$ crystal field parameters for the zircon structure crystals [NU72]. Table 5.11 gives coordination angles and combined coordination factors.

| Compound | $b_2^0 \times 10^4$ cm$^{-1}$ | $B_0^2$ cm$^{-1}$ |
|----------|---------------------------------|---------------------|
| $YVO_4$  | $-445$ | $-102.8$ |
| $ScVO_4$ | $-381$ | $-238.6$ |
| $YPO_4$  | $-729$ | $141.4$  |
| $YAsO_4$ | $-319$ | $-30.6$  |

### 7.4.3 $Gd^{3+}$ with oxygen ligands

The first application of the superposition model to spin-Hamiltonian parameters [NU72, NU75] was carried out in order to seek an understanding of the fact that, as shown in Table 7.7, the rank 2 crystal field parameters show more variation than spin-Hamiltonian parameters in different zircon structure hosts. (The comparison is made with $Er^{3+}$ crystal field parameters because better results are available for this ion and it is similar in size to $Gd^{3+}$). The superposition model analysis reported in Section 5.4.2 showed that the variations in magnitude and sign of the rank 2 crystal field parameters in these host crystals (see Table 7.7) can be understood in terms of reasonable values of the intrinsic crystal field parameter $\overline{B}_2$ and the corresponding power law exponent $t_2$, viz.

$$\overline{B}_2 = 725 \text{ cm}^{-1}, \ t_2 = 7. \tag{7.13}$$

The spin-Hamiltonian parameters quoted in Table 7.7 are negative for all nine hosts. Newman and Urban [NU72] showed that these small variations could be understood in terms of an 'effective' power law exponent $t_2' \leq 1$ and an intrinsic parameter $\overline{b}_2$ of about $-0.1$ cm$^{-1}$. If the more distant ion contributions can really be neglected, such a small power law exponent is only likely to be valid over a very small range of interionic distances. The most likely explanation is that there are two cancelling contributions with different distance dependences giving rise to a minimum in $\overline{b}_2$ very close to the oxygen distance in the zircon structure hosts. In these circumstances the power law approximation is very poor and one must expect the effective power law exponent to vary with oxygen distance.

It is of interest to compare the spin-Hamiltonian intrinsic parameters and power law exponents for $Gd^{3+}$ in yttrium aluminium garnet (YAG) and yttrium gallium garnet (YGG) host crystals [New75, NE76]. There are three rank 4 spin-Hamiltonian parameters, making it possible to determine $t'_4 = 4.2$ for YAG and $t'_4 = 3.5$ for YGG [New75]. If it is assumed that $Gd^{3+}$ ions in the two garnets have effectively the same intrinsic parameters for a given oxygen distance, it is also possible to use the two rank 4 intrinsic parameter values to obtain a third estimate of the power law exponent. This is obtained as $t'_4 = 5.9$, in reasonable agreement with the two values obtained for the separate systems.

### 7.4.4 Divalent europium

Levin and Eriksonas [LE87] carried out a detailed superposition model analysis of the rank 2 spin-Hamiltonian parameters for $Eu^{2+}$ substituted into $LaCl_3$, the fluorites ($CaF_2$, $SrF_2$, $BaF_2$), CaFCl, SrFCl and BaFCl. Allowing for effects of local distortion in the fluorites, they obtained consistent values for the fluorine intrinsic parameters and power law exponents, viz. $\bar{b}_2 = -0.0483$ cm$^{-1}$, $t'_2 = -2.0$. These values are reasonably close to those obtained from uniaxial stress experiments [EN75], viz. $\bar{b}_2 = -0.0616(30)$ cm$^{-1}$, $t'_2 = -0.8(2)$. A rather more complicated analysis of the data for $Eu^{2+}$:$LaCl_3$ gives positive values for both the rank 2 chlorine intrinsic parameter $\bar{b}_2 = 0.0955$ cm$^{-1}$ and power law exponent $t'_2 = 2.0$. Using these values, Levin and Eriksonas [LE87] were able to get quite good agreement with the experimental rank 2 spin-Hamiltonian parameters for $Eu^{2+}$ substituted for the alkaline earths in $BaF_2$, CaFCl, SrFCl and BaFCl.

### 7.4.5 Ground state splittings in Mn²⁺ and Fe³⁺

The successful application of the superposition model to spin-Hamiltonian parameters of ions with $4f^7$ configurations [NU75] led to the idea that a similar approach might be valid for $3d^5$ ions [NS76]. This was supported, at least in part, by the theoretical analysis of Novák and Veltruský [NV76], who showed that important overlap and covalency contributions to the rank 2 spin-Hamiltonian parameters could indeed be calculated using their 'independent bond approach'. The poor numerical results of their calculation, however, made it clear that they had omitted some important contributions to the intrinsic parameters. Previous theoretical derivations of the ground state splittings in $3d^5$ ions ([SDO66, SDO67, SDO68]) had predicted significant contributions to the spin-Hamiltonian parameters that are non-linear in the crystal field parameters, bringing the linearity assumption of the superposition model into question. However, the superposition model works

Table 7.8. Selected intrinsic spin-Hamiltonian parameters $\bar{b}_k$ and power law exponents $t'_k$ for $Mn^{2+}$ and $Fe^{3+}$ from [NS76] and [SM79b]. $R$ is the ligand distance. Square brackets indicate that assumed values were used in the analysis.

| Ion | Host | $R$(Å) | $\bar{b}_2$(cm$^{-1}$) | $t'_2$ | $\bar{b}_4 \times 10^4$(cm$^{-1}$) | $t'_4$ |
|---|---|---|---|---|---|---|
| $Mn^{2+}$ | MgO | 2.101 | $-0.157(5)$ | 7(1) | 2.72 | 8(1) |
| $Mn^{2+}$ | CaO | 2.398 | $-0.050(10)$ | 7(1) | 0.84 | 8(1) |
| $Mn^{2+}$ | Ca(WO)$_4$ | 2.438 | $-0.0241$ | [7] | 0.7(3) | |
| $Mn^{2+}$ | Sr(WO)$_4$ | 2.59 | $-0.0185$ | [7] | | |
| $Mn^{2+}$ | Ca(CO)$_3$ | 2.398 | $-0.026(20)$ | | 0.8(1) | 8(1) |
| $Mn^{2+}$ | ZnS | | | | 1.13 | |
| $Fe^{3+}$ | CaO | 2.398 | $-0.225(20)$ | 5(1) | | |
| $Fe^{3+}$ | MgO | 2.101 | $-0.412(25)$ | 8(1) | | |
| $Fe^{3+}$ | SrTiO$_3$ | 1.952 | $-0.63(6)$ | [8] | | |
| $Fe^{3+}$ | KNbO$_3$ | 1.883 | | | 16.0(3) | [8] |
| $Fe^{3+}$ | TlCdF$_3$ | 2.198 | $-0.095(30)$ | [8] | | |
| $Fe^{3+}$ | RbCdF$_3$ | 2.198 | $-0.084(32)$ | [8] | | |
| $Fe^{3+}$ | RbCaF$_3$ | 2.229 | $-0.076(18)$ | [8] | | |

well in practice, providing reasonably consistent values of intrinsic parameters and power law exponents for both $Mn^{2+}$ and $Fe^{3+}$. The initial results obtained by Newman and Siegel [NS76], with the extensions and corrections of Siegel and Müller [SM79b], are summarized in Table 7.8.

The regularities in the entries of Table 7.8 are evidence for the value of the superposition model in relating experimental results for different systems. These regularities were exploited by Siegel and Müller [SM79a] to show that substituted ferric ions do not follow the cooperative ferroelectric displacements of titanium ions in BaTiO$_4$. Little work has been carried out for $3d^5$ ions with ligands other than oxygen, however, so our comments must necessarily be restricted to the intrinsic parameters and power law exponents for this ligand. In summary, the rank 2 intrinsic parameters for both $Mn^{2+}$ and $Fe^{3+}$ ions are negative, with those for $Fe^{3+}$ being significantly greater than those for $Mn^{2+}$. This sign agrees with that of $\bar{b}_2$ for $Gd^{3+}$ and $Eu^{2+}$, suggesting that the same mechanisms dominate. Rank 2 intrinsic parameters for $Fe^{3+}$ with fluorine ligands are much smaller in magnitude than they are for oxygen ligands. The power law exponents for both $Mn^{2+}$ and $Fe^{3+}$

ions are close to 7 for the rank 2 parameters and close to 8 for the rank 4 parameters.

Given the complexities in the dependence of intrinsic parameters on ligand distance in the case of the lanthanides, it would be surprising if the intrinsic parameters for the $3d^5$ ions followed a simple power law distance dependence for significant displacements of the relative position of the magnetic ion and ligand. This question has been investigated in some detail for ferric ions in perovskites by Donnerberg et al. [DEC93]. They carried out a detailed examination of the behaviour of rank 2 intrinsic parameters, using a shell model analysis to determine the local distortion near $Fe^{3+}$ substituted into the potassium sites of $KTaO_3$. They were led to propose a 'Lennard–Jones-type' distance dependence for $\bar{b}_2(R)$, very similar in form to that which had already been suggested [NU72, NU75] for the distance dependence of $\bar{b}_2$ in $Gd^{3+}$ (see Section 7.4.3) and $Eu^{2+}$. While further experimental evidence is still required, this general form of distance dependence of the $\bar{b}_2$ parameter can now be regarded as well established.

Donnerberg [Don94] also performed a superposition model analysis of the spin-Hamiltonian parameters for $Fe^{3+}$ substituting niobium in $KNbO_3$. Because of the associated oxygen vacancy, a large local distortion is produced. Consistent results were obtained by combining a shell model calculation with the superposition model analysis. The distance dependence of $\bar{b}_2$ was again shown to take the form of a Lennard–Jones potential. The clear message is, therefore, that successful applications of the superposition model to systems with large local distortions depend on the use of realistic distance dependences for the intrinsic parameters. When this is taken into account, the results obtained by Donnerberg et al. [DEC93, Don94] suggest that the superposition model provides a useful tool in the determination of quite large changes in the atomic structure of crystals near substituted sites.

## 7.5 Mechanisms for zero-field splitting

The most detailed analysis has been carried out for the $Gd^{3+}$ ion [NU75], so it is of particular interest to consider the various zero-field splitting mechanisms for this case. Similar arguments can certainly be applied to $Eu^{2+}$ zero-field splittings, and possibly to the zero-field splittings of actinide and $3d$ transition metal ions with half-filled shells.

### 7.5.1 *Relativistic crystal field*

The relativistic crystal field (introduced in Section 6.2.1) is produced by a spin–orbit admixture of the one-electron states $5f$, $6f$, etc. into the $4f$ open-shell states. In consequence, the open-shell one-electron states have radial wavefunctions that depend on whether $j = 5/2$ or $7/2$. The resulting difference in the crystal field splitting is very small, being of the order of 5% of the one-electron crystal field parameters in the lanthanides, and even smaller in the actinides and transition metal ions. Normally such small contributions cannot be separated from the one-electron crystal field. However, in the $S$ ground state of $Gd^{3+}$, the rank 2 relativistic crystal field has non-zero matrix elements between the $^6P_{\frac{7}{2}}$ and $^8S_{\frac{7}{2}}$ states. Because $^6P_{\frac{7}{2}}$ is the dominant higher-lying multiplet admixed into the ground multiplet, this produces relativistic crystal field contributions to the ground state splitting which are similar in magnitude to those produced by the one-electron crystal field.

### 7.5.2 *Contributions to the ground state splitting in $Gd^{3+}$*

Apart from the relativistic crystal field (RCF) contributions just described, significant contributions to the rank 2 spin-Hamiltonian parameters arise from crystal field (CF) contributions to the cross matrix element between the states $^6P_{\frac{7}{2}}$ and $^6D_{\frac{7}{2}}$ and correlation crystal field (CCF) contributions to the $^6P_{\frac{7}{2}}$ diagonal matrix elements (see Chapter 6 and [NU75]). It is convenient, therefore, to express the rank 2 intrinsic spin-Hamiltonian parameter as the sum of these three contributions, viz.

$$\bar{b}_2 = \bar{b}_2(\text{CF}) + \bar{b}_2(\text{RCF}) + \bar{b}_2(\text{CCF}). \tag{7.14}$$

As shown in [NU75], the first contribution in this equation can be expressed directly in terms of the crystal field rank 2 intrinsic parameter $\overline{B}_2$, as follows

$$\bar{b}_2(\text{CF}) = \frac{2\sqrt{5}}{105}pd\overline{B}_2. \tag{7.15}$$

Formulae giving additional contributions due to the admixture of other excited multiplets into the $f^7$ ground multiplet are given in [LBH93], but these are not important in the case of $Gd^{3+}$. Inserting the values of the coefficients $p, d$ for $Gd^{3+}$ (given in (7.6)) into (7.15) produces the numerical result

$$\bar{b}_2(\text{CF}) = -1.70 \times 10^{-4}\overline{B}_2, \tag{7.16}$$

which is very close to the empirical result given in equation (7.12). However, not too much should be made of this, because it does not provide an understanding of the considerable differences in power law exponents determined for the rank 2 spin-Hamiltonian and crystal field parameters.

The relativistic crystal field contribution has been calculated [NU75] to be

$$\bar{b}_2(\text{RCF}) = \frac{4}{35\sqrt{14}} sp\left\{\overline{B}_2\left(j = \frac{7}{2}\right) - \overline{B}_2\left(j = \frac{5}{2}\right)\right\}, \qquad (7.17)$$

where the intrinsic crystal field parameters are differentiated by the $j$ values which label the corresponding wavefunctions. Using the estimate of this difference derived by Newman and Urban [NU75], (7.17) gives

$$\bar{b}_2(\text{RCF}) = -(5.3 \times 10^{-4})\overline{B}_2, \qquad (7.18)$$

where $\overline{B}_2$ is the intrinsic parameter derived from *all* multiplets. Note that this has the same sign, but is much greater in magnitude, than the crystal field contribution. As it can be expressed in terms of the same intrinsic parameter, it should depend on ligand distance in much the same way as $\bar{b}_2(\text{CF})$.

Theoretical estimates of the magnitude of correlation crystal field contributions to the ground state splitting are difficult to obtain. However, empirical determinations show the ratio $\bar{b}_2/\overline{B}_2$ to be negative, in agreement with the above calculation, but much smaller in magnitude than is obtained by summing the crystal field and relativistic crystal field contributions. This leads to the hypothesis that the correlation crystal field contributions have the opposite sign to the crystal field and relativistic crystal field contributions. Using the experimentally determined result for several systems with oxygen ligands, viz. $\bar{b}_2 = -(1.4 \times 10^{-4})\overline{B}_2$, and the crystal field and correlation crystal field contributions quoted above, the correlation crystal field contribution can be estimated to be

$$\bar{b}_2(\text{CCF}) = (5.6 \times 10^{-4})\overline{B}_2. \qquad (7.19)$$

This should be regarded as an estimate of the magnitude of the correlation crystal field contribution at the nearest ligand distance: it should not be taken to imply that the distance dependence is the same for the correlation crystal field contribution as it is for the crystal field and relativistic crystal field contributions.

Such a strong cancellation between the different contributions is consistent with the suggestion, made on empirical grounds in Section 7.4.3, that the low dependence of spin-Hamiltonian intrinsic parameter $\bar{b}_2$ on ligand distance is produced by the cancellation of two contributions with opposite signs and large power law exponents. Moreover, a two-power law model of the distance dependence (see [NU75]) shows the positive contribution, due to the correlation crystal field, to have a smaller power law exponent than the negative contribution, due to the crystal field and relativistic crystal field.

## 7.6 Outlook

Superposition model analyses of spin-Hamiltonian parameters, especially when combined with shell model calculations, provide a useful tool in the determination of crystal structures in the neighbourhood of substituted magnetic ions. While it is clear that spin-Hamiltonian parameters also contain useful information about ligand polarizability and the correlation crystal field, their potential use in obtaining a better understanding of the electronic structure of magnetic ions and their ligands has yet to be developed.

# 8

# Invariants and moments

## Y. Y. YEUNG

*Hong Kong Institute of Education*

The coordinate frame dependence of conventional crystal field parameters $B_q^k$ makes it possible that quite different sets of values of $B_q^k$ represent the same crystal field. This is of especial significance in the case of low-symmetry sites, where there are several alternative directions of the principal axis. In order to make direct comparisons between different sets of parameters it is necessary to carry out transformations between the various (implicit) coordinate frames (see Appendix 4).

In the case of low-symmetry sites, the least-squares fitting procedure is often ill-conditioned, allowing many or even continuous sets of equivalent minima. It may not be apparent, therefore, just what coordinate frame corresponds to a given set of phenomenological parameters. A frequently used approach has been to use an approximate higher symmetry in carrying out the crystal field analysis. For example, in the LaF$_3$ host crystal, the La$^{3+}$ site symmetry is C$_2$. As indicated in Chapter 2, the $C_2$ axis may be aligned in either the $z$ or $y$ directions, determining respectively whether the odd $q$ or negative $q$ parameters are identically zero. In either case, 15 non-zero crystal field parameters are required (see Table 2.3). Non-linear least-squares fits to as many as 15 parameters are notoriously unreliable unless good approximate values are known which can be used to initiate the fitting procedure. Two higher symmetries, D$_{3h}$ and C$_{2v}$ have been used in spectroscopic investigations (e.g. see [SN71a], [CCC77], [ML79], [CC83] and [CGRR89]). A comparison of the five independent fits for Er$^{3+}$:LaF$_3$ given in Table 8.1 shows that a consensus is far from being obtained due to the coordinate dependence of the parameters, even though there is rough agreement in signs and orders of magnitude. It is therefore highly desirable to seek an alternative characterization of crystal fields in terms of *crystal field* or *rotational invariants* ([Lea82, YN85a, YN86a]), which are independent of coordinate frame and/or site symmetry.

160

Table 8.1. Fitted crystal field parameters (cm$^{-1}$) for Er$^{3+}$:LaF$_3$ with assumed site symmetries C$_2$, C$_{2v}$ and D$_{3h}$. Values in brackets denote errors and [0] indicates that the parameter has been set at zero. Part of this table is reproduced (with the permission of the American Institute of Physics) from table I of [YN85a].

| | C$_2$ | C$_2$ | C$_{2v}$ | C$_{2v}$ | D$_{3h}$ |
|---|---|---|---|---|---|
| $B_0^2$ | −66 | −228 | −209 | −238(17) | 282 |
| $B_2^2$ | −149 | −119 | −99 | −91(14) | [0] |
| $B_{-2}^2$ | 20 | [0] | [0] | [0] | [0] |
| $B_0^4$ | 232 | 545 | 492 | 453(90) | 1160 |
| $B_2^4$ | 195 | 301 | 380 | 308(60) | [0] |
| $B_{-2}^4$ | −188 | 108 | [0] | [0] | [0] |
| $B_4^4$ | 675 | 358 | 404 | 417(56) | [0] |
| $B_{-4}^4$ | 158 | 219 | [0] | [0] | [0] |
| $B_0^6$ | 80 | 275 | 423 | 373(83) | 773 |
| $B_2^6$ | −642 | −520 | −473 | −489(51) | [0] |
| $B_{-2}^6$ | [0] | −73 | [0] | [0] | [0] |
| $B_4^6$ | −23 | 56 | −238 | −240(51) | [0] |
| $B_{-4}^6$ | 122 | −305 | [0] | [0] | [0] |
| $B_6^6$ | −403 | −307 | −529 | −536(49) | [0] |
| $B_{-6}^6$ | −52 | −368 | [0] | [0] | 453 |
| Source | [SN71a] | [ML79] | [CC83] | [CGRR89] | [CCC77] |

Leavitt [Lea82] obtained a relationship between crystal field parameters and crystal field invariants, which are quadratic in the parameters. He also showed that values of these invariants could be obtained directly from the experimental spectra by means of the so-called 'quadratic crystal field splitting moments'. In Section 8.1, we give the general relations between the quadratic and higher order moments and their corresponding rotational invariants. This approach is further developed in Section 8.2 by relating the quadratic invariants to the superposition model intrinsic parameters (described in Chapter 5).

The possibility of using quadratic invariants in conjunction with the superposition model to determine crystal field parameters for cubic and low-symmetry systems is investigated in Section 8.3. In the same section it is shown that the rotational invariants also provide a viable method [YN86b] for determining the spin-correlated crystal field parameters (introduced in Chapter 6).

### 8.1 Relations between moments and rotational invariants

Apart from analysing the individual $J$ multiplet splitting levels, it is possible to study the spread of the energy levels in the $J$ multiplets by means of statistics such as their mean value (first order moment), standard deviation (quadratic or second order moment), cubic (third order) moment and quartic (fourth order) moment, etc. These moments are clearly invariant with respect to any rotation of the (implicit) coordinate frames used to describe the crystal field in terms of the parameters $B_q^k$.

### 8.1.1 Definition of the crystal field splitting moments

The crystal field energies of the levels $\Gamma_i$ in an $\alpha LSJ$ multiplet can be written in terms of the corresponding eigenstates $|(\alpha LSJ)\Gamma_i\rangle$ as

$$E_i(\alpha LSJ) = E^0(\alpha LSJ) + \langle(\alpha LSJ)\Gamma_i|V_{\mathrm{CF}}|(\alpha LSJ)\Gamma_i\rangle, \qquad (8.1)$$

where $E^0$ is the mean energy of the multiplet. The $n$th order moment $(\sigma_n(\alpha LSJ))$ is defined as

$$(\sigma_n(\alpha LSJ))^n = \frac{1}{2J+1}\sum_i [E_i(\alpha LSJ) - E^0(\alpha LSJ)]^n. \qquad (8.2)$$

If $H_{\mathrm{CF}}$ denotes the matrix of the crystal field operator $V_{\mathrm{CF}}$, referred to an arbitrary basis spanning the multiplet $\alpha LSJ$, the $n$th order moment can also be expressed as

$$(\sigma_n(\alpha LSJ))^n = \frac{1}{2J+1}\mathrm{Tr}(H_{\mathrm{CF}})^n, \qquad (8.3)$$

where the trace operation Tr is taken over the $\alpha LSJ$ basis. The right hand side of equation (8.3) can be evaluated by using diagrammatic angular momentum theory (e.g. see [LM86] or [BS68]).

Substituting the crystal field expansion (2.1)

$$H_{\mathrm{CF}} = \sum_{k,q} \hat{B}_q^k C_q^{(k)}(\alpha LSJ),$$

into (8.3), where $C_q^{(k)}(\alpha LSJ)$ is to be interpreted as the matrix of the tensor operator $C_q^{(k)}$ for the $\alpha LSJ$ basis, it is possible to express the $n$-th order moment in terms of the complex parameters $\hat{B}_q^k$. The results of applying this for $n = 2$, $3$ and $4$ are discussed next.

### 8.1.2 Moments in terms of crystal field parameters

Quadratic, or $n = 2$, rotational invariants can be defined in terms of the crystal field parameters as

$$s_k^2 = \frac{1}{2k+1} \sum_{q=-k}^{k} |\hat{B}_q^k|^2. \tag{8.4}$$

In the case of lanthanides and actinides there are three of these invariants, corresponding to the ranks $k = 2$, $k = 4$ and $k = 6$. The quadratic, or second order $(n = 2)$, moments of the multiplets can be expressed in terms of rotational invariants as

$$(\sigma_2(\alpha LSJ))^2 = \frac{1}{2J+1} \sum_{k=2,4,6} s_k^2 \langle \alpha LSJ || C^{(k)} || \alpha LSJ \rangle^2, \tag{8.5}$$

where $\langle \alpha LSJ || C^{(k)} || \alpha LSJ \rangle$ is a reduced matrix element (see Chapter 3 and Appendix 1).

Cubic, or $n = 3$, rotational invariants are defined as

$$\nu_3(k_1 k_2 k_3) = \sum_{q_1,q_2,q_3} \begin{pmatrix} k_1 & k_2 & k_3 \\ q_1 & q_2 & q_3 \end{pmatrix} \hat{B}_{q_1}^{k_1} \hat{B}_{q_2}^{k_2} \hat{B}_{q_3}^{k_3}. \tag{8.6}$$

The definition of the (bracketed) $3j$ symbol can be found in Appendix 1. In the case of the lanthanides and actinides there are nine such invariants corresponding to the nine inequivalent sets of $k$ values, viz.

$$(k_1 k_2 k_3) = (222), (224), (246), (244), (444), (446), (266), (466), (666). \tag{8.7}$$

In high symmetry systems, the invariants $\nu_3(k_1 k_2 k_3)$ may not all be independent. Their inter-dependence is determined by their relationships to crystal field parameters given in (8.6).

The third order $(n = 3)$ moments can be expressed in terms of the cubic rotational invariants as

$$(\sigma_3(\alpha LSJ))^3 = \frac{1}{2J+1} \sum_{k_1,k_2,k_3=2,4,6} (-1)^{2J} \nu_3(k_1, k_2, k_3) \begin{Bmatrix} k_1 & k_2 & k_3 \\ J & J & J \end{Bmatrix}$$
$$\times \prod_{k_i; i=1,2,3} \langle \alpha LSJ || C^{(k_i)} || \alpha LSJ \rangle. \tag{8.8}$$

The fourth order $(n = 4)$ moments are most conveniently written as

$$(\sigma_4(\alpha LSJ))^4 = \frac{1}{2J+1} \sum_{k_1,k_2,k_3,k_4} \prod_j \langle \alpha LSJ || C^{(k_j)} || \alpha LSJ \rangle$$

$$\times \sum_k (-1)^k (2k+1) \times \begin{Bmatrix} k_1 & k_2 & k \\ J & J & J \end{Bmatrix} \begin{Bmatrix} k & k_3 & k_4 \\ J & J & J \end{Bmatrix} \nu_4(k_1 k_2 k_3 k_4; k),$$

$$(8.9)$$

where the quartic rotational invariants are given by

$$\nu_4(k_1 k_2 k_3 k_4; k) = \sum_{q,q_1,q_2,q_3,q_4} (-1)^{k+q} \begin{pmatrix} k_2 & k_1 & k \\ q_2 & q_1 & -q \end{pmatrix} \begin{pmatrix} k & k_4 & k_3 \\ q & q_4 & q_3 \end{pmatrix} \prod_j \hat{B}_{q_j}^{k_j}.$$

$$(8.10)$$

There are, in general, 21 inequivalent sets of such invariants, corresponding to all the distinct ways of assigning $k_1, k_2, k_3, k_4$ the values 2, 4, 6 without ordering.

### 8.1.3 Special results

The general expressions for rotational invariants derived above can be simplified considerably for site symmetries in which only a few parameters are non-vanishing. Derived expressions for hexagonal and cubic sites are given below in terms of the *real* parameters in Wybourne normalization, $B_q^k$.

#### 8.1.3.1 Rotational invariants for hexagonal sites

For $f$-electron systems, i.e. lanthanides and actinides, the quadratic rotational invariants for hexagonal sites (i.e. with symmetries $C_6$, $C_{3h}$, $C_{6h}$, $D_6$, $C_{6v}$, $D_{3h}$ and $D_{6h}$) take the form

$$(s_2)^2 = (B_0^2)^2/5, (s_4)^2 = (B_0^4)^2/9, (s_6)^2 = [(B_0^6)^2 + 2(B_6^6)^2]/13. \quad (8.11)$$

Cubic rotational invariants for hexagonal sites can be expressed in terms of crystal field parameters as

$$\nu_3(222) = -\sqrt{\frac{2}{35}} (B_0^2)^3,$$

$$\nu_3(224) = \sqrt{\frac{2}{35}} (B_0^2)^2 B_0^4,$$

$$\nu_3(244) = -\frac{2}{3}\sqrt{\frac{5}{77}} B_0^2 (B_0^4)^2,$$

$$\nu_3(246) = \sqrt{\frac{5}{143}} B_0^2 B_0^4 B_0^6,$$

$$\nu_3(444) = 3\sqrt{\frac{2}{1001}} (B_0^4)^3,$$

$$\nu_3(446) = -\frac{2}{3}\sqrt{\frac{5}{143}} (B_0^4)^2 B_0^6,$$

$$\nu_3(266) = -\sqrt{\frac{14}{715}} B_0^2 (B_0^6)^2 + 2\sqrt{\frac{22}{455}} B_0^2 (B_6^6)^2,$$

$$\nu_3(466) = 2\sqrt{\frac{7}{2431}} B_0^4 (B_0^6)^2 + 3\sqrt{\frac{11}{1547}} B_0^4 (B_6^6)^2,$$

$$\nu_3(666) = \frac{-20}{\sqrt{46\,189}} (B_0^6)^3 + 6\sqrt{\frac{11}{4199}} B_0^6 (B_6^6)^2.$$

*8.1.3.2 Rotational invariants for cubic sites*

The quadratic moments for ions in cubic sites, i.e. T, $T_d$, $T_h$, O and $O_h$, are given by

$$(s_2)^2 = 0,\ (s_4)^2 = 4(B_0^4)^2/21,\ (s_6)^2 = 8(B_0^6)^2/13. \tag{8.12}$$

Rotational moments for cubic sites are

$$\nu_3(444) = 8\sqrt{\frac{2}{1001}} (B_0^4)^3,$$

$$\nu_3(446) = -\frac{32}{7}\sqrt{\frac{5}{143}} (B_0^4)^2 B_0^6,$$

$$\nu_3(466) = -24\sqrt{\frac{7}{2431}} B_0^4 (B_0^6)^2,$$

$$\nu_3(666) = \frac{64}{\sqrt{46\,189}} (B_0^6)^3,$$

and the other $\nu_3(k_1 k_2 k_3)$ all vanish.

In order to obtain closed form expressions for simplifying calculations of cubic sites, octahedral ($O_h$) group coupling coefficients (or $3j$ factors) tabulated by Butler [But81] can be used to relate the fourth order invariants explicitly with products of the crystal field parameters $B_0^4$ and $B_0^6$. The resulting expressions are

$$\nu_4(k_1 k_2 k_3 k_4; k) = d_k(k_1 k_2 k_3 k_4)(B_0^4)^m (B_0^6)^n, \tag{8.13}$$

where the parameters $d_k(k_1 k_2 k_3 k_4)$ are given in Table 8.2. There are only

Table 8.2. Factors $d_k(k_1k_2k_3k_4)$, in (8.13), relating quartic invariants to the cubic crystal field parameters. (Closed expressions for the entries in this table are given in table II of [YN86a]. In the expression for $(k; k_1k_2k_3k_4) = (8; 4466)$, the denominator 211 should have been 221.)

| $k_1k_2k_3k_4=$<br>$m, n=$ | 4444<br>4,0 | 4446<br>3,1 | 4466<br>2,2 | 4646<br>2,2 | 4666<br>1,3 | 6666<br>0,4 |
|---|---|---|---|---|---|---|
| $k = 0$ | 0.327 | 0 | 1.27 | 0 | 0 | 4.92 |
| 2 | 0 | 0 | 0 | 0 | 0 | 0 |
| 4 | 0.0746 | −0.178 | −0.269 | 0.426 | 0.642 | 0.968 |
| 6 | 0.0913 | 0.138 | −0.0318 | 0.207 | −0.0479 | 0.0111 |
| 8 | 0.0443 | −0.0561 | 0.253 | 0.0711 | −0.320 | 1.44 |
| 10 | 0 | 0 | 0 | 0.0849 | 0.226 | 0.603 |
| 12 | 0 | 0 | 0 | 0 | 0 | 0.523 |

six inequivalent sets of $(k_1k_2k_3k_4)$ for cubic sites corresponding to the $m, n$ values shown in Table 8.2. These determine 25 non-vanishing invariants $\nu_4(k_1k_2k_3k_4; k)$.

The rotational invariant parameters $s_k^2$, $\nu_3(k_1k_2k_3)$, $\nu_4(k_1k_2k_3k_4; k)$, etc. provide an alternative way of describing the crystal field. In particular, a field of cubic symmetry, *apart from the signs of the parameters*, is completely described by $(s_4)^2$ and $(s_6)^2$ alone. Lower symmetry crystal fields require higher order invariants to provide a complete description.

### 8.1.4 Corrections to the free-ion Hamiltonian

It has been shown ([Lea82]) that the mean energy of a $J$ multiplet can be corrected by the following formula to take account of the crystal field $J$-mixing effects

$$\bar{E}(\alpha J) = E^0(\alpha J) + \frac{1}{2J+1}\sum_k s_k^2 \sum_{\beta J' \neq \alpha J} \frac{\langle \alpha J||C^{(k)}||\beta J'\rangle^2}{E^0(\alpha J) - E^0(\beta J')}, \qquad (8.14)$$

where $\alpha, \beta$ absorb all the unspecified state labels. This equation is a simple linear function of the invariants $s_k^2$ and does not explicitly involve the crystal field parameters $B_q^k$. Hence, to a first approximation, crystal field $J$-mixing effects can be incorporated into the fits to free-ion parameters via equation (8.14) without resorting to the much more complicated simultaneous diagonalization of the free-ion and crystal field Hamiltonians.

## 8.2 Quadratic rotational invariants and the superposition model

As shown in (8.5), the three quadratic rotational invariants are linearly related to the standard deviation (or second order moment) of the crystal field splitting of the $J$ multiplets, allowing the direct determination of the values of $s_k^2$ independently of the values of $B_q^k$ themselves. It follows that $s_k^2$ values obtained from the crystal field parameters using (8.4) can be compared with directly determined values to provide a check on fitted sets of $B_q^k$ parameters.

A superposition model (Chapter 5) expression for $s_k^2$ can be obtained by direct substitution of the relations given in (5.4) into (8.4), giving

$$s_k^2 = \frac{1}{2k+1} \sum_{i,j} \overline{B}_k(R_i)\overline{B}_k(R_j)C_0^{(k)}(\omega_{ij}). \tag{8.15}$$

Note that the functions $C_q^{(k)}(\theta, \phi)$, defined in (2.6), only depend on single angle ($\theta$) when $q = 0$. The angle $\omega_{ij}$, between the direction $\mathbf{R}_i = (R_i, \theta_i, \phi_i)$ of ligand $i$ and the direction $\mathbf{R}_i = (R_j, \theta_j, \phi_j)$ of ligand $j$, is given by

$$\cos \omega_{ij} = \sin \theta_i \sin \theta_j \cos(\phi_i - \phi_j) + \cos \theta_i \cos \theta_j.$$

In most applications of (8.15) the intrinsic parameters are expressed in terms of their values at some specific distance and the power law exponents $t_k$ (see (5.5)). Equation (8.15) shows the $s_k^2$ to be functions of both the intrinsic parameters *and* the angles $\omega_{ij}$. Hence, although the $s_k^2$ values are rotational invariants, they cannot be interpreted simply as representing the *strength* of the ligand interactions as is frequently done in the literature. Unlike the intrinsic parameters $\overline{B}_k(R_i)$, they do not have a simple physical significance.

The main value of (8.15) is that it enables values of the intrinsic parameters to be obtained directly from the standard deviation of the crystal field splittings, without resorting to a detailed crystal field fit for $B_q^k$. Given the site structure, this only requires some reasonable assumption to be made about the distance dependence of the intrinsic parameters $\overline{B}_k(R_i)$.

Auzel (see [Auz84] and references therein) introduced the so-called *crystal field strength parameters*, $S \equiv \sqrt{\sum_k s_k^2}$. These are almost linearly correlated with the maximum crystal field splittings of $J$ multiplets, and are sometimes used for comparative purposes (e.g. [CLWR91]). The crystal field strength parameters are also believed to be of importance for the understanding of energy transfers between ions.

Table 8.3. Quadratic rotational invariants $s_k$ and $S$ as calculated from the crystal field parameters for $Er^{3+}$:$LaF_3$ given in Table 8.1 ($cm^{-1}$). Errors are shown in brackets.

|        | $C_2$    | $C_2$    | $C_{2v}$ | $C_{2v}$  | $D_{3h}$ |
|--------|----------|----------|----------|-----------|----------|
| $s_2$  | 100      | 127      | 112      | 121(11)   | 126      |
| $s_4$  | 359      | 308      | 309      | 287(48)   | 387      |
| $s_6$  | 303      | 314      | 316      | 317(38)   | 278      |
| $S$    | 480      | 457      | 456      | 445(61)   | 493      |
|        | [SN71a]  | [ML79]   | [CC83]   | [CGRR89]  | [CCC77]  |

## 8.3 Examples of application

This section highlights three different types of application of the theory outlined in the previous sections. The aim is to show that the use of crystal field invariants and moments produces significant simplifications in various types of calculation.

### 8.3.1 Estimating crystal field parameters for low-symmetry sites

For low-symmetry systems, where there are many minima in the fitting space, it is necessary to use starting values of the crystal field parameters which are very close to the final fitted values. Several alternative strategies that can be employed to determine starting sets of crystal field parameters are described in Section 3.2.3. Once a reliable set of fitted values has been obtained for one lanthanide ion, these parameters can, of course, be used as starting values in fits to other ions in the same host crystal. The expected smooth variation of crystal field parameters down the lanthanide series normally provides the most reliable check that significant results are being obtained. Unfortunately, for low-symmetry systems like $Ln^{3+}$:$LaF_3$, where the coordinate frame is not uniquely defined, individual crystal field parameters $B_q^k$ can be drastically different even for the same $Ln^{3+}$ ion (see Table 8.1). Nevertheless, the corresponding rotational invariants, shown in Table 8.3, are fairly consistent whatever site symmetry was assumed in the fit.

Table 8.4 lists the crystal field invariants $s_k$ and $S$, for the $Ln^{3+}$:$LaF_3$ series, calculated from the crystal field parameters obtained by Morrison and Leavitt [ML79] for $C_2$ symmetry and by Carnall et al. [CGRR89] for

Table 8.4. *Crystal field invariants (in cm$^{-1}$) for trivalent lanthanide ions in LaF$_3$. Values (a) and (b) are determined, respectively, from parameters obtained in [CGRR89] and [ML79]. Uncertainties, based on possible errors in the parameters, are of the order of 10%.*

|        | Pr  | Nd  | Sm  | Eu  | Tb  | Dy  | Ho  | Er  | Tm  |
|--------|-----|-----|-----|-----|-----|-----|-----|-----|-----|
| $s_2$(a) | 124 | 118 | 105 | 102 | 121 | 117 | 127 | 121 | 130 |
| $s_2$(b) | 82  | 99  | 83  | 111 | —   | 106 | 104 | 127 | 65  |
| $s_4$(a) | 431 | 398 | 373 | 368 | 334 | 332 | 311 | 287 | 294 |
| $s_4$(b) | 398 | 375 | 343 | 328 | —   | 365 | 295 | 308 | 270 |
| $s_6$(a) | 529 | 521 | 472 | 463 | 364 | 349 | 340 | 317 | 296 |
| $s_6$(b) | 544 | 528 | 451 | 418 | —   | 358 | 340 | 314 | 322 |
| $S$(a)   | 694 | 666 | 610 | 600 | 508 | 495 | 478 | 445 | 437 |
| $S$(b)   | 679 | 655 | 573 | 543 | —   | 522 | 462 | 457 | 425 |

C$_{2v}$ approximate symmetry. The rank 2 and 4 values of $s_k$ demonstrate a fairly smooth variation across the series but the $s_6$ invariant shows a sudden drop in its value for ions with more than half-filled shells. This reflects the behaviour of the crystal field of lanthanides in anhydrous chloride host crystals, which has been attributed to a large rank 6 spin-correlated crystal field contribution in lanthanides (see Chapter 6).

The properties of rotational invariants suggest a different way of using the superposition model to estimate starting values of crystal field parameter fits. Given values of the invariants $s_k$, (8.15) provides a relationship between each $\overline{B}_k(R_i)$ and its corresponding $t_k$. Taken together with prior knowledge of the expected values of these parameters, this can provide sufficient information to estimate an initial set of crystal field parameters.

This method can be illustrated by listing the essential steps required to obtain starting crystal field parameters to fit the optical spectrum of Er$^{3+}$:LaF$_3$.

(i) Use (8.2) to evaluate the quadratic moments $\sigma_2$ for the crystal field splitting of the 10 low-lying $J$ multiplets (specified in Table 8.5) of the optical spectrum of Er$^{3+}$:LaF$_3$ [ML79, ML82]. In doing this it is necessary to take into account the degeneracy of each energy level in the summation over states. $E^0$ is the mean value of the energies within a given $J$ multiplet. Overlapping $J$ multiplets have been omitted from this analysis. The calculated $\sigma_2$ values are shown in Table 8.5 as $\sigma_2$(exp).

Table 8.5. Reduced matrix elements and quadratic moments for 10 low-lying multiplets of $Er^{3+}$:$LaF_3$, based on experimental results given in [ML82]. All quantities are given in $cm^{-1}$.

| Term | $\langle J||C^{(2)}||J\rangle$ | $\langle J||C^{(4)}||J\rangle$ | $\langle J||C^{(6)}||J\rangle$ | Centroid | $\sigma_2(\text{exp})$ | $\sigma_2(\text{fit})$ |
|------|------|------|------|------|------|------|
| $^4I_{15/2}$ | 0.6796 | 0.6962 | 1.7402 | 217 | 147.4 | 146.4 |
| $^4I_{13/2}$ | 0.5677 | 0.4696 | 0.6120 | 6701 | 69.3 | 67.5 |
| $^4I_{11/2}$ | 0.3820 | 0.2139 | 0.2120 | 10 340 | 31.2 | 30.4 |
| $^4I_{9/2}$ | 0.0866 | 0.3157 | 1.1373 | 12 597 | 115.5 | 114.9 |
| $^4F_{9/2}$ | −0.5080 | 0.3096 | −0.2918 | 15 453 | 45.4 | 46.7 |
| $^4S_{3/2}$ | −0.2605 | 0 | 0 | 18 573 | 15.5 | 13.5 |
| $^2H_{11/2}$ | 0.0648 | −0.3122 | 0.4241 | 19 337 | 48.3 | 48.4 |
| $^4F_{7/2}$ | −0.5365 | −0.1147 | 0.4041 | 20 720 | 47.3 | 49.4 |
| $^2H_{9/2}$ | −0.5266 | −0.1645 | −0.9907 | 24 748 | 97.4 | 98.7 |
| $^4G_{11/2}$ | −0.0961 | −0.5566 | 0.3243 | 26 612 | 61.0 | 61.8 |

(ii) Using published free-ion parameters for $Er^{3+}$:$LaF_3$ [ML79], calculate reduced matrix elements $\langle \alpha SLJ||C^{(k)}||\alpha SLJ\rangle$ for ranks $k = 2, 4, 6$ and each of the 10 selected $J$ multiplets.

(iii) Using (8.5), perform a linear least-squares fit of the rotational invariants $s_k^2$ to the 10 calculated quadratic moments $\sigma_2^2(J)$. This yields

$$s_2 = 104 \text{ cm}^{-1}, \ s_4 = 341 \text{ cm}^{-1}, \ s_6 = 305 \text{ cm}^{-1}.$$

(Weighting factors equal to the reciprocals of $\sigma_2^2(J)$ were used to obtain these results.) These values of $s_k$ are in good agreement with values obtained from fitted crystal field parameters shown in Table 8.3.

(iv) Estimate ranges of values for the intrinsic parameters $\overline{B}_k$ as follows:

(a) Make use of the neutron scattering results of Cheetham *et al.* [CFFW76] for $LaF_3$ to obtain the polar coordinates $(R_i, \theta_i, \phi_i)$ of the 11 nearest neighbour fluoride ligands $i$ (see table II of [YN85a]).

(b) Using (8.15) and (5.5), plot $|\overline{B}_k/s_k|$ versus the power law exponents $t_k$. The resulting graphs are given in figures 1–3 of [YN85a].

(c) Assuming $t_k$ to be in the range 5–15, the following bounds can be placed on the values of the intrinsic parameters at

$R = 2.421$ Å:

$$\overline{B}_2 = 830 \pm 100 \ \text{cm}^{-1},$$
$$\overline{B}_4 = 560 \pm 40 \ \text{cm}^{-1}, \qquad (8.16)$$
$$\overline{B}_6 = 310 \pm 40 \ \text{cm}^{-1}.$$

This provides a direct way to estimate the values of intrinsic parameters, but does not determine the power law exponents.

(v) Finally, employ the superposition model expression (5.4) to estimate starting values of the crystal field parameters. Several alternative sets of starting values determined from intrinsic parameters within the ranges allowed by (8.16) should be tried in order to test for the existence of multiple minima in the fitting space.

### 8.3.2 Cubic crystal field parameters

To illustrate the use of higher order crystal field splitting moments, we consider the crystal field splittings of the ground multiplet of $Nd^{3+}$ in the cubic monopnictides. For the $^4I_{9/2}$ ground multiplet of $Nd^{3+}$, the cubic crystal field splits the $J = 9/2$ degenerate free-ion state into three levels $(\Gamma_6 + 2\Gamma_8)$ so that there are only two independent crystal field splitting energies. These are just sufficient to determine the two cubic crystal field parameters, namely, $B_0^4$ and $B_0^6$ in a conventional crystal field fit.

An alternative approach involves using the quadratic and quartic moments of the $^4I_{9/2}$ ground multiplet of $Nd^{3+}$. Just as in conventional crystal field fits, it is necessary to have reduced matrix elements for the free-ion $^4I_{9/2}$ ground multiplet. In the following calculation the reduced matrix elements for $LS$ coupling are employed, following [YN86a]. Two simultaneous equations must be solved, the first of which is obtained by substituting the expressions (8.12) into (8.5):

$$\sigma_2^2 = (4.58(B_0^4)^2 + 25.6(B_0^6)^2) \times 10^{-3}.$$

The second of the simultaneous equations is obtained from (8.9) and (8.13) together with Table 8.2, viz.

$$\sigma_4^4 = (0.279(B_0^4)^4 - 1.05(B_0^4)^3 B_0^6 + 17.68(B_0^4)^2(B_0^6)^2$$
$$- 23.65 B_0^4(B_0^6)^3 + 84.05(B_0^6)^4) \times 10^{-3}.$$

Note that only the *relative* sign of $B_0^4$ and $B_0^6$ can be obtained from the quadratic and quartic moments.

Results obtained by Yeung and Newman [YN86a] for two neodymium

Table 8.6. Moment analysis of $Nd^{3+}$ monopnictides, reproduced from table III of [YN86a], with the permission of the American Institute of Physics. All quantities are in meV. Brackets indicate experimental or fitting uncertainties.

| Crystal | | $\Gamma_8^{(1)}$ | $\Gamma_6$ | $\sigma_2$ | $\sigma_4$ | $B_0^4$ | $B_0^6$ |
|---------|------|---------|---------|---------|---------|---------|---------|
| NdP | Exp | 14.6(3) | 2.8(1) | 6.8(2) | 7.1(2) | | |
| | CF fit | 14.54 | 2.41 | 6.79 | 7.09 | 91(2) | 10.2(5) |
| | $\sigma$ fit | 14.62 | 2.91 | 6.77 | 7.08 | 90(2) | 10.9(12) |
| NdAs | Exp | 13.5(1) | 2.6(1) | 6.26(8) | 6.55(8) | | |
| | CF fit | 13.48 | 2.19 | 6.30 | 6.58 | 85(2) | 9.4(3) |
| | $\sigma$ fit | 13.55 | 2.83 | 6.26 | 6.55 | 83(1) | 10.2(4) |

monopnictides are summarized in Table 8.6. Experimental values of the energy levels are from Furrer *et al.* [FKV72]. The results of a direct crystal field fit to the ground multiplet energy levels [FKV72] are shown as the 'CF fit' in Table 8.6. These may be compared with the results of calculating $\sigma_2$ and $\sigma_4$ and solving the simultaneous equations given above for $B_0^4$ and $B_0^6$, which are shown as the '$\sigma$ fit' in Table 8.6.

### 8.3.3 Spin-correlated crystal field parameters

Certain features of the crystal field splittings of trivalent lanthanide ions doped in crystals can be better understood when the spin-correlated crystal field is included in the crystal field Hamiltonian (see Sections 6.2.2 and 6.3.3). However, the standard fitting procedure used, for example, in [CN84c] and [RR85], is only applicable in particularly favourable cases where many spectroscopic energies are known and the site symmetry is high. It requires the simultaneous determination of a large number of parameters, including many free-ion parameters.

In order to bypass the problems encountered when using the standard approach, Yeung and Newman [YN86b] developed a method for obtaining spin-correlated crystal field parameters $c_k$ based on the use of quadratic invariants. One advantage of this approach is that it does not require the standard one-electron crystal field parameters to be determined. This is particularly important in the case of paramagnetic ions in sites of low symmetry. A second advantage is that it is not necessary to carry out simultaneous fits to the free-ion and crystal field parameters. Nevertheless, a free-ion calculation must first be carried out to determine the reduced matrix elements in intermediate coupling. In common with other applications of the rotational invariants, it is necessary to assume that there is no crystal field $J$-mixing. This assumption makes the approach potentially less useful for the analysis of the actinide spectra.

The spin-correlated crystal field is introduced by substituting the operator $(1 + c_k \mathbf{S} \cdot \mathbf{s}_i C_q^{(k)}(i))$ for the one electron tensor operator $C_q^{(k)}(i)$ in the crystal field Hamiltonian (see Section 6.3.3). Spin correlated effects are thus represented by the three fitted parameters $c_k$ ($k = 2$, 4, 6). The validity of assuming this parametrization to be $q$ independent has already been discussed in Section 6.3.3. Some empirical confirmation of the $q$ independence of the $c_k$ parameters has been provided by Crosswhite and Newman [CN84c].

The operator substitution introduced above is equivalent to replacing the reduced matrix element $\langle J||C^{(k)}||J\rangle$ in (8.5) by the expression

$$\langle J||C^{(k)}||J\rangle + c_k\langle J||V^{(k)}||J\rangle$$

[Jud77a]. Here the '$V$' operators are defined in terms of their single electron components

$$V_q^{(k)}(i) = \mathbf{S} \cdot \mathbf{s}_i C_q^{(k)}(i), \tag{8.17}$$

and $J$ is used as an abbreviation for the complete multiplet labels $\alpha LSJ$. The '$V$' operators can be shown to have the reduced matrix elements

$$\langle J||V^{(k)}||J\rangle = \langle l||C^{(k)}||l\rangle\left\langle J\left|\left|\sqrt{\frac{S(S+1)}{2S+1}}V^{(1k)}\right|\right|J\right\rangle. \tag{8.18}$$

These can be evaluated using the reduced matrix elements of the operator $V^{(1k)}$, which have been tabulated by Nielson and Koster [NK63]. With the above substitution, the quadratic moments $\sigma_2$ for a given $\alpha LSJ$ multiplet generalize to

$$\sigma_2^2(J) = \frac{1}{2J+1}\sum_k(\langle J||C^{(k)}||J\rangle + c_k\langle J||V^{(k)}||J\rangle)^2 s_k^2. \tag{8.19}$$

When the values of $\sigma_2$ have been determined for multiplets in which all the crystal field energy levels are known, (8.19) can be used to fit both the quadratic invariants $s_k$ and the spin-correlated crystal field parameters $c_k$. Because of the small values of $c_k$ it is possible to neglect the term in (8.19) which depends on $c_k^2$. In order to obtain good results it is necessary that the reduced matrix elements of $C^{(k)}$ and $V^{(k)}$ are linearly independent for the set

Table 8.7. Spin-correlated crystal field analysis for $Ho^{3+}$:$LaCl_3$, using quadratic invariants, adapted from table 1 of [YN86b] (with the permission of the Institute of Physics). The symbol [0] indicates parameters which were constrained to zero in fits (a)–(c). In (e) all entries were calculated using the crystal field parameter and $c_k$ values determined in [CN84c].

| | $s_2$ | $s_4$ | $s_6$ | $c_2$ | $c_4$ | $c_6$ | $\Delta\sigma_2$ | $\Delta\sigma_4$ |
|---|---|---|---|---|---|---|---|---|
| (a) | 98.4 | 104.0 | 170.1 | [0] | [0] | [0] | 4.38 | 4.99 |
| (b) | 99.9 | 107.6 | 213.9 | [0] | [0] | 0.150 | 1.17 | 1.71 |
| (c) | 65.5 | 102.2 | 214.6 | −0.390 | [0] | 0.149 | 0.85 | 1.28 |
| (d) | 73.6 | 145.3 | 211.7 | −0.225 | 0.159 | 0.143 | 0.84 | 1.23 |
| (e) | 105.1 | 89.5 | 217.2 | 0.04 | −0.05 | 0.16 | 1.57 | 2.10 |

of multiplets used in the fitting. A formal check of linear independence can be carried out using a method described in [YN86b]. A check on the results of fitting the quadratic moments to the $s_k$ and $c_k$ parameters is provided by comparing the quartic moments $\sigma_4$ with their calculated values, as described in [YN86b].

Although it would be very complicated to include these directly in a fit to a system with lower symmetry than cubic, they do provide a useful test of the parameters obtained using the quadratic invariants. In order to assess its accuracy, Yeung and Newman [YN86b] applied the method described above to $Ho^{3+}$:$LaCl_3$, which had already been studied by Crosswhite and Newman [CN84c]. The free-ion eigenstates used to calculate the intermediate coupling reduced matrix elements for $Ho^{3+}$ were obtained from the work of Rajnak and Krupke ([RK67] and [RK68]). Experimental results were taken from Crosswhite *et al.* [CCER77]. Using the energy levels determined by Crosswhite *et al.*, quadratic moments were determined for the 13 multiplets $^5I_8$, $^5I_7$, $^5I_6$, $^5I_5$, $^5I_4$, $^5F_5$, $^5F_4$, $^5F_3$ $^3K_8$, $^5G_6$, $^5G_5$, $^5G_4$ and $^5D_4$. A least-squares fitting procedure was used to fit to the rotational invariants $s_k$ and spin-correlated crystal field parameters $s_k$ using (8.19).

The results of four fits of the data to the parameters $s_k$ and $c_k$, with various combinations of the $c_k$ held to zero, are reported in Table 8.7. The final row gives the corresponding results obtained from the fit to crystal field parameters [CN84c]. It can be seen from Table 8.7 that the introduction of $c_6$ results in a considerable improvement in the fit to the quadratic moments. $c_2$ produces a smaller, but quite definite reduction to the mean square deviation, while the effect of including $c_4$ is negligible.

Deviations in the quartic moments are also given in Table 8.7. These are based on a comparison between values determined directly from the experimental data and values calculated using fitted crystal field parameters. The root mean square deviations of the quartic moments supply a useful independent check on the physical significance of the fitted parameters. The relative quality of the fits, as determined by the quartic moments, is the same as that determined by the quadratic moments.

These fits reported in Table 8.7 determine essentially the same positive values for $c_4$ and $c_6$ as were obtained by Crosswhite and Newman, but a significantly larger negative value of $c_2$. Further details of this analysis can be found in Yeung and Newman [YN86b], which also reported similar analyses for $Ho^{3+}:YPO_4$, $Er^{3+}:YAlO_3$ and $Nd^{3+}:PbMoO_4$.

## 8.4 Prospects

Crystal field invariants have useful applications in fitting optical spectra where the normal crystal field fitting procedures, described in Chapters 3 and 4, become difficult. This occurs when the number of crystal field parameters is large, as is the case for low-symmetry sites, and when the values of additional types of crystal field parameters are required. The example of the determination of spin-correlated crystal field parameters suggests that moment methods could also be used to test other models of correlation crystal field effects. One step in this direction has been taken in [YN87], where the possible effects of *orbitally* correlated crystal fields were studied.

# 9

# Semiclassical model

## K. S. CHAN

*City University of Hong Kong*

The semiclassical, or tunnelling, model of crystal fields was introduced by Trammel [Tra63] as an heuristic aid in understanding the magnetic ordering in lanthanide phosphides. Pytte and Stevens [PS71] subsequently used this model to describe Jahn–Teller ordering in lanthanide vanadates.

The semiclassical model also provides an explanation of the eightfold *clusters* of energy levels observed in the ground multiplets of holmium and erbium in both garnets [JDR69, OH69] and superconducting cuprates [FBU88a, FBU88b, SLK92] (see Section 9.3). The usual reason for the clustering of enery levels, viz. the existence of an approximate higher symmetry, cannot be relevant, as no site symmetry higher than cubic is possible (see Appendix 1).

Trammel [Tra63] explained clustering in terms of the preferred alignment of a classical angular momentum vector along the directions of minima in the crystal field potential. A quantitative version of the semiclassical model was subsequently developed by Harter and Patterson [HP79] to explain the even more pronounced clustering that is observed in the rotational energy levels of polyatomic molecules. The semiclassical model is very accurate for molecular rotational energies, mainly because the states of interest have very large angular momenta. While it is necessary to be selective in applying the model to crystal field splittings, where the angular momenta are relatively small, it can occasionally be very useful indeed, as this chapter seeks to demonstrate.

In the semiclassical model the states of the lowest-lying cluster of energy levels are described by superpositions of several angular momentum eigenstates $|J, M_J = J\rangle$ with the quantization axis aligned in different directions. Each of these states is interpreted as the angular momentum vector $\mathbf{J}$ localized in one of the degenerate minima in the effective crystal field potential. The number of minima determines the overall degeneracy of the cluster.

In nearly cubic systems this degeneracy will normally be either sixfold or eightfold. In some, but not all, cases the number of minima corresponds to the number of nearest neighbour ligands. The ground state degeneracy produced in this way is subject to splitting due to the tunnelling of the angular momentum vector between the equivalent potential minima. The matrix elements between states localized in the different potential minima are interpreted as *tunnelling amplitudes*. Within a cluster, the energy level splittings can be expressed in terms of tunnelling, as described in Section 9.2.

One indication that the semiclassical model is likely to be of value is the separation of the observed spectrum of a ground multiplet into two or more well-defined 'clusters' of energy levels, such that the splittings within the clusters are much smaller than the splittings between them. Clustering of this type is most often observed in lanthanide ion spectra when the ground multiplet has a large value of $J$.

Another property of the ground state can also provide an indicator of the applicability of the semiclassical model. When the observed spectroscopic splitting factors (also known as the $g$-factors) for ground states are found to be near their maximum possible (free-ion) values, it is likely that they correspond to a superposition of states with $M_J = J$. Writing the ground states in this way can be useful in describing thermal and magnetic behaviour involving independent or coupled ions, as these phenomena only involve excitations to the low-lying energy levels. An example is Jahn–Teller ordering in vanadates, discussed in Section 9.1.3 [PS71, EHHS72].

The semiclassical model can be used in a variety of ways. In particular, it provides simple analytic expressions for the states which correspond to a cluster of low-lying energy levels. This is of particular value when only a limited amount of information is available, as is sometimes the case when crystal field splittings are determined using electron spin resonance, far-infrared spectroscopy or inelastic neutron scattering. It may also be possible to use analytic expressions obtained for a site of high symmetry (say cubic) as the first step in investigating crystal fields of lower symmetry sites with similar structure. If the individual energy levels are very broad, so that the energy levels within the clusters are merged, the semiclassical model can be used to relate intercluster splitting to the crystal field parameters. An application of this type is given in Section 9.1.2.

A more formal application of the semiclassical model is to provide 'unique symmetry labels' for crystal field split energy levels [New83b, YN85b]. However, this is mainly of group theoretical interest, and is not an appropriate topic to pursue in this book.

The organization of this chapter is as follows. Section 9.1 gives three in-

troductory examples of the application of the semiclassical model. Section
9.2 is devoted to a particular example of the construction and diagonalization of tunnelling Hamiltonians for lanthanide ions in approximately cubic
sites. The overall aim is to assess the types of application that are most
suitable for the semiclassical approach.

## 9.1 Introductory examples

The first example uses results generated by the program LLWDIAG (see
Appendix 3), or the corresponding results given in the paper by Lea, Leask
and Wolf [LLW62], to determine circumstances when ground multiplet clustering is likely to occur for lanthanides and actinides in cubic crystal fields.
The second example is concerned with the interpretation of the spectra of
certain lanthanide ions in glasses in terms of intercluster splittings, which
provides a potentially important, but previously unexplored, area of application of the semiclassical model. The final example concerns the application
of the semiclassical model to describe Jahn–Teller ordering in dysprosium
and terbium vanadates [PS71].

### 9.1.1 *Clustering in cubic and nearly cubic crystal fields*

Trammel [Tra63] pointed out that clustering is predicted by several of the
Lea, Leask and Wolf [LLW62] energy level diagrams† for restricted ranges
of the $x$ parameter (defined in Section 2.2.5, see (2.13)). This is particularly
apparent for the higher values of $J$, especially $J = 8$, $J = 15/2$ and $J = 6$,
all of which correspond to ground state multiplets of certain lanthanide and
actinide ions. The relevance of this observation to real systems depends on
whether or not actual values of the ratio of rank 4 and rank 6 cubic crystal
parameters correspond to the $x$-values for which clustering occurs.

Realistic values of $x$ for ionic crystals can be estimated from the ratio of
intrinsic parameters $\mu = \overline{B}_6/\overline{B}_4$. Values of the intrinsic parameters given in
Chapter 5 show that, for most negatively charged ligands, $0.67 < \mu < 1.0$.
Using the formulae for eight and sixfold coordinated cubic sites (i.e. $c = 8$
or 6) given in Section 5.3, this enables the physically significant ranges of
$x$ shown in Table 9.1 to be determined. Reference to the Lea, Leask and
Wolf [LLW62] diagrams then shows in which cases *strong* (s), *weak* (w) or
*no* (n) clustering can be expected. As shown in Table 9.1, strong clustering
is only likely to occur in ionic host crystals for the ground multiplets of

---

† These diagrams can be generated using the Mathematica program LLWDIAG described in
Appendix 3.

Table 9.1. Clustering predictions for lanthanide ion ground state splittings
in cubic sites. Cubic parameters $W$ and $x$ are defined by (2.13). The
ranges of $x$ shown correspond to $\mu = \overline{B}_6/\overline{B}_4$ between 2/3 and 1. The sign
of $W$ determines the position of the low-lying clusters in the Lea, Leask
and Wolf figures [LLW62]. The labels s,w,n denote strong, weak and no
clustering, respectively.

| Ion | $J$ | $W$ | $\beta F(4)/\gamma F(6)$ | $x$ for $c = 6$ | $x$ for $c = 8$ |
|-----|-----|-----|------------------|-----------------|-----------------|
| $Tb^{3+}$ | 6 | $-$ | $-0.867$ | $-0.89$ to $-0.92$: w | 0.75 to 0.82: s |
| $Dy^{3+}$ | 15/2 | $+$ | $-0.0825$ | $-0.46$ to $-0.54$: s | 0.22 to 0.30: n |
| $Ho^{3+}$ | 8 | $-$ | 0.111 | 0.51 to 0.61: s | $-0.28$ to $-0.37$: s |
| $Er^{3+}$ | 15/2 | $+$ | 0.0929 | 0.46 to 0.57: w | $-0.25$ to $-0.33$: w |
| $Tm^{3+}$ | 6 | $-$ | $-0.231$ | $-0.68$ to $-0.76$: n | 0.45 to 0.55: n |

holmium (in both coordinations), terbium (with eightfold coordination) and
dysprosium (with sixfold coordination). These four cases provide the most
likely applications for the semiclassical approach.

Reference to the Lea, Leask and Wolf diagrams [LLW62] shows in which
cases the potential minima are in the ligand directions, i.e. with sixfold
clustering in the case $c = 6$ or eightfold clustering for the case $c = 8$. When
the minima lie between the ligand directions there is eightfold clustering for
$c = 6$ or sixfold clustering for $c = 8$. These indicators show that, in the case
of the holmium ground multiplet, the potential minima are in the ligand
directions, whereas both dysprosium and terbium ground multiplets have
potential minima located between the ligand directions.

### 9.1.2 Low-lying states of lanthanide ions in glasses

Neutron scattering spectroscopy provides an effective means of probing the
energy levels of the lowest-lying multiplets of lanthanide ions in glasses.
Well-defined spectra are obtained, but these do not have the sharp lines
characteristic of the lanthanides in crystalline environments. However, given
the disordered nature of glasses, it is rather unexpected to find well-defined
spectra at all. There are two possible reasons for their occurrence.

(i) There is a well-defined *average* crystal field potential in glasses. (Random orientations of this potential do not affect the energy levels.)

Table 9.2. Lowest-lying cluster energies, as factors of $\overline{A}_4 = \overline{B}_4/8$, for $Ho^{3+}$ and $Tb^{3+}$ in eightfold and sixfold coordination, for several values of $\mu = \overline{B}_4/\overline{B}_6$.

| Ion | Coord | $M_J$ | Axis | $\mu = 0.75$ | $\mu = 1$ | $\mu = 1.5$ | $\mu = 3$ |
|-----|-------|-------|------|-------------|-----------|-------------|-----------|
| $Ho^{3+}$ | eightfold | 8 | [111] | $-5.44$ | $-4.46$ | $-3.47$ | $-2.49$ |
|  |  | 7 |  | 6.78 | 5.18 | 3.58 | 1.99 |
|  | sixfold | 8 | [100] | $-3.48$ | $-3.24$ | $-3.01$ | $-2.78$ |
|  |  | 7 |  | 2.17 | 1.79 | 1.42 | 1.04 |
| $Tb^{3+}$ | eightfold | 6 | [100] | $-2.45$ | $-2.40$ | $-2.34$ | $-2.29$ |
|  |  | 5 |  | 2.06 | 1.93 | 1.79 | 1.65 |
|  | sixfold | 6 | [111] | $-1.87$ | $-1.83$ | $-1.78$ | $-1.74$ |
|  |  | 5 |  | 1.54 | 1.44 | 1.34 | 1.23 |

(ii) The semiclassical approximation is good, giving energy differences between localized states with $M_J = J$ and $M_J = J - 1$, which are effectively independent of the details of the site structure. It is conjectured that this situation would be most likely to occur when the potential minima are located in the ligand directions.

An interesting example is provided by the inelastic neutron scattering spectra of various lanthanide ions in the phosphate glasses (B. Rainford and R. Langan, private communication). The $Ho^{3+}$ ground multiplet spectrum consists of a broad, but well defined, single line at 41 meV. This is consistent with a clear separation of the two lowest-lying clusters. $Tb^{3+}$ also has a single line ground multiplet spectrum, but this appears as a broad shoulder from zero to 20 meV. This is consistent with a weak separation of the two lowest clusters. Other lanthanide ions have more complicated, and generally better defined, spectra.

These observations suggest that the $Ho^{3+}$, and possibly the $Tb^{3+}$, inelastic neutron scattering spectra arise from transitions between (approximate) semiclassical localized states. Magnetic dipole transition intensities between the localized states forming the two lowest-lying semiclassical clusters should be strong in neutron scattering spectroscopy, as they are produced by transitions between localized $M = J$ and $M = J - 1$ states.

In order to carry out the semiclassical calculation, several explicit simplifying assumptions must be made.

(i) The semiclassical model can be employed for the ground multiplets of $Ho^{3+}$ and $Tb^{3+}$. This is consistent with the results shown in Table 9.1.

(ii) Repulsion between the ligands ensures that they are spaced around the paramagnetic ion in a regular way, tending to produce a high site symmetry, *on average*. More specifically, the average local point symmetry at the lanthanide ions in the phosphate glasses is assumed to be cubic, with six or eight oxygen ligands. In this approximation only rank 4 and rank 6 crystal field parameters have to be considered.

(iii) Given their similar size, it is assumed that $Ho^{3+}$ and $Tb^{3+}$ both have the same number of ligands.

(iv) The superposition model (see Chapter 5) is assumed to hold, with the same intrinsic parameters for $Ho^{3+}$ and $Tb^{3+}$.

The intrinsic parameters $\overline{B}_k(R_L)$, for a similar ligand distance $R_L$, can be expected to have a similar magnitude to those in a crystalline environment. In cubic symmetry only the intrinsic parameters $\overline{B}_4$ and $\overline{B}_6$ need to be considered. As has been shown in Chapter 5, these can be related, to the parameters $W$ and $x$ introduced by Lea, Leask and Wolf [LLW62]. The advantage of characterizing the crystal field by means of (one-electron) intrinsic parameters, rather than $W$ and $x$, is that they take very similar values for all the lanthanide ions.

Table 9.2 shows the low-lying energy levels calculated for $Ho^{3+}$ and $Tb^{3+}$ using both eightfold and sixfold coordination, for a range of intrinsic parameter ratios $\mu = \overline{B}_4/\overline{B}_6$. The axis, corresponding to the **J**-vector direction, is chosen in each case to give the lowest-lying ground states. The best agreement with the experimental ratio (2.0) of cluster energy differences for the two ions is obtained with eightfold coordination and $\mu = 1$, corresponding to the ratio $9.64/4.33 = 2.2$. The experimental inter-cluster splittings then give the intrinsic parameters $\overline{B}_4 = \overline{B}_6 = 272$ cm$^{-1}$. This solution corresponds to there being eight potential energy minima for $Ho^{3+}$, with the **J**-vectors pointing towards the oxygen ligands, and six potential energy minima for $Tb^{3+}$, with the **J**-vectors pointing between the ligands towards the faces of the cube.

Using the relationships

$$A_4\langle r^4 \rangle = -(7/18)\overline{B}_4, \ A_6\langle r^6 \rangle = (1/9)\overline{B}_6, \tag{9.1}$$

the Lea, Leask and Wolf parameters for $Ho^{3+}$ are determined as $x = -0.28$ and $W = -0.094$ meV. Figure 1 of [LLW62] (for $J = 8$) confirms that the value $x = -0.28$ corresponds to a large gap (approx. $430W$) between two clusters of energy levels. Using the observed inter-cluster splitting for $Ho^{3+}$, viz. 41 meV = 330 $cm^{-1}$, gives $W = -0.095$ meV, in agreement with the semiclassical result quoted above.

In the case of $Tb^{3+}$ the LLW parameters are calculated as $x = 0.75$ and $W = -0.129$ meV. Figure 5 of [LLW62] shows strong clustering for this value of $x$ with an inter-cluster splitting of about $160W$. Comparing this with the experimental shoulder at 20 meV gives $W = -0.125$ meV, in good agreement with the semiclassical result.

In summary, the semiclassical approximation for cubic statistical symmetry with eightfold coordination has been shown to give consistent results for both $Ho^{3+}$ and $Tb^{3+}$ in phosphate glasses. The values of the intrinsic parameters have a similar ratio, but are smaller than those obtained in superposition model analyses of the crystal spectra for $Er^{3+}$ in $YVO_4$, viz. $\overline{B}_4 = 441$ $cm^{-1}$ and $\overline{B}_6 = 378$ $cm^{-1}$ (see Table 5.2). This is consistent with the expectation that ligand distances will, on average, be slightly greater in phosphate glasses than they are in phosphate crystals.

### 9.1.3 Cooperative Jahn–Teller effects in the vanadates

Pytte and Stevens [PS71] used the semiclassical model to determine approximate low-lying wavefunctions which form the basis of their explanation of the cooperative Jahn–Teller effect in the $DyVO_4$ and $TbVO_4$. In both of these materials the relevance of the semiclassical description can be established from the high values of the observed spectroscopic splitting (or $g$-)factors [EHHS72]. For example, the observed $g$-factors for $DyVO_4$, in its ordered state below the transition temperature, are $g_x = 19$, $g_y < 1$ and $g_z = 0.5$ for the ground state, and $g_y = 19$, $g_x \simeq g_z \simeq 0$ for the lowest-lying excited state. The values may be compared with a maximum possible $g$-value of 20 for a $J = \frac{15}{2}$ state, corresponding to $M_J = J$. In the semiclassical approximation these two doublets are degenerate, the overall fourfold degeneracy corresponding to the existence of four directions in which the energy of the $\mathbf{J}$-vector is a minimum. The measured $g$-factors suggest that these minima form a square in the $x$–$y$ plane. It can been shown that, for $Dy^{3+}$, the $\mathbf{J}$-vector tends to point towards the ligands.

The $D_{2d}$ site structure of vanadates has already been described in Section 5.3.2. The four nearest neighbour oxygens lie about 12 degrees away from the $x$–$y$ plane. However, taking the inversion operator into account increases the effective symmetry to $D_{4h}$, which does have four equivalent coplanar directions. These directions bisect the directions of each nearest neighbour oxygen and its inverted image at $\pm 12°$ relative to the $x$–$y$ plane.

According to the semiclassical model the low-lying crystal field states take the form [PS71]

$$
\begin{aligned}
|\omega\rangle = N^{-1/2}(&|J_x = 15/2\rangle + \omega|J_y = 15/2\rangle \\
&+ \omega^2|J_x = -15/2\rangle + \omega^3|J_y = -15/2\rangle),
\end{aligned} \tag{9.2}
$$

where $\omega$ is one of the four solutions of $\omega^4 = -1$ and $N$ is chosen to ensure normalization. Given the degeneracy of the pairs of states with $J_x = \pm 15/2$ and $J_y = \pm 15/2$, it is to be expected that tunnelling will reduce the degeneracy to two (Kramers) doublets.

At low temperatures the higher energy doublet will be depopulated, producing an electronic state with lower symmetry than that of the site. In the vanadates this is stabilized by elastic distortions of the crystal, leading to an explanation of the observed Jahn–Teller ordering. Further details are given in [PS71, EHHS72].

## 9.2 Tunnelling in a cubic site with eightfold coordination

This section is concerned with the explicit parametrization of the tunnelling matrix for single electrons, and the relationship between the tunnelling parameters and the crystal field parameters introduced in Chapter 2. In order to simplify the discussion, the ground cluster of $Ho^{3+}$ in cubic sites with eightfold coordination will be used as a running example. The approach described in this section can, however, be readily extended to other magnetic ions, other site symmetries and other coordinations.

The chosen example is of particular interest because of the strong clustering that is observed in the ground multiplet energy levels of $Ho^{3+}$ in garnets and superconducting cuprates, which both have approximate cubic symmetry. For example, the neutron scattering spectrum for holmium in both orthorhombic and tetragonal forms of the high-temperature superconductor $HoBa_2Cu_3O_{7-\delta}$ has a cluster of observed levels up to approximately 100 cm$^{-1}$ (see Table 3.5), the next highest level being about 500 cm$^{-1}$ [FBU88b, AFB$^+$89, GLS91].

The application of the semiclassical model to the description of crystal field split energy levels is complicated by the fact that, with angular momenta $J \leq 8$, there are significant overlaps between the localized wavefunctions. It is therefore necessary to extend the formalism developed by Harter and Patterson [HP79] to take account of these overlaps (see Section 9.2.2). In practice, there is an added benefit in calculating overlaps because their magnitudes provide a useful estimate of the relative importance of tunnelling between the various energy minima.

### 9.2.1 The tunnelling Hamiltonian

In order to find an expression for the tunnelling Hamiltonian for a given cluster it is necessary to:

(i) determine the directions of potential minima for the **J**-vector;
(ii) find expressions for states localized at these minima;
(iii) use these expressions to construct the tunnelling Hamiltonian and overlap matrices; and
(iv) calculate the elements of the overlap matrix.

The tunnelling Hamiltonian is then expressed in terms of one or more tunnelling amplitudes, which have to be fitted to observed energy levels using a method similar to that used to fit crystal field parameters (described in Chapter 3). The eigenvalues of this Hamiltonian correspond to the energy levels of the states in the cluster.

#### 9.2.1.1 Extrema of a cubic potential

Consider a cubic potential function with the quantization axis along the $z$-direction. Following Harter and Patterson [HP79], this may be expressed in the form

$$V_{\mathrm{CF}}(\theta, \phi) = Y_4(\theta, \phi) \cos \alpha + Y_6(\theta, \phi) \sin \alpha, \qquad (9.3)$$

where the angle $\alpha$ is a parameter and $Y_4(\theta, \phi)$ and $Y_6(\theta, \phi)$ are, respectively, the rank 4 and rank 6 cubic crystal field potential functions. These functions are related to the spherical harmonic functions by

$$\begin{aligned}
Y_4(\theta, \phi) &= \sqrt{1/24} \left( \sqrt{14}\, Y_{4,0}(\theta, \phi) + \sqrt{5}\, (Y_{4,4}(\theta, \phi) + Y_{4,-4}(\theta, \phi)) \right), \\
Y_6(\theta, \phi) &= (1/4) \left( \sqrt{2}\, Y_{6,0}(\theta, \phi) - \sqrt{7}\, (Y_{6,4}(\theta, \phi) + Y_{6,-4}(\theta, \phi)) \right),
\end{aligned} \qquad (9.4)$$

when the $z$-direction is along a fourfold axis. In the coordinate system used in (9.3), one of the (equivalent) fourfold symmetry axes is along $\theta = 0$, and one of the threefold symmetry axes is along $\theta = \pi/4$ and $\phi = \pi/4$. The angle $\alpha$ is related to the cubic crystal field parameters as follows:

$$\tan \alpha = \frac{\sqrt{42} B_0^6}{\sqrt{13} B_0^4}.$$

Harter and Patterson [HP79] provide a graphical representation of the way the potential function in (9.3) changes with $\alpha$. Table 9.3 gives the value of $\alpha$ for the maxima and minima of this function. The minimum energy for a given **J**-vector can correspond to either of these potential function extrema.

Table 9.3. Values of $\alpha$ for potential function extrema along various symmetry axes.

| Symmetry axis | $\alpha$ (min.) | $\alpha$ (max.) |
|---|---|---|
| Fourfold $(\theta = 0, \phi = 0)$ | $\pi$ | $\pi/6$ |
| Threefold $(\theta = \cos^{-1}(\sqrt{(2/3)}), \phi = 0)$ | $0$ | $4\pi/6$ |
| Twofold $(\theta = \cos^{-1}(\sqrt{(1/2)}), \phi = \pi/4)$ | $5\pi/12$ | $\pi$ |

### 9.2.1.2 Localized basis set

Consider the case when the potential energy minima of the **J**-vectors are along the eight threefold symmetry axes. According to the semiclassical picture [Tra63, HP79], the approximate eigenfunctions of the localized crystal field potentials should have angular momentum wavefunctions with $M_J = J, J - 1, \ldots$ with their quantization axes $z$ aligned along each of the threefold symmetry axes. Eigenfunctions for the energy levels in a given cluster can be constructed as linear superpositions of *localized* angular momentum states with **J** pointing along the direction of all the equivalent potential extrema. In particular, the lowest-lying cluster of energy levels correspond to linear superpositions of the localized states $|M_J = J\rangle$. Similarly, the eigenstates of the next highest-lying cluster are composed of linear superpositions of the localized states $|M_J = J - 1\rangle$. In the case of $Ho^{3+}$, the approximate eigenfunctions of the lowest-lying cluster are formed from superpositions of eight states of the form $|M_J = J\rangle$, with their quantization axes aligned along each of the threefold symmetry axes.

Coupling between these localized states by the crystal field potential splits the eightfold degeneracy of the cluster. There are two different types of coupling parameters; those which correspond to tunnelling between the localized states which make up the given cluster, and those which couple localized states in different clusters. The semiclassical model is only concerned with the first type of coupling parameters, which are contained in 'intra-cluster' Hamiltonians.

States localized in a potential minimum correspond to irreducible representations of one of the eight $C_{3v}$ subgroups of the site symmetry group $O_h$. The first localized state to be considered is $|M_J = J\rangle$, for which the quantization direction defines the $z$-axis. This state is denoted $|1\rangle$. The other seven states belonging to the same cluster are obtained by rotating

Table 9.4. Overlap matrix elements for eight potential energy minima.

| $J$ | $M_J$ | $o_2$ | $o_6$ |
|------|-------|---------|---------|
| 8 | 8 | 0.0390 | 0.0001 |
| 8 | 7 | 0.2536 | 0.0044 |
| 15/2 | 15/2 | i0.0478 | i0.0003 |
| 15/2 | 13/2 | i0.2867 | i0.0071 |
| 6 | 6 | 0.0878 | 0.0014 |
| 6 | 5 | 0.3951 | 0.0288 |

this localized state into the other threefold symmetry directions. They are denoted $|\kappa\rangle$, with $\kappa = 2, 3, 4$, etc. In terms of the Euler angle $(\alpha, \beta, \gamma)$ rotations used to obtain the states $|\kappa\rangle$ from $|1\rangle$, they are given by

$$
\begin{aligned}
|1\rangle &= |0, 0, 0\rangle \\
|2\rangle &= |0, 2\theta_1, \pi\rangle \\
|3\rangle &= |2\pi/3, 2\theta_1, \pi\rangle \\
|4\rangle &= |4\pi/3, 2\theta_1, \pi\rangle \\
|5\rangle &= |0, \pi, 0\rangle \\
|6\rangle &= |0, -2\theta_2, \pi\rangle \\
|7\rangle &= |2\pi/3, -2\theta_2, \pi\rangle \\
|8\rangle &= |4\pi/3, -2\theta_2, \pi\rangle,
\end{aligned}
\tag{9.5}
$$

where $\theta_1 = \cos^{-1}(\sqrt{2/3})$ and $\theta_2 = \sin^{-1}(\sqrt{2/3})$. It should be noted that while these localized states are linearly independent, they are *not* all mutually orthogonal. Their overlaps, which depend on the angular momentum vector $\mathbf{J}$, are given in Table 9.4.

In order to obtain an explicit expression for the intra-cluster, or tunnelling, Hamiltonian, it is first necessary to express the localized wavefunctions (9.5) in terms of the angular momentum wavefunctions for a single quantization axis, which is taken to be the direction defined by $|1\rangle$. The other localized wavefunctions can be expressed in terms of angular momentum wavefunctions along this direction as follows

$$
|\kappa\rangle = \sum_{M_J} d_{M_J J}(\kappa) |M_J\rangle,
\tag{9.6}
$$

where $d_{M_J J}(\kappa)$ denotes the matrix which rotates $|1\rangle$ into $|\kappa\rangle$.

### 9.2.1.3 The Hamiltonian and overlap matrices

Using the set of eight states defined by (9.5) as a non-orthogonal basis set, the matrix of the tunnelling Hamiltonian $\overline{H}$ for the ground cluster, takes the form:

$$
\overline{H} = \begin{pmatrix}
h_1 & h_2 & h_2 a & h_2 a^2 & 0 & h_6 & h_6 a & h_6 a^2 \\
h_2 & h_1 & h_6 a^2 & h_6 a & h_6 & 0 & h_2 & h_2 \\
h_2 a^2 & h_6 a & h_1 & h_6 a^2 & h_6 & h_2 & 0 & h_2 \\
h_2 a & h_6 a^2 & h_6 a & h_1 & h_6 a^2 & h_2 & h_2 & 0 \\
0 & h_6 & h_6 a^2 & h_6 a & h_1 & h_2 & h_2 a^2 & h_2 a \\
h_6 & 0 & h_2 & h_2 & h_2 & h_1 & h_6 a & h_6 a^2 \\
h_6 a^2 & h_2 & 0 & h_2 & h_2 a & h_2 a^2 & h_1 & h_6 a \\
h_6 a & h_2 & h_2 & 0 & h_2 a^2 & h_6 a & h_6 a^2 & h_1
\end{pmatrix}. \tag{9.7}
$$

All the matrix elements are defined in terms of the phase factor $a = \exp(-2\pi i/3)$:

$h_1 = \langle 1|\overline{H}|1\rangle$, the average energy of the cluster,

$h_2 = \langle 1|\overline{H}|2\rangle$, the magnitude of the tunnelling amplitude between adjacent minima, and

$h_6 = \langle 1|\overline{H}|6\rangle$, the tunnelling amplitude between next nearest neighbour minima.

Energy levels within the cluster can be obtained by solving

$$
\overline{H}\Phi = E\overline{O}\Phi, \tag{9.8}
$$

where the overlap matrix $\overline{O}$ is obtained by substituting the corresponding overlap matrix elements $o_1 = 1$, $o_2 = \langle 1|2\rangle$ and $o_6 = \langle 1|6\rangle$ for $h_1$, $h_2$ and $h_6$, respectively. In the case of the $J = 8$ ground multiplet of $Ho^{3+}$, the overlap $o_6$ is very much smaller than $o_2$. Hence the semiclassical model allows us to predict that the tunnelling amplitude $h_6$ is significantly smaller than $h_2$.

### 9.2.2 Symmetry adapted states and analytic expressions for eigenvalues

Symmetry adapted states for a general value of the total angular momentum $J$ can be obtained by applying irreducible representation projection operators [LN69] to the localized states. In semiclassical model applications, the derivation of these operators involves the theory of induced representations [Led77, Alt77], which is beyond the scope of this book. However, in order to proceed with the present example, it is only necessary to quote results for the $M_J = J$ cluster of the $J = 8$ multiplet of $Ho^{3+}$. In this case, the energy minima are aligned with the threefold symmetry axes of the cube ($O_h$ symmetry), and the eight symmetry adapted functions form a doublet, labelled $E$, and two triplets, labelled $T_1$ and $T_2$ [New83b]†. Just one of the components for each case is given below:

$$|E\rangle = N(E)(|2\rangle + a|3\rangle + a^2|4\rangle + |5\rangle)$$
$$|T_1\rangle = N(T_1)(|2\rangle + |3\rangle + |4\rangle - |6\rangle - |7\rangle - |8\rangle) \tag{9.9}$$
$$|T_2\rangle = N(T_2)(|2\rangle + |3\rangle + |4\rangle + |6\rangle + |7\rangle + |8\rangle).$$

The normalization constants, which allow for the non-orthogonality of the localized states, are given by

$$N(E) = (4 + 12o_6)^{-1/2}$$
$$N(T_1) = [6(1 - 2o_2 - o_6)]^{-1/2} \tag{9.10}$$
$$N(T_2) = [6(1 + 2o_2 - o_6)]^{-1/2}.$$

With this symmetry adapted basis, solving equation (9.8) gives the following eigenenergies

$$E(E) = (h_1 + 3h_6)/(1 + 3o_6)$$
$$E(T_1) = (h_1 - 2h_2 - h_6)/(1 - 2o_2 - o_6) \tag{9.11}$$
$$E(T_2) = (h_1 + 2h_2 - h_6)/(1 + 2o_2 - o_6).$$

The overlap integrals in these energy expressions are given in Table 9.4. The parameters $h_1$, $h_2$ and $h_6$ can be fitted directly to the three observed energy levels.

The energy levels in the $^5I_8$ ground multiplet of $Ho^{3+}$ in the high-temperature superconductor $HoBa_2Cu_3O_{7-\delta}$ have been determined by neutron scattering [FBU88b, GLS91]. The eight nearest neighbour oxygen ligands of the $Ho^{3+}$ ions are at the corners of an almost perfect cube. Hence the eight non-degenerate energy levels that make up the lowest-lying cluster (see Table 3.5) can be associated with eight localized states, showing that the localized **J**-vectors point towards the ligands. Table 3.5 shows that, while the energies of the eight lowest-lying states have been determined experimentally, only a few of the higher-lying levels of the ground multiplet have been determined. This makes the semiclassical model particularly appropriate in this case.

In order to use the theory developed above, it is first necessary to group the eight energy levels (corresponding to the true $D_{2h}$ site symmetry) according to the symmetry labels of the approximate cubic symmetry. Equation (9.11), taken together with the expected dominance of $h_2$, leads to the

† Irreducible representation labels are defined in Appendix 1.

prediction that the three lowest-lying levels, with mean energy 6.2 cm$^{-1}$, correspond to the irreducible representation label $T_1$; the three highest levels, with mean energy 82.0 cm$^{-1}$, correspond to the irreducible representation label $T_2$; and the two remaining levels, with mean energy 32.7 cm$^{-1}$, correspond to the irreducible representation label $E$. Using group theoretical methods (see Appendix 1), these assignments can be shown to be in agreement with the $D_{2h}$ assignments shown in Table 3.5. Solving (9.11), as simultaneous equations for the tunnelling amplitudes, then gives $h_1 = 43.5$ cm$^{-1}$, $h_2 = 20.7$ cm$^{-1}$ and $h_6 = -3.6$ cm$^{-1}$. These results must, of course, be interpreted as mean values of the tunnelling amplitudes for the $D_{2h}$ site, and may contain significant contributions from the way this is distorted from $O_h$ symmetry. A similar calculation, using energy levels obtained by Allenspach *et al.* [AFB$^{+}$89] for the tetragonal ($D_{4h}$) form of the same superconductor, gives $h_1 = 43.2$ cm$^{-1}$, $h_2 = 19.5$ cm$^{-1}$ and $h_6 = -1.2$ cm$^{-1}$, confirming the expectation that the tunnelling amplitude $h_6$ is significantly smaller in magnitude than $h_2$. At the same time, however, it becomes clear that the mean value of $h_6$ is very sensitive to local distortions from cubic symmetry.

In principle, the tunnelling parameters $h_2$ and $h_6$ provide sufficient information to determine the two cubic crystal field parameters and hence the two intrinsic parameters $\overline{B}_4$ and $\overline{B}_6$. However, in practice the sensitivity of the mean value of the tunnelling parameter $h_6$ to symmetry changes would introduce considerable uncertainties into such a calculation. This problem is the subject of current research.

## 9.3 Outlook

The semiclassical model has limited applications in the analysis of crystal field split energy levels, and has generally been ignored. However, the examples given in this chapter demonstrate that, on occasion, it can provide a very useful first step in the analysis of ground multiplet spectra. In order to develop the full power of this approach, some elaboration and generalization of the formal procedures will be necessary.

# 10

## Transition intensities

### M. F. REID

*University of Canterbury*

The intensities of transitions within the $4f^N$ configurations of lanthanide ions are strongly dependent on the environment of the ion. The electric dipole transitions that dominate the solid state and solution spectra are forbidden for an isolated ion. They only become allowed when the symmetry is reduced from the full rotational symmetry of a free ion to the point group symmetry at an ion in a condensed matter environment.

The phenomenological and *ab initio* descriptions of lanthanide intensities are much less developed than the description of crystal field split energy levels. This is due to various technical problems, both experimental and theoretical, that will be alluded to in this chapter. As a result, the detailed fitting of transitions between crystal field split levels is still relatively uncommon, over 35 years since it became possible.

The methods described here are relevant to the sharp line spectra of lanthanide and actinide ions in crystals, though the examples refer mostly to lanthanides. A detailed analysis of the broadband spectra typical of transition metal systems generally requires the addition of extra theoretical machinery to take account of vibronic progressions, Franck–Condon factors, etc. Analyses of the intense $4f^N \leftrightarrow 4f^{N-1}5d$ transitions of lanthanide ions also require extensions.

The history of the field, particularly that of fitting total transition intensities between $J$ multiplets, has been covered in a number of reviews [Pea75, Hüf78, GWB98]. The derivation of selection rules for optical transitions and the relationships between line strengths and physical observables: oscillator strengths, cross-sections, Einstein coefficients, etc. are also well known. The present discussion concentrates on how to fit crystal field level intensity data, and how one might make sense of the resulting parameters in terms of the physics of the interactions between the lanthanide ion, its ligands, and the radiation field.

The 1962 papers of Judd [Jud62] and Ofelt [Ofe62] provided the first practical parametrization of electric dipole transition intensities within the $4f^N$ configurations of lanthanide ions in solids and solutions. However, as in the crystal field case, the success of this parametrization scheme does not prove the correctness of the assumptions inherent in the model on which it is based. Unfortunately, many papers on $4f^N$ transition intensities have placed too great an emphasis on the original mechanisms invoked by Judd and Ofelt, i.e. contributions to the transition intensities due to the mixing of $4f^{N-1}nd$ and $4f^{N-1}ng$ configurations into the $4f^N$ configuration by an external crystal field. Subsequent work by Judd and others [JJ64, MPS74, HFC76, Jud79, PN84, RN89] widened the scope of possible mechanisms to include various forms of mixing and interaction of the $4f^N$ states with ligand states. Even for the case of mixing of $4f$ orbitals with $nd$ and $ng$ orbitals, naive point charge calculations run into the logical difficulty that the $nd$ and $ng$ orbitals extend beyond the ligands. The sources of the crystal field are, therefore, hardly 'external' to the lanthanide ion.

With these cautions in mind, in this chapter we develop a general one-electron parametrization scheme for transitions within the $4f^N$ configuration, being careful not to rely too much on guidance from the original Judd–Ofelt model. The approach presented here originates with Newman and Balasubramanian [NB75], who were the first to point out that the Judd–Ofelt approach did not provide the most general one-electron spin-independent parametrization of transition intensities.

Two phenomenological parametrizations were described by Newman and Balasubramanian. One, an extension of Judd's [Jud62] parametrization, was adapted by Reid and Richardson [RR83, RR84b] and has been widely used. However, Burdick *et al.* [BCR99] have shown this parametrization to have the disadvantage that several very different parameter sets give the same calculated intensities. In the other parametrization, which has only recently been used in parameter fits, alternative sets of parameters that give the same intensities are very clearly related.

Having established a general parametrization scheme (at the one-electron, spin-independent level) the implications of the superposition model are discussed. In contrast to the one-electron crystal field case, the superposition model restricts the number of intensity parameters. We discuss data analysis methods for both total multiplet-to-multiplet transitions, and for transitions between crystal field levels. Since the fitting of multiplet-to-multiplet transitions has recently been extensively reviewed [GWB98], particular attention is paid to transitions between crystal field levels, and an example where the superposition approximation breaks down is given.

After considering the phenomenology we briefly discuss first principles calculations. The obvious mechanism of crystal field mixing of opposite parity configurations is unable to explain the experimental data. The relative signs of the intensity parameters can only be reproduced when ligand excitations are included.

Extensions of the parametrization schemes to include two-body and spin-dependent effects are briefly considered. Finally, we discuss some related topics: vibronic transitions, circular dichroism, two-photon absorption and Raman scattering.

Readers who are interested in analysing transition intensity data, without getting involved with the formal aspects of the theory, can go straight to Section 10.2.2 (which describes the parametrization) and Section 10.4 (which describes phenomenological analyses of experimental results). A program package, which includes the means to fit intensity parameters, has been developed by the author of this chapter. A brief description of this package is given in Appendix 3.

## 10.1 General aspects

Experimental measurements of absorption or emission may be expressed in terms of various observables, such as oscillator strengths, cross-sections, and Einstein $A$ and $B$ coefficients. In calculating these observables the electromagnetic interaction between the atom and the radiation is decomposed in a multipolar expansion leading to electric dipole, magnetic dipole, electric quadrupole, and higher multipole operators (e.g. [Wei78, HI89]). Note that in forming the multipolar expansion there are some subtleties regarding the choice of gauge for the electromagnetic field [Rei88, WS93, CTDRG89]. We do not discuss the electric quadrupole operator here, though it is important for some circular dichroism transitions (Section 10.7.2) and some transitions in centrosymmetric compounds [TS92].

### *10.1.1 Electric and magnetic dipole operators*

The electric dipole operator is given by

$$-eD_q^{(1)} = -erC_q^{(1)},\tag{10.1}$$

and the magnetic dipole operator by

$$M_q^{(1)} = \frac{-e\hbar}{2mc}\left(L_q^{(1)} + 2S_q^{(1)}\right),\tag{10.2}$$

where the symbols have their usual significance.

Since there are a variety of physical observables it is convenient to focus attention on one specific calculated quantity, the line strength (or dipole strength). In order to define the line strength we consider a set of $g_I$ degenerate initial states $\{|Ii\rangle\}$ and a set of $g_F$ degenerate final states $\{|Ff\rangle\}$. The electric dipole line strength for a transition between $I$ and $F$ with polarization $q$ (either spherical, $q = 0, \pm1$, or Cartesian, $q = x, y, z$) is defined to be

$$S_{FI,q}^{\mathrm{ED}} = \sum_i \sum_f e^2 \left| \langle Ff | D_q^{(1)} | Ii \rangle \right|^2 \tag{10.3}$$

and the magnetic dipole line strength is defined to be

$$S_{FI,q}^{\mathrm{MD}} = \sum_i \sum_f \left| \langle Ff | M_q^{(1)} | Ii \rangle \right|^2 . \tag{10.4}$$

For linear polarized light the electric and magnetic dipole moment interactions do not interfere, and the electric and magnetic dipole strengths may be added (after multiplying by appropriate factors). If the light is circularly polarized then interference between electric and magnetic dipole moments can lead to circular dichroism in certain symmetries (see Section 10.7.2).

To relate the dipole strengths to oscillator strengths, or $A$ or $B$ coefficients, they must be divided by the degeneracy of the initial state, $g_I$, and multiplied by appropriate physical constants, including refractive index corrections (various powers of $n$) and Boltzmann factors for the populations of the initial states.

For a full discussion of the relation between line strengths and physical observables, and the determination of these physical observables from experimental measurements, we refer readers to the literature [HI89, GWB98]. Henderson and Imbusch [HI89] give a modern treatment that uses SI units and takes careful account of corrections due to the refractive index of the medium (see also [SIB89]). Here we only quote the results for oscillator strengths and leave the $A$ and $B$ coefficients to the literature, but with the caution that extra refractive index corrections arise in those cases.

The electric dipole and magnetic dipole oscillator strengths are defined by

$$f_{FI,q}^{\mathrm{ED}} = \frac{2m\omega}{\hbar e^2} \frac{\chi_{\mathrm{L}}}{n} \frac{1}{g_I} S_{FI,q}^{\mathrm{ED}}, \tag{10.5}$$

$$f_{FI,q}^{\mathrm{MD}} = \frac{2m\omega}{\hbar e^2} n \frac{1}{g_I} S_{FI,q}^{\mathrm{MD}}, \tag{10.6}$$

where $\omega$ is the angular frequency of the radiation and the other symbols have their usual meanings. The refractive index $n$ appears in these expressions because the electric and magnetic fields have values in the medium different from their values in free space. The bulk correction factors are supplemented by a local correction to the electric field, $\chi_L$, to account for the fact that the ion is less polarizable than the bulk medium. This local electric field correction is often taken to be

$$\chi_L = \left(\frac{n^2 + 2}{3}\right)^2.$$                      (10.7)

This equation is an approximation. It is only strictly correct for high-symmetry sites [HI89]. In addition, it is based on the simplistic assumption that the lanthanide ion is an isolated entity surrounded by a dielectric medium. The transition intensity mechanisms discussed in Section 10.5 consider the lanthanide plus its ligands as a unit, making it difficult to disentangle the refractive index corrections, particularly the local correction, from the *ab initio* calculations.

Many authors define a $\chi$ that is a product of $\chi_L$, given in (10.7) and the bulk refractive index correction $1/n$ (e.g. [MRR87b]). Since the exact form of the local correction has a much more tenuous justification than the bulk correction it seems sensible to keep them separate.

### 10.1.2 Polarization and selection rules

Most experiments are performed with unpolarized light, linearly polarized light, or circularly polarized light. We can use equations (10.3) and (10.4), in either a spherical basis ($q = 0, \pm1$), or a Cartesian ($q = x, y, z$) basis to calculate dipole strengths for the polarizations of interest.

If the lanthanide ion sites are randomly oriented (as in a powder or solution) or embedded in a cubic crystal (such as a garnet), and the light is unpolarized or linearly polarized, then the physical observables are proportional to an average over the polarization components and we may define an isotropic dipole strength

$$\bar{S}_{FI}^{ED} = \frac{1}{3}\sum_q S_{FI,q}^{ED}.$$                      (10.8)

For polarized light and oriented crystals the situation is more complex. To perform the calculations we require the transformations between spherical ($q = 0, \pm1$) and Cartesian ($q = x, y, z$) bases. The electric dipole moment has odd parity, so the transformation involves the odd-parity irreducible representation $D^{(1)-}$ of the full rotation group $O_3$ (defined in Appendix 1):

$$|1^-0\rangle = |1^-z\rangle, \qquad |1^- \pm 1\rangle = \mp\frac{1}{\sqrt{2}}(|1^-x\rangle \pm i|1^-y\rangle). \qquad (10.9)$$

The magnetic dipole operator has even parity, so the transformation involves the irreducible representation $D^{(1)+}$:

$$|1^+0\rangle = |1^+z\rangle, \qquad |1^+ \pm 1\rangle = \mp\frac{1}{\sqrt{2}}(|1^+x\rangle \pm i|1^+y\rangle). \qquad (10.10)$$

In some low-symmetry situations it is necessary to measure not only intensities along the $x$-, $y$- and $z$-axes, but also at intermediate angles [Ste85, Ste90]. However, in the following discussion we generally restrict ourselves to the case of unpolarized radiation or uniaxial crystal systems (i.e. crystals containing sites with parallel axes of at least threefold symmetry). The traditional experimental arrangements are orthoaxial (propagation vector perpendicular to the high-symmetry axis) and axial (propagation vector parallel to the axis).

Consider first the case of an orthoaxial measurement using linearly polarized light with the electric field along the $z$-axis ($\pi$ polarization). Equation (10.9) indicates that the expression for the electric dipole line strength is given by the spherical expression (10.3) with $q = 0$, so that

$$S^{ED}_{FI,\pi} = S^{ED}_{FI,0}. \qquad (10.11)$$

In the case of the magnetic dipole line strength it is necessary to take into account the fact that the magnetic field of the radiation is perpendicular to the electric field, i.e. in the $x$–$y$ plane. Without loss of generality the magnetic field may be chosen to be along the $x$-axis. Then, from (10.10), we may derive an expression containing the square moduli of the matrix elements of the operator $\left(-M^{(1)}_1 + M^{(1)}_{-1}\right)/\sqrt{2}$. For uniaxial symmetries the cross terms cancel, giving

$$S^{MD}_{FI,\pi} = \frac{1}{2}\left(S^{MD}_{FI,1} + S^{MD}_{FI,-1}\right). \qquad (10.12)$$

For linearly polarized light with the electric field perpendicular to the $z$-axis ($\sigma$ polarization)

$$S^{ED}_{FI,\sigma} = \frac{1}{2}\left(S^{ED}_{FI,1} + S^{ED}_{FI,-1}\right), \qquad (10.13)$$

and

$$S^{MD}_{FI,\pi} = S^{MD}_{FI,0}. \qquad (10.14)$$

For axial spectra the polarization is irrelevant (as long as it is linear), so that

$$S^{\text{ED}}_{FI,\text{axial}} = S^{\text{ED}}_{FI,\sigma},$$
(10.15)

$$S^{\text{MD}}_{FI,\text{axial}} = S^{\text{MD}}_{FI,\pi}.$$
(10.16)

Circularly polarized measurements are generally performed on isotropic samples or in an axial configuration so that the dichroism of the bulk medium does not obscure the circular dichroism of the lanthanide sites. In the convention used by most chemists (where the incoming radiation is presumed to be viewed from the sample) the $q = \pm 1$ components of the dipole moment operators correspond to the absorption of left and right circularly polarized light [PS83]. Most physicists use the opposite convention [HI89].

Derivations of the selection rules for optical transitions may be found in many sources [Wyb65a, Hüf78, PS83, HI89, Ste90, GWB98]. For uniaxial crystals the operators corresponding to $\pi$ and $\sigma$ polarization transform as different irreducible representations of the point group of the site. The transformation properties may be determined from appropriate tables [KDWS63, But81, AH94]. For example, the transition will be electric dipole allowed if $\Gamma_I \times \Gamma^{\text{ED}}_T$ contains $\Gamma^*_F$, where the polarization $T$ may be $\sigma$ or $\pi$ in the examples considered here.

In determining the appropriate irreducible representations (see Appendix 1) it is necessary to consider the *odd-parity* irreducible representation $D^{(1)-}$ of $O_3$ in the case of the electric dipole operator, and the *even-parity* irreducible representation $D^{(1)+}$ of $O_3$ in the case of the magnetic dipole operator (equations (10.9) and (10.10)). It is also necessary to remember that the electric and magnetic fields of the radiation are perpendicular.

As an example, consider a site of $C_{4v}$ symmetry. The branchings from the $O_3$ irreducible representations $D^{(1)+}$ and $D^{(1)-}$ to the subgroup $C_{4v}$ are [KDWS63, But81, AH94]:

$$D^{(1)+} \to A_2 + E, \qquad D^{(1)-} \to A_1 + E.$$

The $A_2$ and $A_1$ irreducible representations are associated with the functions transforming as $S_z$ or $z$, i.e. $|1^\pm 0\rangle$. The $E$ irreducible representations are associated with the functions $|1^\pm \pm 1\rangle$. Because $\pi$-polarized light has the electric vector along the $z$-axis, in the electric dipole case the appropriate operator transforms as $A_1$. The operator appropriate for $\sigma$-polarized electric dipole radiation transforms as $E$. On the other hand, for $\pi$-polarized light the magnetic field is perpendicular to the $z$-axis, so the magnetic dipole operator corresponding to $\pi$ polarization transforms as $E$. The operator appropriate for $\sigma$-polarized magnetic dipole radiation transforms as $A_2$.

## 10.2 Parity-forbidden transitions

Once the $4f^N$ crystal field eigenfunctions have been calculated, as described in Chapters 3 and 4, it is possible to calculate the magnetic dipole moments by simply evaluating the magnetic dipole moment operator between crystal field eigenfunctions. However, this is not the case for the electric dipole moments, since electric dipole transitions within a pure $f^N$ configuration are forbidden, as the odd-parity operator $D_q^{(1)}$ cannot connect two states of the same parity. In order for electric dipole transitions to occur there must be an admixture of other ion or ligand states of opposite parity into $4f^N$. This admixture will not be possible, and the transitions will remain forbidden, if the site symmetry group contains the inversion operation.

### 10.2.1 Effective transition operators

To include all the states of opposite parity in a calculation would be impossible. The approach followed here is to write the transition amplitude as an *effective* operator that acts between $4f^N$ eigenstates of the phenomenological Hamiltonian (see Section 4.1). This is the same approach that has been used for the description of the crystal field (in Chapters 1–4) and the correlation crystal field (in Chapter 6). A discussion of the theoretical basis of the effective operator approach can be found in the literature [HF93, HF94, BR98a]. It is assumed that the time dependent perturbation expansion that leads to Fermi's golden rule (recall that the one-photon transition rate is proportional to the line strength) is carried out separately from the time independent perturbation expansion of the eigenstates of the time independent Hamiltonian. Thus, the task is to find a time independent expansion for the dipole moment operator. This expansion can then be substituted into the expression (10.3) for the dipole strength.

The first step in the time independent perturbation theory calculation is to partition the Hamiltonian into a zero-order part $H_0$ and a perturbation $V$. The total Hamiltonian is

$$H = H_0 + V. \tag{10.17}$$

The eigenstates of $H_0$ are labelled with Greek letters,

$$H_0|\alpha\rangle = E_\alpha^{(0)}|\alpha\rangle, \tag{10.18}$$

and the eigenstates of $H$ are labelled with Roman letters,

$$H|a\rangle = E_a|a\rangle. \tag{10.19}$$

An effective (crystal field) Hamiltonian $H_{\text{eff}}$ is now constructed to act within a 'model space' M. The equations are simpler if all members of M have identical $H_0$ eigenvalues, $E_0$, i.e. the model space is 'degenerate'. This simplification will be used in what follows.

In the case of interest here the model space is the $4f^N$ configuration and a typical choice for $H_0$ would be the free-ion Hamiltonian (see Chapters 1 and 4). The perturbation $V$ is a sum of several terms:

$$V = V_{\text{ee}} + V_{\text{so}} + V_{\text{CF}} + \cdots. \tag{10.20}$$

This expression includes that part of the coulomb repulsion $V_{\text{ee}}$ between $f$ electrons which is not included in $H_0$, the spin–orbit interaction $V_{\text{so}}$, and the crystal field potential $V_{\text{CF}}$ arising from the interaction of the magnetic ion with the ligands.

For eigenstates $|a_0\rangle$ of $H_{\text{eff}}$,

$$H_{\text{eff}}|a_0\rangle = E_a|a_0\rangle, \tag{10.21}$$

where the $E_a$ in this equation is identical to the $E_a$ in (10.19). In our applications the $|a_0\rangle$ are linear combinations of the $|4f^N\alpha SLJM\rangle$ states.

$H_{\text{eff}}$ may be constructed via a standard Rayleigh–Schrödinger perturbation expansion as:

$$H_{\text{eff}} = H_0 + V + \sum_{\beta\notin\text{M}} \frac{V|\beta\rangle\langle\beta|V}{E_0 - E_\beta^{(0)}} + \cdots, \tag{10.22}$$

where M refers to the model space, and the exact states are given by

$$|a\rangle = |a_0\rangle + \sum_{\beta\notin\text{M}} \frac{|\beta\rangle\langle\beta|V|a_0\rangle}{E_0 - E_\beta^{(0)}} + \cdots. \tag{10.23}$$

Note that if we intend to go to higher than first order in $V$ we must be careful to consider the orthogonality and normalization of the states [HF94].

We can now use equation (10.23) to write the matrix elements of the dipole moment operator to first order in $V$ as

$$\langle f|D_q^{(1)}|i\rangle = \langle f_0|D_q^{(1)}|i_0\rangle$$
$$+ \left\langle f_0\left|D_q^{(1)} \sum_{\beta\notin\text{M}} \frac{|\beta\rangle\langle\beta|V}{E_0 - E_\beta^{(0)}}\right|i_0\right\rangle$$

$$+ \left\langle f_0 \left| \sum_{\beta \notin M} \frac{V |\beta\rangle\langle\beta|}{E_0 - E_\beta^{(0)}} D_q^{(1)} \right| i_0 \right\rangle$$

$$+ \cdots \tag{10.24}$$

(recalling that $E_i^{(0)} = E_f^{(0)} \equiv E_0$). The matrix element $\langle f_0 | D_q^{(1)} | i_0 \rangle$ is zero for our case of interest ($|i_0\rangle$ and $|f_0\rangle$ in the $f^N$ configuration) but we include it here for completeness.

We require an effective operator $D_{\text{eff},q}$ such that the matrix elements of $D_{\text{eff},q}$ between eigenstates of $H_{\text{eff}}$ are identical to the matrix elements of $D_q^{(1)}$ between the exact eigenstates (if the expansion is carried to infinite order), i.e.

$$\langle f_0 | D_{\text{eff},q} | i_0 \rangle = \langle f | D_q^{(1)} | i \rangle. \tag{10.25}$$

The required operator may be taken from (10.24) as

$$D_{\text{eff},q} = D_q^{(1)} + D_q^{(1)} \sum_{\beta \notin M} \frac{|\beta\rangle\langle\beta| V}{E_0 - E_\beta^{(0)}} + \sum_{\beta \notin M} \frac{V |\beta\rangle\langle\beta|}{E_0 - E_\beta^{(0)}} D_q^{(1)} + \cdots. \tag{10.26}$$

Again, the first term, $D_q^{(1)}$, will not contribute in the case of interest here.

The denominators in the two summations in (10.26) are equal term by term and the effective operator is Hermitian. We refer readers to Hurtubise and Freed [HF93, HF94] for a careful discussion showing that for Hermitian operators (such as the dipole moment operator) it is always possible to construct *effective* operators that are Hermitian [HF93] and perturbation expansions that are Hermitian order by order in $V$ [HF94].

The time-reversal plus Hermiticity symmetry argument that restricts the one-electron phenomenological crystal field to even rank operators (see Appendix 1) may also be applied to our effective dipole moment operator. It is only necessary to consider even rank operators in order to parametrize the one-electron, spin-independent, part of the dipole moment effective operator. Since they are one-electron operators the rank of the effective operators is restricted to be less than or equal to 6 (for $f$ electrons).

The denominators in (10.26) are eigenvalues of $H_0$ and do not vary for different states within the $4f^N$ configuration or within each excited configuration. This is often referred to as the 'closure approximation' (though this terminology sometimes refers to closure only over the excited configuration). Our development emphasizes that this approximation arises naturally when we restrict the Rayleigh–Schrödinger expansion to first order in $V$. At higher orders there will be additional terms. In a Rayleigh–Schrödinger expansion the denominators will still only contain eigenvalues of $H_0$. However, it is straightforward to recast some of these higher order contributions as corrections to the denominators appearing in lower order terms (for example, see Brandow's [Bra67] transformations between Brillouin–Wigner and Rayleigh–Schrödinger expansions).

The perturbation $V$ is a sum of one- and two-electron operators (crystal field, coulomb interaction, spin–orbit interaction, etc.). At first order in $V$ only one-electron spin-independent operators can occur in $D_{\text{eff},q}$. This is because the only part of $V$ that can contribute is the odd-parity part of the operator $V_{\text{CF}}$, which is a one-electron operator. $V_{\text{CF}}$ can connect the $4f^N$ configuration to lanthanide or ligand states of opposite parity. The two-electron Coulomb operator $V_{\text{ee}}$ or the spin–orbit operator $V_{\text{so}}$ (acting within the lanthanide ion) cannot, by themselves, generate non-zero contributions to $D_{\text{eff},q}$, since they cannot connect the $4f^N$ configuration to states of opposite parity. At higher orders two-body and spin-dependent operators can arise because $V$ appears more than once in the perturbation expansion and so the perturbation can involve $V_{\text{ee}}$ or $V_{\text{so}}$, in addition to $V_{\text{CF}}$ (see Section 10.6).

### 10.2.2 Parametrization

The theory of effective transition operators can now be used to derive a parametrization analogous to the crystal field Hamiltonian. Because of its invariance under site symmetry operations, the effective operator for the crystal field Hamiltonian transforms as the identity irreducible representation of the site symmetry group (i.e. as a scalar). In the case of the effective dipole moment operator (10.26) the dipole moment operator $D_q^{(1)}$ is coupled to the perturbation operator $V$. Each part of $V$ is an invariant of the site symmetry group (see Chapter 2 and Appendix 1) so the effective operator must transform as the irreducible representations of the dipole moment operator, $D_q^{(1)}$ (e.g. $\Gamma_\pi^{\text{ED}}$ or $\Gamma_\sigma^{\text{ED}}$ for uniaxial crystals). As discussed above, the effective operator for a one-electron spin-independent parametrization must be a combination of tensor operators of even rank, and less than or equal to 6. A completely scalar operator (rank 0 in spin and orbital spaces) cannot contribute to transitions between different states, so we may write an expression for the effective dipole moment operator as

$$D_{\text{eff},q} = \sum_{\lambda,l} B_{lq}^\lambda U_l^{(\lambda)}, \tag{10.27}$$

with $\lambda = 2, 4, 6$. In this equation we use the notation of Burdick *et al.*

[BCR99]. However, the ideas date back to Newman and Balasubramanian [NB75], who referred to expression (10.27) as the *vector crystal field*.

The non-zero $B_{lq}^{\lambda}$ will yield all linear combinations of $U_l^{(\lambda)}$ with $\lambda = 2$, $4$, $6$ that transform as $\Gamma_{\pi}^{\mathrm{ED}}$ or $\Gamma_{\sigma}^{\mathrm{ED}}$. We may define parameters $B_{l\sigma}^{\lambda}$ and $B_{l\pi}^{\lambda}$, and use equations (10.9) and (10.27), to derive expressions for the effective dipole moment operators transforming as $\Gamma_{\pi}^{\mathrm{ED}}$ or $\Gamma_{\sigma}^{\mathrm{ED}}$:

$$D_{\mathrm{eff},\pi} \equiv D_{\mathrm{eff},0} = \sum_{\lambda,l} B_{l\pi}^{\lambda} U_l^{(\lambda)}, \tag{10.28}$$

$$D_{\mathrm{eff},\sigma} \equiv D_{\mathrm{eff},x} = (-D_{\mathrm{eff},1} + D_{\mathrm{eff},-1})/\sqrt{2} = \sum_{\lambda,l} B_{l\sigma}^{\lambda} U_l^{(\lambda)}. \tag{10.29}$$

Until recently (10.27) had not been used in any data analyses. Newman and Balasubramanian [NB75] emphasized the advantages of using a different parametrization, based on the concept of vector spherical harmonics, that is more closely related to the original papers of Judd and Ofelt [Jud62, Ofe62], and the perturbation theory calculations of Section 10.2.1. An adaptation of this parametrization by Reid and Richardson [RR83, RR84b] is now in common use. In this adaptation the effective electric dipole moment operator is written as

$$D_{\mathrm{eff},q} = \sum_{\lambda,t,p} A_{tp}^{\lambda} U_{p+q}^{(\lambda)} (-1)^q \langle \lambda(p+q), 1 - q | tp \rangle, \tag{10.30}$$

with $\lambda = 2$, $4$, $6$, $t = \lambda - 1$, $\lambda$, $\lambda + 1$ and $p$ restricted by the site symmetry. The Clebsch–Gordan coefficient, $\langle \lambda(p+q), 1 - q | tp \rangle$, is defined in equation (A1.5). It can be seen from (10.26) that the effective dipole moment arises from coupling the dipole moment operator $D_q^{(1)}$ with the perturbation operator $V$. Since the perturbation is a scalar of the site symmetry the $A_{tp}^{\lambda}$ are non-zero only if $tp$ (or linear combinations of $tp$) transform as the identity irreducible representation of the site symmetry group. Equation (10.30) emphasizes the coupling between the perturbation (transforming as $tp$) and the dipole moment operator (transforming as $1q$), to give an effective operator (transforming as $\lambda(p+q)$).

The two parametrizations, (10.27) and (10.30), must give identical results and have the same number of independent parameters. We may use the relationship

$$B_{(p+q)q}^{\lambda} = \sum_{t} A_{tp}^{\lambda} (-1)^q \langle \lambda(p+q), 1 - q | tp \rangle, \tag{10.31}$$

together with the spherical–Cartesian transformation (10.9), to derive relationships between the $A_{tp}^{\lambda}$ and the $B_{l\sigma}^{\lambda}$, $B_{l\pi}^{\lambda}$ parameter sets.

The parametrization (10.30) has been preferred because it is not only more closely related to the earlier work of Judd [Jud62], but it is also more amenable to superposition model analyses (Section 10.3). However, Burdick *et al.* [BCR99] pointed out that there are inherent ambiguities with this parametrization. These ambiguities have their origin in the nature of intensities. Because intensities involve squares of dipole moments, they are invariant under a sign change of the dipole moments.

It is clear from (10.30) that there are always at least two parameter sets that give identical calculated intensities, related by a change in the sign of all of the $A_{tp}^\lambda$ parameters. However, the parametrization (10.27) shows that there is often more freedom, since the signs of the parameters associated with each distinct polarization may be changed independently. Thus, in the uniaxial case, the parameter sets $B_{l\sigma}^\lambda$ and $B_{l\pi}^\lambda$ may be changed independently, giving not just two but four distinct sets of parameters. Transforming to the $A_{tp}^\lambda$ parameter set using (10.31) and (10.9) will, in general, give $A_{tp}^\lambda$ parameter sets that differ not only in *sign*, but also in *magnitude*. An example of this multiplicity will be given in Section 10.4.4: it has obvious implications for attempts to interpret fitted parameters in terms of *ab initio* calculations.

As well as the restrictions on the $A_{tp}^\lambda$ parameters that arise from the necessity to have combinations of $tp$ that are scalar under the site symmetry group there are restrictions arising from the requirement that the effective operator is Hermitian. This leads to a complex conjugation symmetry similar to that for the phenomenological crystal field parameters

$$\left(A_{tp}^\lambda\right)^* = (-1)^{t+1+p} A_{t-p}^\lambda. \tag{10.32}$$

This symmetry restricts parameters with $p = 0$ to be either purely real or purely imaginary for $t$ odd or even, respectively.

For the restricted case $t = \lambda \pm 1$ the $A_{tp}^\lambda$ parameters are related to the product $A_{tp}\Xi(t, \lambda)$ from Judd's 1962 paper [Jud62] by

$$A_{tp}^\lambda = -A_{tp}\Xi(t, \lambda)\frac{(2\lambda + 1)}{(2t + 1)^{1/2}}. \tag{10.33}$$

The products $A_{tp}\Xi(t, \lambda)$ have been used as phenomenological parameters by various authors, starting with Axe [Axe63], and are sometimes referred to as $B_{\lambda tp}$ [PC78].

To complete the calculation of dipole moments it is necessary to evaluate the matrix elements of the effective dipole operators between crystal field eigenstates. If the crystal field eigenstates are given in terms of $JM$ states as

$$|i_0\rangle = \sum_{\alpha,S,L,J,M} |4f^N \alpha SLJM\rangle C_{i,\alpha,S,L,J,M},$$

$$|f_0\rangle = \sum_{\alpha',S',L',J',M'} |4f^N \alpha'S'L'J'M'\rangle C_{f,\alpha',S',L',J',M'},$$

(10.34)

where $C$ denote eigenvector coefficients, then the matrix elements of the effective dipole operator (10.30) are given by

$$\langle f_0|D_{\mathrm{eff},q}|i_0\rangle = \sum_{\lambda,t,p} A^\lambda_{tp}(-1)^q \langle \lambda(p+q), 1-q|tp\rangle$$

$$\times \sum_{\alpha',S',L',J',M'} \sum_{\alpha,S,L,J,M} C^*_{f,\alpha',S',L',J',M'} C_{i,\alpha,S,L,J,M}$$

$$\times \langle 4f^N \alpha'S'L'J'M'|U^{(\lambda)}_{(p+q)}|4f^N \alpha SLJM\rangle.$$

(10.35)

This expression will be squared, summed over degenerate states, and multiplied by $e^2$ to give the electric dipole line strength (10.3). Computer programs for this purpose are described in Appendix 3.

In this section we have used the $JM$ basis. Many aspects of the derivation are clearer if point group coupling coefficients and point group basis functions are used instead. This has been discussed in [RR84b] and by Kibler and Gâcon [KG89].

### 10.2.3 Summation over multiplets, $\Omega_\lambda$ parameters

Judd [Jud62] showed that the intensities for total $J$ multiplet to $J'$ multiplet transitions may be fitted to a three-parameter linear model. The standard parameters are now labelled $\Omega_\lambda$, with $\lambda = 2, 4, 6$. The derivation is discussed in detail by [Pea75, GWB98]. Consider transitions from an initial multiplet $|\alpha_I J_I\rangle$ to a final multiplet $|\alpha_F J_F\rangle$. If the (rather drastic) assumption is made that all components of the initial multiplet are equally populated, averaging over all polarizations, summing the dipole strength over the $M$ components of the multiplets and using the orthogonality of the $3j$ symbols leads to an expression for the isotropic dipole strength, viz.

$$\bar{S}^{\mathrm{ED}}_{\alpha_F J_F, \alpha_I J_I} = \frac{1}{3}e^2 \sum_\lambda \Omega_\lambda \langle \alpha_F J_F \| U^{(\lambda)} \| \alpha_I J_I \rangle^2,$$

(10.36)

with

$$\Omega_\lambda = \sum_{t,p} \frac{1}{2\lambda+1} \left| A^\lambda_{tp} \right|^2.$$

(10.37)

This equation is analogous to equation (8.4), relating crystal field invariants to crystal field parameters. Note that the reduced matrix elements in equation (10.36) are for intermediate coupling (see Chapter 3). Equation (10.36) may be combined with the oscillator strength expression (10.5), with $1/g_I = 1/(2J_I+1)$, to recover the familiar expressions given by, for example, [Pea75, GWB98].

Equation (10.36) has the virtue of being linear in the $\Omega_\lambda$ parameters. This makes fitting to experimental data straightforward. However, in forming the sum and reducing the parametrization to just three parameters a considerable amount of information is lost, since each $\Omega_\lambda$ parameter is a combination of $A_{tp}^\lambda$ parameters with $t = \lambda - 1$, $\lambda$, $\lambda + 1$ and vibronic intensity is absorbed into the $\Omega_\lambda$ parametrization. Thus tests of models, such as the superposition model (Section 10.3), or detailed comparisons with *ab initio* calculations are not possible, since the $\Omega_\lambda$ parameters contain many contributions and furthermore are merely magnitudes with no sign information. In addition, the assumption of equal population of states of the initial multiplet may be wildly inaccurate if the crystal field splitting is several hundred wave numbers. Nevertheless, the relative simplicity of the measurements and calculations have permitted extremely useful analyses of large amounts of experimental data, as discussed in Section 10.4.

## 10.3 Superposition model

The superposition model may be developed for intensity parameters in analogy to its development for crystal field parameters (see Chapter 5). We begin by considering the hypothetical case of a single ligand at $R_0$ on the $z$-axis. In this case the only non-zero parameters will be those with $p = 0$. We define *intrinsic* intensity parameters to be the intensity parameters for this situation, viz.

$$\bar{A}_t^\lambda \equiv A_{t0}^\lambda. \tag{10.38}$$

The intrinsic parameters are restricted to odd $t$, since the identity irreducible representation of $C_{\infty v}$ does not occur in the $O_3$ irreducible representation $t^-$ unless $t$ is odd. Therefore, if the superposition model is applicable then the $A_{tp}^\lambda$ parameters are restricted to $t = \lambda \pm 1$ and the superposition model gives an important prediction about the allowed parameters.

For a set of (identical) ligands we may write

$$A_{tp}^\lambda = \bar{A}_t^\lambda \sum_L (-1)^p C_{-p}^t(\theta_L, \phi_L) \left(\frac{R_0}{R_L}\right)^{\tau_t^\lambda}. \tag{10.39}$$

Just as in the crystal field case, discussed in Chapter 5, the geometric factors $(-1)^p C^t_{-p}(\theta_L, \phi_L)$ are referred to as *coordination factors*. Each intensity parameter $\bar{A}^\lambda_t$ is associated with a corresponding power law exponent $\tau^\lambda_t$.

Equation (10.39) may be used to calculate the intrinsic $\bar{A}^\lambda_t$ parameters from the phenomenological parameters $A^\lambda_{tp}$, evaluating the coordination factors either with computer programs (Appendix 3) or from tables [Rud87b]. We will discuss the $\bar{A}^\lambda_t$ parameters derived from various experimental analyses in Section 10.4.3.

Most model calculations assume that the ligands may be represented by point charges and isotropic dipole moments. These calculations therefore implicitly contain the superposition model assumptions, restricting $t$ to be odd. Before the work of Newman and Balasubramanian [NB75], it was assumed that the equivalent of the $A^\lambda_{tp}$ parametrization with $\lambda = 2$, 4, 6 (i.e. $\lambda$ even) and $t = \lambda \pm 1$ (i.e. $t$ odd) formed the most general one-electron parametrization. However, in some circumstances, especially when the ligands are complex organic molecules, the $t = \lambda$ parameters are essential to the phenomenological fits (see Section 10.4.4).

## 10.4 Phenomenological treatment

The one-electron multiplet-to-multiplet $\Omega_\lambda$ or crystal field level $A^\lambda_{tp}$ or $B^\lambda_{lq}$ effective operator parametrizations, plus the $M_q^{(1)}$ magnetic dipole operator, can explain most aspects of lanthanide transition intensities within the $4f^N$ configuration.

The $\Omega_\lambda$ parametrization involves the drastic assumption that all states in the initial multiplet are equally populated, and some of the inaccuracies which occur in fitting these parameters may be attributed to this assumption. In the case of the $A^\lambda_{tp}$ parametrization there are other problems, such as the non-linearity of the fit, that will be discussed in Section 10.4.2.

In using either parametrization it is necessary to choose the experimental quantities to which the parameters should be fitted. Measurements may include absorption coefficients that are related to oscillator strengths and emission branching ratios and lifetimes that are related to the Einstein $A$ coefficients. It is generally most convenient to transform the experimental data into line strengths. Since the refractive index correction factors for electric and magnetic dipole interactions are different, and the refractive index may vary with frequency, an absolutely accurate extraction is impossible for transitions with mixed electric and magnetic dipole character.

Nevertheless, these correction factors may be absorbed into the parameters [MRR87b].

$D_{\text{eff},q}$ contains one-electron spin-independent operators so we expect the selection rule $\Delta S = 0$. For $M_q^{(1)}$ we expect $\Delta S = 0$ and $\Delta L = 0$. These selection rules are rather weak since the large spin–orbit coupling in the $4f^N$ configuration mixes states of different $S$ and $L$ (see Chapter 2) and $\Delta S = \pm 1$, $\Delta L = \pm 1$ transitions such as $^7F_J \leftrightarrow {}^5D_{J'}$ in $Eu^{3+}$ are routinely observed.

The parametrization also yields various selection rules on $J$. In the absence of crystal field mixing of the multiplets magnetic dipole transitions are restricted to $\Delta J \leq 1$ and not $J = 0 \leftrightarrow J = 0$, and electric dipole to $\Delta J \leq 6$ and $0 \leftrightarrow J = 2, 4, 6$ only. This selection rule is commonly violated. For $Eu^{3+}$ in some crystals $J = 0 \leftrightarrow J = 0$ and $J = 0 \leftrightarrow J = 3$ transitions are routinely observed. In most cases these can be explained by crystal field mixing of the $J = 0$ or $J = 3$ states with other states. The $\Omega_\lambda$ parametrization is unable to take this into account. To quantitatively explain some transitions it is necessary to consider spin-dependent or two-electron effective operators (see Section 10.6).

### 10.4.1 Phenomenology of multiplet-to-multiplet transitions

The fitting of experimental data using (10.36) is relatively straightforward, since the expression is linear in the three $\Omega_\lambda$ parameters. Experimental measurements are commonly made at room temperature, and the spectra are broad band. Such absorption spectra are relatively easy to measure and calibrate using standard spectrometers.

A large number of fits with $\Omega_\lambda$ parameters have appeared in the literature [Pea75, GWB98]. The focus of many of these studies has been on the 'hypersensitive' transitions, with $\Delta J = 2$, that are most strongly affected by the $\Omega_2$ parameters. Strong hypersensitive transitions and large $\Omega_2$ parameters appear to be correlated with the polarizability of the ligands.

We present a typical example by reproducing part of the calculation of Krupke [Kru71] for $Nd^{3+}$:YAG. In Table 10.1 we give a few of Krupke's measured and calculated line strengths. Note that the $^4G_{5/2}$ and $^4G_{7/2}$ multiplets are not resolved in the room temperature measurements, so we must add the line strengths for absorption from the ground multiplet $(^4I_{9/2})$ to both of these excited multiplets.

Squares of the relevant reduced matrix elements are given in Table 10.2. By multiplying these by the $\Omega_\lambda$ parameters given in Table 10.3, we can re-

Table 10.1. Calculation of line strengths for selected absorption transitions from the ground multiplet $^4I_{9/2}$ of $Nd^{3+}$:YAG (from [Kru71]). The line strengths are in units of $10^{-20}\,cm^2/e^2$.

| Excited state | Experiment | Calculated |
|---|---|---|
| $^2K_{11/2}$ | 0.054 | 0.060 |
| $^4G_{5/2} + {}^4G_{7/2}$ | 2.17 | 2.17 |

Table 10.2. Squared unit tensor intermediate-coupled reduced matrix elements $U(\lambda)$ between the $^4I_{9/2}$ ground state and excited states of $Nd^{3+}$ [CFR68].

| Excited state | $U(2)$ | $U(4)$ | $U(6)$ |
|---|---|---|---|
| $^2K_{11/2}$ | 0.0001 | 0.0027 | 0.0104 |
| $^4G_{5/2}$ | 0.8979 | 0.4093 | 0.0359 |
| $^4G_{7/2}$ | 0.0757 | 0.1848 | 0.0314 |

Table 10.3. Phenomenological $\Omega_\lambda$ parameters for various $Nd^{3+}$ systems (in units of $10^{-20}\,cm^2$).

| System | $\Omega_2$ | $\Omega_4$ | $\Omega_6$ |
|---|---|---|---|
| $Nd^{3+}$:YAG [Kru71] | 0.2 | 2.7 | 5.0 |
| $Nd^{3+}$:$Y_2O_3$ [Kru66] | 8.6 | 5.3 | 2.9 |
| $Nd^{3+}$:$LaF_3$ [Kru66] | 0.35 | 2.6 | 2.5 |
| $Nd^{3+}$:$LiYF_4$ [RB92] | 0.36 | 4.0 | 4.8 |

produce the calculation. For example, in units of $10^{-20}\,cm^2/e^2$, the electric dipole strength for the $^4I_{9/2}$ to $^2K_{11/2}$ transition is given by

$$0.2 \times 0.0001 + 2.7 \times 0.0027 + 5.0 \times 0.0104 = 0.06.$$

Tables of relevant matrix elements may be found in various reports and papers [CCC77, CGRR88, CGRR89] or generated from computer programs (see Appendix 3). In this example we have used the matrix elements of [CFR68] in order to be consistent with Krupke [Kru71].

Because the $\mathbf{U}^{(2)}$ reduced matrix element is particularly large the transition to the $^4G_{5/2}$ multiplet is very sensitive to the value of the $\Omega_2$ parameter. Since $\Delta J = 2$, this is a classic *hypersensitive* transition [Pea75, GWB98].

A few parameter sets for $Nd^{3+}$ in different host crystals are given in Table 10.3. Many more may be found in the literature (e.g. [GWB98]). From Table 10.3 we see that the $\Omega_2$ parameter tends to vary more than the others. There does appear to be a correlation between the value of $\Omega_2$ and the ligand polarizability [Pea75, GWB98]. However, we would expect the geometry of the site to also have an effect. For example, the $C_2$ site symmetry in $Y_2O_3$ admits an intensity parameter $A_{10}^2$, whereas the $D_2$ site symmetry in YAG does not. This is presumably an important contributing factor to the very different values of $\Omega_2$ for $Y_2O_3$ and YAG.

The $\Omega_\lambda$ parametrization has been useful in laser and phosphor design (e.g. [Kam96]). However, the approach does have limitations. The assumption that the parametrization is based on, i.e. that there is an equal occupation of all the states of the ground multiplet, is a poor approximation, even at room temperature. Since measurements are performed at room temperature, vibronic processes add to the zero-phonon processes, making comparison with *ab initio* calculations difficult.

### 10.4.2 Phenomenology of crystal field level transitions

To extract the maximum information from lanthanide spectra it is necessary to measure transition intensities between crystal field energy levels. This requires low temperatures, careful calibration, and sometimes deconvolution of instrumental broadening. Emission spectra can be particularly troublesome, since absolute measurements are difficult. It is possible to carry out fits using relative intensities within multiplets, but this is not as satisfactory as using absolute intensities. In some cases it is possible to calibrate the emission data by assuming that calculated magnetic dipole intensities are accurate [PC78]. Physical processes may also conspire to create extra difficulties. Lines may be broadened or overlap, making measurement difficult.

In contrast to the $\Omega_\lambda$ parametrization, fits using the $A_{tp}^\lambda$ parameters are highly non-linear, since the line strengths contain squares of sums of products of $A_{tp}^\lambda$ parameters with matrix elements. This leads to many local minima and it is difficult and time consuming to exhaustively search the parameter space for the global minimum. Furthermore, small uncertainties in the crystal field parameters (and therefore the eigenvectors) or measured intensities lead to large uncertainties in the fitted parameters. The values of fitted parameters given in the literature should therefore be treated with some caution.

Many workers (the author of this chapter included) have carried out least-squares fits that minimize

$$\sum_{i=1}^{n} \frac{1}{N-M} \left| \frac{e_i - c_i}{e_i + c_i} \right|^2, \tag{10.40}$$

where $e_i$ and $c_i$ are the experimental and calculated dipole strengths respectively, $N$ is the number of data points, and $M$ is the number of parameters. The reason for using this function is to avoid having the fit dominated by the most intense transitions. However, by effectively giving identical weight to all measurements, it is possible that very weak transitions, which may have large uncertainties, are overemphasized. Also, (10.40) is not a standard statistical function. The more usual definition of the *variance*,

$$\sum_{i=1}^{n} \frac{1}{N-M} \left| \frac{e_i - c_i}{\sigma_i} \right|^2, \tag{10.41}$$

where the $\sigma_i$ are the uncertainties of the measurements, is generally a more appropriate function to minimize, giving several advantages, including a smaller number of local minima. If equal weighting of all transitions is desired, it may be accomplished by setting the $\sigma_i$ proportional to the $e_i$.

The first fits to crystal field level intensity measurement were carried out by Axe [Axe63]. During the 1960s and early 1970s there were only a small number of phenomenological analyses of crystal field level intensity measurements. Since the late 1970s there have been a number of fits by Porcher and various coworkers [PC78], many on emission transitions in $Eu^{3+}$, and Richardson and coworkers [DRR84, BJRR94]. The Richardson group has carried out extensive analyses of the spectra of lanthanide oxydiacetate (ODA) crystals that display the interesting property of circular dichroism [MRR87a, HMR98]. These crystals also provide a model system for which the superposition model appears to be invalid [DRR84].

Since in many cases the superposition model does appear to hold, we discuss two categories of crystal.

    (i) Crystals for which the superposition model is a good approximation.
    (ii) Crystals containing complex ligands for which the axial symmetry assumption of the superposition model is invalid.

Restricting the parameter set to the odd $t$ parameters, allowed by the superposition model, removes the problem of multiple solutions for the $A_{tp}^{\lambda}$ parameters. In this case the $A_{tp}^{\lambda}$ parameters are the most appropriate.

Table 10.4. Intensity parameters, $A_{tp}^\lambda$ (units $i \times 10^{-13}\,cm$) for $Nd^{3+}$:YAG [BJRR94]. Signs are chosen for consistency with Table 10.15.

| Parameter | Value |
|-----------|-------|
| $A_{32}^2$ | $-^a$ |
| $A_{32}^4$ | 1700 |
| $A_{52}^4$ | $-4150$ |
| $A_{54}^4$ | 4000 |
| $A_{52}^6$ | 1150 |
| $A_{54}^6$ | $-7490$ |
| $A_{72}^6$ | 1900 |
| $A_{74}^6$ | $-^a$ |
| $A_{76}^6$ | $-^a$ |

$^a$ Parameter value was statistically insignificant, so it was omitted from the fit.

However, this is not the case for the second category of crystal, where it is essential to use the full parameter set. In this situation the $B_{lq}^\lambda$ parameter set has some advantages.

### 10.4.3 Crystals for which the superposition model is applicable

In crystals with simple ionic ligands, such as oxide, fluoride, or chloride, the superposition model is expected to provide a good approximation. Superposition model assumptions can be tested by carrying out fits with and without the $t = \lambda$ parameters forbidden by the superposition model. In most cases these 'non-superposition' parameters have been shown to have little effect on the fit [CR89, BJRR94].

One of the most extensive analyses of this kind is for $Nd^{3+}$ ions in YAG [BJRR94]. Table 10.4 gives parameter values obtained for one of the fits reported in that paper. This fit does not include parameters forbidden in the superposition approximation, but does use eigenfunctions including correlation crystal field effects (Chapter 6). The addition of parameters not allowed by the superposition model (such as $A_{20}^2$) had only a small effect on the quality of the fit, reducing the standard deviation by 4%, whereas using correlation crystal field eigenfunctions instead of one-electron crystal field eigenfunctions reduces the deviation by 6%. Burdick *et al.* [BJRR94] therefore concluded that the superposition model is a good approximation for this system. The fit was to 97 absorption measurements. In Table 10.5 results are given for a small subset: the transitions to the $^2H_{11/2}$ multiplet.

Table 10.5. Experimental ($e$) and fitted ($c$) dipole strengths (in units of $10^{-20}$ cm$^2/e^2$) for absorption transitions from the ground state of Nd$^{3+}$:YAG to the $^2H_{11/2}$ multiplet [BJRR94].

| Energy (cm$^{-1}$) | $e$ | $c$ | $\dfrac{e-c}{e+c}$ |
|---|---|---|---|
| 15 741 | 69 | 78 | −0.065 |
| 15 831 | 217 | 182 | 0.088 |
| 15 865 | 290 | 251 | 0.073 |
| 15 950 | 325 | 295 | 0.052 |
| 16 088 | 325 | 186 | 0.275 |
| 16 104 | 61 | 104 | −0.273 |

This typifies the quality of fit, with most data being well fitted by the model, but some fitted values differing from the experimental data by as much as a factor of two.

We may calculate effective $\Omega_\lambda$ parameters from the parameters in Table 10.4 and equation (10.37). For example, summing over $t$ and $p$ (recalling that parameters for $\pm p$ are not independent) we have

$$\Omega_4 = \frac{2}{9}\left(\left(A_{32}^4\right)^2 + \left(A_{52}^4\right)^2 + \left(A_{54}^4\right)^2\right) = 8.0 \times 10^{-20}\,\text{cm}^2. \qquad (10.42)$$

This is completely different from Krupke's (room temperature) determination of $2.7 \times 10^{-20}$ cm$^2$ (Table 10.3). Clearly, it is not straightforward to compare room temperature with low temperature analyses. The room temperature analyses are based on the assumption of equal population of all states of the ground multiplet. This is a rather poor approximation for Nd$^{3+}$:YAG, where the ground multiplet spans more than 800 cm$^{-1}$.

Intrinsic intensity parameters may be determined with the aid of (10.39). For example, the intrinsic parameter $\bar{A}_3^4$ may be calculated from $A_{32}^4$. Burdick *et al.* [BJRR94] calculate the combined coordination factors for the four O$^{2-}$ ligands at 2.303 Å as 0.891i and the sum for the four O$^{2-}$ ligands at 2.4 323 Å as −0.634i. Assuming a power law exponent $\tau_3^4$ of 5, we can write the total contribution from the ligands as $0.891 - 0.634 \times (2.303/2.4\,323)^5 = 0.409$. The intrinsic parameter $\bar{A}_3^4$ is therefore calculated to be $1700 \times 10^{-13}$ cm$/0.409 = 4160 \times 10^{-13}$ cm, for $R_0 = 2.303$ Å.

When using the full set of $A_{tp}^\lambda$ parameters, even with the restriction to $t = \lambda \pm 1$, Burdick *et al.* [BJRR94] were unable to obtain meaningful values for certain parameters, in particular $A_{32}^2$. Consequently, they carried out

Table 10.6. Phenomenological intrinsic intensity parameters $\bar{A}_t^\lambda$ (in units $10^{-13}$ cm) for various systems. Signs have been chosen for consistency with Table 10.15.

| System | Source | $\bar{A}_1^2$ | $\bar{A}_3^2$ | $\bar{A}_3^4$ | $\bar{A}_5^4$ | $\bar{A}_5^6$ | $\bar{A}_7^6$ |
|---|---|---|---|---|---|---|---|
| $Pr^{3+}$:LiYF$_4$ | [RR84a] | — | −950 | 1050 | 150 | −1790 | −260 |
| $Pr^{3+}$:LaAlO$_3$ | [RDR83] | — | −310 | 1870 | 1620 | −3900 | −8000 |
| $Nd^{3+}$:YAG | [BJRR94] | — | −1100 | 1920 | 2710 | −4290 | −1030 |
| $Eu^{3+}$:KY$_3$F$_{10}$ | [RDR83] | −1600 | −3000 | 710 | 1700 | 1900 | 370 |
| $Eu^{3+}$:LiYF$_4$ | [Rei87b] | — | −3000 | 1000 | 1200 | −400 | 17 800 |

additional fits using the superposition model to reduce the number of parameters to five (there being no parameters with $t = 1$ in $D_2$ symmetry), using (10.39) to fix ratios of parameters with the same $\lambda$ and $t$ but different $p$. The intrinsic parameters quoted in Table 10.6 were obtained by this method.

Intrinsic intensity parameters for a selection of systems are summarized in Table 10.6. More examples may be found in the literature [RDR83, Rei87b, Rei93]. The overall sign of the parameter sets are not determined by the data and the signs in Table 10.6 have been chosen to agree with the calculations discussed in Section 10.5. We see that the relative signs of the parameters for $\lambda = 2$ and $\lambda = 4$ are quite consistent, but the $\lambda = 6$ parameters are less consistent. In the case of the $Eu^{3+}$ parameters this is most likely because those fits do not include any transitions that are strongly dependent on the $\lambda = 6$ parameters. Note that the ratio of $\bar{A}_3^2$ and $\bar{A}_3^4$ is always negative. We return to this in Section 10.5.

The number of intrinsic parameter determinations is so small that it is difficult to make systematic comparisons. It is particularly unfortunate that only one of the site symmetries admits an $A_{1p}^2$ parameter so that only one determination of the intrinsic parameter $\bar{A}_1^2$ has been made. Comparing the magnitudes of the parameters in Table 10.6, it can be seen that the determinations of $\lambda = 2$, 4 intrinsic parameters for $Eu^{3+}$ ions with $F^-$ ligands, in KY$_3$F$_{10}$ and LiYF$_4$, are reasonably consistent (both experimental sets were obtained by Porcher and coworkers [PC78, GWBP+85]). However, other parameters in the table vary widely. Not only have these parameters been derived from analyses by several different experimental groups, using very different techniques, but for practical reasons intensity measurements only sample a small number of the possible $4f^N$ transitions, often from only

Table 10.7. Transformation properties of selected kets with $D_3$ symmetry labels [But81].

| $D_3$ | $JM$ |
|---|---|
| $\|0^\pm A_1\rangle$ | $\|0^\pm 0\rangle$ |
| $\|1^\pm A_2\rangle$ | $\|1^\pm 0\rangle$ |
| $\|1^\pm E_\pm\rangle$ | $-\|1^\pm \pm 1\rangle$ |
| $\|2^\pm A_1\rangle$ | $-\|2^\pm 0\rangle$ |
| $\|2^\pm E_\pm(1)\rangle$ | $\mp\sqrt{2/3}\|2^\pm \pm 1\rangle + 1/\sqrt{3}\|2^\pm \mp 2\rangle$ |
| $\|2^\pm E_\pm(2)\rangle$ | $\mp 1/\sqrt{3}\|2^\pm \pm 1\rangle - \sqrt{2/3}\|2^\pm \mp 2\rangle$ |
| $\|3^\pm A_1\rangle$ | $-(\|3^\pm 3\rangle + \|3^\pm -3\rangle)/\sqrt{2}$ |

one initial state (typically the ground state). This restriction to a small subset of the possible transitions may explain some of the discrepancies in Table 10.6. Another source of uncertainty is that, for some parameters, the contributions to the sum in (10.39) cancel strongly. Unlike the crystal field case (Chapter 5) there are no systems for which the power law exponents $\tau_t^\lambda$ can be reliably determined. To derive the value of $\bar{A}_t^\lambda$ it is necessary to assume values for $\tau_t^\lambda$.

### 10.4.4 Crystals for which the superposition model is invalid

Kuroda *et al.* [KMR80] pointed out that complex ligands can produce contributions to intensities that, in the context of their model, arise from the anisotropic polarizability of the ligands. Reid and Richardson [RR83, RR84b] have recast these ideas in terms of the $A_{tp}^\lambda$ intensity parametrization and the breakdown of the superposition model. If the lanthanide–ligand interaction is not cylindrically symmetric, which is the case for complex ligands, then the restriction to odd $t$ does not apply.

The analysis of transition intensities for $Eu^{3+}$ ions in europium oxydiacetate (EuODA) provides a useful example [DRR84]. The site symmetry for these crystals is $D_3$. Table 10.7 shows the $JM$ decomposition of some $D_3$ irreducible representations. For $D_3$ symmetry the even- and odd-parity states have the same branching rules. Note that for $J = 2$ there are two $E$ irreducible representations, labelled $E(1)$ and $E(2)$ in the table.

From the definitions of $\pi$ and $\sigma$ polarizations given in Section 10.1.2 we may work out the transformation properties of the electric and magnetic dipole moment operators (Table 10.8). For a $\pi$ transition the operator $D_{\text{eff},0}$

Table 10.8. Electric and magnetic dipole operators in $D_3$ symmetry.

|  | Polarization | Symmetry | $JM$ labels |
|---|---|---|---|
| Electric dipole | $\pi$ | $A_2$ | $\|1^+0\rangle$ |
|  | $\sigma$, axial | $E_\pm$ | $\|1^+\pm1\rangle$ |
| Magnetic dipole | $\pi$, axial | $E_\pm$ | $\|1^-\pm1\rangle$ |
|  | $\sigma$ | $A_2$ | $\|1^-0\rangle$ |

Table 10.9. $JM$ composition of crystal field eigenstates of EuODA.

| State | Energy (cm$^{-1}$) | $JM$ composition |
|---|---|---|
| $^7F_0(A_1)$ | 0 | $\|0^+0\rangle$ |
| $^5D_0(A_1)$ | 17 526 | $\|0^+0\rangle$ |
| $^5D_1(E)$ | 19 038 | $\|1^+\pm1\rangle$ |
| $^5D_1(A_2)$ | 19 042 | $\|1^+0\rangle$ |
| $^5D_2(A_1)$ | 21 549 | $\|2^+0\rangle$ |
| $^5D_2(E(a))$ | 21 561 | $\pm0.46\|2^+\mp1\rangle - 0.88\|2^+\pm2\rangle$ |
| $^5D_2(E(b))$ | 21 607 | $0.88\|2^+\mp1\rangle \pm 0.46\|2^+\pm2\rangle$ |

is required and from Table 10.8 this operator transforms as the irreducible representation $A_2$. The operator for $\sigma$ or axial polarization transforms as the irreducible representation $E$.

Electric dipole transitions from the ground state of Eu$^{3+}$, $^7F_0$, to the $^5D_2$ multiplet will now be discussed in greater detail. These transitions only involve operators with $\lambda = 2$. The eigenstates of the relevant energy levels are given in Table 10.9. Note that the two $^5D_2$ eigenstates that transform as $E$ of $D_3$, labelled $E(a)$ and $E(b)$, are mixtures of the functions $E(1)$ and $E(2)$ of Table 10.7.

According to the discussion in Section 10.2.2 we may use $B^\lambda_{l\pi}$ and $B^\lambda_{l\sigma}$ to parametrize the electric dipole moments ((10.28) and (10.29)). For $\lambda = 2$ there are no $A_2$ irreducible representations, but there are two $E$ irreducible representations (Table 10.7). Therefore we have two $\lambda = 2$ parameters and expect all electric dipole transitions to have $\sigma$ (or axial) polarization. The $\lambda = 2$ part of the effective dipole moment operator may be written as

$$D_{\text{eff},\sigma} = B^2_{1\sigma}\left(U_1^{(2)} + U_{-1}^{(2)}\right) + B^2_{2\sigma}\left(U_2^{(2)} - U_{-2}^{(2)}\right). \tag{10.43}$$

(The relative signs of the $U_{\pm l}^{(2)}$ are determined by complex conjugation symmetry.)

For the other parametrization, equation (10.30), all values of $A_{tp}^{\lambda}$ for which the $tp$ combinations transform as $A_1$ are used. From Table 10.7 it can be seen that there are two possibilities, one each for $t = 2$ and $t = 3$. The parameters we require are $A_{20}^2$ and $A_{33}^2$. The $\lambda = 2$ part of the effective dipole moment operator can be written as

$$
\begin{aligned}
D_{\text{eff},\sigma} &= (-D_{\text{eff},1} + D_{\text{eff},-1})/\sqrt{2} \\
&= A_{20}^2 \left( -U_1^{(2)}(-1)^1 \langle 21, 1-1|20 \rangle + U_{-1}^{(2)}(-1)^1 \langle 2-1, 11|20 \rangle \right)/\sqrt{2} \\
&\quad - A_{3-3}^2 U_{-2}^{(2)}(-1)^1 \langle 2-2, 1-1|3-3 \rangle/\sqrt{2} \\
&\quad + A_{33}^2 U_2^{(2)}(-1)^1 \langle 22, 11|33 \rangle/\sqrt{2}\,.
\end{aligned}
\tag{10.44}
$$

From (10.32), $A_{3-3}^2 = -(A_{33}^2)^*$ and again there are only two independent parameters.

Clearly the two parametrizations, shown in equations (10.43) and (10.44), are equivalent. In fact there is a very simple relationship between the $B_{1\sigma}^2$, $B_{2\sigma}^2$ and $A_{20}^2$, $A_{33}^2$ parameter sets in this particular case (Table 10.12). The $A_{20}^2$ parameter would not appear if the superposition model were valid. If only the parameter $A_{33}^2$ (allowed by the superposition model) were nonzero, equation (10.35) and the eigenvectors in Table 10.9 could be used to calculate the relative intensity of the transitions from $^7F_0$ to $^5D_2$ as

$$
\frac{0.46^2}{0.88^2} = 0.27,
$$

since all the extra factors cancel. This compares poorly to the measured ratio of 4.43. The experimental values (as well as other measurements) may be quite well fitted by using the parameter values [DRR84]

$$
A_{20}^2 = -\text{i} \times 1580 \times 10^{-13}\,\text{cm}^{-1}, \qquad A_{33}^2 = \text{i} \times 1560 \times 10^{-13}\,\text{cm}^{-1}.
$$

(Note that the signs of both of these parameters have been changed from [DRR84] for consistency with the calculations described in Section 10.5.) We can understand the need for the factors of $\text{i} = \sqrt{-1}$ as follows. If the $A_{20}^2$ parameter was not imaginary then the complex conjugation symmetry (equation (10.32)) would lead to a contradiction. This will not be affected by rotating the system about the $z$-axis. A similar argument applies to the $A_{3\pm3}^2$ parameter. However, a rotation about $z$ could make this parameter real, or change its sign. At the same time the crystal field parameters would also change (see table VI of [DRR84], Chapter 2 and Appendices 1 and 4).

Table 10.10. Two equivalent $A_{tp}^{\lambda}$ intensity parameter sets for EuODA. Set 1 is from [BSR88], with signs changed for consistency with the calculations described in Section 10.5. Set 2 gives identical calculated intensities and is obtained from Set 1 by transforming to the $B_{li}^{\lambda}$ parameter set (Table 10.11), changing the signs of the $B_{l\pi}^{\lambda}$ parameters, and transforming back. All values have units $i \times 10^{-13}$ cm.

|          | Set 1  | Set 2  |
|----------|--------|--------|
| $A_{20}^{2}$ | $-1100$ | $-1100$ |
| $A_{33}^{2}$ | $2070$  | $2070$  |
| $A_{33}^{4}$ | $-510$  | $1274$  |
| $A_{40}^{4}$ | $-370$  | $-370$  |
| $A_{43}^{4}$ | $-110$  | $-2824$ |
| $A_{53}^{4}$ | $3140$  | $726$   |
| $A_{53}^{6}$ | $-5420$ | $-1135$ |
| $A_{60}^{6}$ | $-430$  | $-430$  |
| $A_{63}^{6}$ | $-340$  | $-3711$ |
| $A_{66}^{6}$ | $-360$  | $-883$  |
| $A_{73}^{6}$ | $920$   | $-3908$ |
| $A_{76}^{6}$ | $1630$  | $1416$  |

Table 10.11. Two equivalent $B_{li}^{\lambda}$ intensity parameter sets for EuODA corresponding to the $A_{tp}^{\lambda}$ parameter sets of Table 10.10. All values have units $i \times 10^{-13}$ cm.

|          | Set 1  | Set 2  |
|----------|--------|--------|
| $B_{1\sigma}^{2}$ | $-550$  | $-550$  |
| $B_{2\sigma}^{2}$ | $-1463$ | $-1463$ |
| $B_{1\sigma}^{4}$ | $-185$  | $-185$  |
| $B_{2\sigma}^{4}$ | $-1737$ | $-1737$ |
| $B_{3\pi}^{4}$ | $2023$  | $-2023$ |
| $B_{4\sigma}^{4}$ | $-21$   | $-21$   |
| $B_{1\sigma}^{6}$ | $-215$  | $-215$  |
| $B_{2\sigma}^{6}$ | $448$   | $448$   |
| $B_{3\pi}^{6}$ | $3641$  | $-3641$ |
| $B_{4\sigma}^{6}$ | $-2887$ | $-2887$ |
| $B_{5\sigma}^{6}$ | $-1163$ | $-1163$ |
| $B_{6\pi}^{6}$ | $282$   | $-282$  |

Table 10.12. Transformations between $A_{tp}^{\lambda}$ and $B_{l\sigma}^{\lambda}$ parameters for $\lambda = 2$ in $D_3$ symmetry.

|  | $A_{20}^2$ | $A_{33}^2$ |
|---|---|---|
| $B_{1\sigma}^2$ | $1/2$ | $0$ |
| $B_{2\sigma}^2$ | $0$ | $-1/\sqrt{2}$ |

The $A_{20}^2$ parameter transforms as $2^-0$, i.e. with angular momentum quantum number 2 and odd parity. It is important to note that this is completely different from a $2^+0$ function, proportional to $3z^2 - r^2$, that arises in the crystal field (energy level) parametrization. The $2^+0$ function is 'superposable'. The $2^-0$ function is not superposable and the $A_{20}^2$ parameter vanishes under the superposition approximation. Hence the superposition approximation (Section 5.1.1) appears to be invalid for the physical system under discussion here.

More extensive analyses of lanthanide oxydiacetate data confirm that the non-superposition model parameters are crucial to explaining the transition intensities and the circular dichroism of these systems [MRR87a, BSR88]. In Table 10.10 we give the parameters for $Eu^{3+}$ from Berry *et al.* [BSR88] (column labelled Set 1).

As noted in Section 10.2.2 it is possible to find another set of $A_{tp}^{\lambda}$ parameters in which some of the parameters have completely different magnitudes. The transformation between the $A_{tp}^{\lambda}$ and $B_{l\sigma/\pi}^{\lambda}$ parameter sets is given in Tables 10.12–10.14. In Table 10.11 we show two sets of $B_{l\sigma/\pi}^{\lambda}$ parameters and in Table 10.10 the two equivalent sets of $A_{tp}^{\lambda}$ parameters. The only difference between the $B_{l\sigma/\pi}^{\lambda}$ parameter sets is the signs of the $B_{l\pi}^{\lambda}$ parameters. However, for the $A_{tp}^{\lambda}$ parameters there are differences of magnitude (except for $A_{20}^2$, $A_{33}^2$, $A_{40}^4$, and $A_{60}^6$, where there is a one-to-one correspondence to $B_{l\sigma}^{\lambda}$ parameters). Clearly we must be very careful when comparing $A_{tp}^{\lambda}$ parameter sets and comparing the experimental parameters to *ab initio* calculations. It is also important to note that the full set of $A_{tp}^{\lambda}$ parameters *cannot* be fitted to experimental data for a single polarization. In that case the $B_{l\sigma}^{\lambda}$ or $B_{l\pi}^{\lambda}$ sets are the natural choice.

The magnetic dipole intensity for absorption to the $^5D_1$ multiplet may be calculated relatively simply. From the Wigner–Eckart theorem (A1.1) the matrix elements of $M_q^{(1)}$ between $^7F_0$ and $^5D_1$ all have the same magnitude. It is predicted, therefore, that the intensity ratio of the $E(\pi = \text{axial})$ to

Table 10.13. Transformations between $A_{tp}^\lambda$ and $B_{l\sigma/\pi}^\lambda$ parameters for $\lambda = 4$ in D$_3$ symmetry.

|            | $A_{33}^4$     | $A_{40}^4$ | $A_{43}^4$    | $A_{53}^4$       |
|------------|----------------|------------|---------------|------------------|
| $B_{1\sigma}^4$ | 0              | $1/2$      | 0             | 0                |
| $B_{2\sigma}^4$ | $-1/\sqrt{72}$ | 0          | $\sqrt{7/40}$ | $-\sqrt{14/45}$  |
| $B_{3\pi}^4$    | $-\sqrt{7/36}$ | 0          | $3/\sqrt{20}$ | $4/\sqrt{45}$    |
| $B_{4\sigma}^4$ | $\sqrt{7/18}$  | 0          | $1/\sqrt{10}$ | $1/\sqrt{90}$    |

Table 10.14. Transformations between $A_{tp}^\lambda$ and $B_{l\sigma/\pi}^\lambda$ parameters for $\lambda = 6$ in D$_3$ symmetry.

|            | $A_{53}^6$     | $A_{60}^6$ | $A_{63}^6$    | $A_{66}^6$   | $A_{73}^6$       | $A_{76}^6$     |
|------------|----------------|------------|---------------|--------------|------------------|----------------|
| $B_{1\sigma}^6$ | 0              | $1/2$      | 0             | 0            | 0                | 0              |
| $B_{2\sigma}^6$ | $-1/\sqrt{26}$ | 0          | $\sqrt{3/14}$ | 0            | $-\sqrt{45/182}$ | 0              |
| $B_{3\pi}^6$    | $-3/\sqrt{26}$ | 0          | $\sqrt{3/14}$ | 0            | $\sqrt{40/91}$   | 0              |
| $B_{4\sigma}^6$ | $\sqrt{15/52}$ | 0          | $\sqrt{5/28}$ | 0            | $\sqrt{3/91}$    | 0              |
| $B_{5\sigma}^6$ | 0              | 0          | 0             | $1/\sqrt{14}$ | 0                | $-\sqrt{3/7}$  |
| $B_{6\pi}^6$    | 0              | 0          | 0             | $\sqrt{6/7}$ | 0                | $1/\sqrt{7}$   |

$A_1(\sigma)$ transitions is 2. This factor comes entirely from the degeneracies of the final states. The experimental ratio is 2.6. This difference suggests that there are physical processes affecting the magnetic dipole (and presumably electric dipole) intensities which have not been included in our models.

## 10.5   *Ab initio* calculations

The phenomenological analyses discussed above provide a useful start to the rationalization of transition intensity data for transitions within the $4f^N$ configuration. Though this phenomenology is important in itself, our aim is to understand these parameters semiquantitatively in terms of the mechanisms which contribute in *ab initio* calculations, as has been done in the case of the crystal field parameters (see Chapter 1).

Most calculations of intensity parameters have employed point charge crystal field mixing with excited lanthanide states and dipolar interactions involving the ligand ('ligand polarization' or 'dynamic coupling' mechanism). Only a few attempts have been made to treat the ligands in a realistic manner to include covalent effects [HFC76, PN84, RN89]. Experience with crystal field calculations suggests that point charge and dipole calculations will give a poor quantitative description of the parameters. Nevertheless, we begin with a discussion of these calculations. No attempt is made to reproduce details of the angular momentum algebra; the discussion concentrates on physical principles.

Given an *ab initio* calculation of the effective dipole moments, the $A_{tp}^\lambda$ parameters may be calculated by comparing (10.26) with (10.30). Most of the calculations discussed below are from Reid and Ng [RN89], for a single Cl$^-$ ligand. These calculations will be compared to experimental intrinsic intensity parameters from Table 10.6. As discussed above, intrinsic parameters for a variety of systems are broadly similar, at least for $\lambda = 2$ and 4, so the fact that Table 10.6 contains no data for Cl$^-$ ligands should not be seen as a difficulty. Note that there was an overall sign error in earlier calculations (see [XR93]) and all the parameters given here have had their signs changed. This has no effect on calculated intensities.

### 10.5.1 Crystal field mixing

This is the mechanism considered in the original work of Judd [Jud62] and Ofelt [Ofe62]. The odd-parity crystal field mixes $4f^{N-1}nd$ and $4f^{N-1}ng$ configurations into $4f^N$, leading to the contribution

$$A_{tp}^\lambda = -A_{tp}\Xi(t,\lambda)\frac{(2\lambda+1)}{(2t+1)^{1/2}}, \qquad (10.45)$$

for $\lambda = 2$, 4, 6 and $t = \lambda \pm 1$. The $A_{tp}$ are *interconfigurational* (e.g. $f$–$d$) crystal field parameters and the $\Xi(t,\lambda)$ factors, defined by Judd [Jud62], contain radial integrals between the $f$ and $d$ orbitals. Experience with crystal field parameters (Chapter 1) suggests that a more elaborate calculation, taking into account the overlap of ligand orbitals, will give contribution of the same sign, but different magnitudes, and this is indeed the case.

The crystal field mechanism can contribute to parameters with $t = 1$ (see [Jud66]). If the crystal field is thought of as a purely external electrostatic potential then this would imply a dipolar potential that would move the ion to a different equilibrium position. Since the crystal field is not merely an external potential (as explained in Chapter 1) this poses no difficulty in a realistic calculation.

Equation (10.45) may be used to predict ratios between parameters with the same $t$ and $p$ but different $\lambda$. Crystal field contributions to intensity parameters for a single $Cl^-$ ligand are given in Table 10.15, in the row labelled the 'crude point charge'. From the earliest fits using the $A_{tp}^\lambda$ (or equivalent) parametrizations [Axe63] calculations of this sort have been incapable of explaining the relative signs of the intensity parameters. In all cases shown in Table 10.6 $\bar{A}_3^2/\bar{A}_3^4$ is negative, whereas the crude point charge calculation (Table 10.15) gives a positive ratio.

### 10.5.2 Ligand polarization

This mechanism (also called 'dynamic coupling' [MPS74]) may be viewed in two different ways [Jud79]. The first is that the radiation field dynamically polarizes the ligand, and this excitation is transferred to the lanthanide via the coulomb interaction between the $4f$ and ligand electrons. The other view is that the $4f$ electrons polarize the ligand and this induced dipole interacts with the radiation. The latter view leads us to expect that the ligand polarization mechanism will give contributions that are generally of opposite sign to the contribution from the crystal field mechanism (in which the ligand electrons polarize the $4f$ electrons.)

For a ligand with isotropic polarizability the ligand polarization mechanism (line 'crude ligand polarization' of Table 10.15) only gives contributions to parameters for which $t = \lambda + 1$. These contributions are, indeed, of opposite sign to the contributions from the crystal field mechanism, for those parameters that the ligand polarization mechanism contributes to. If the ligand polarization mechanism is assumed to dominate for parameters with $t = \lambda + 1$ then the signs of the parameters agree with the experimental values for $\lambda = 2$ and 4 (Table 10.6).

In the case of complex organic ligands, such as the oxydiacetate systems discussed in Section 10.4.4, *anisotropic* ligand polarizability is important. Calculations based on known bond and atom polarizabilities give a qualitative explanation of the non-superposition parameters discussed in Section 10.4.4 [DRR84].

### 10.5.3 Realistic calculations – overlap and covalency

The most extensive *ab initio* calculations of the $A_{tp}^\lambda$ parameters using realistic ligand states are those of Reid and Ng [RN89]. This work was based on the Ng and Newman [NN87b] calculations of crystal field and correlation crystal field parameters described in Chapters 1 and 6. Results are given in lines a–e of Table 10.15. There were various technical difficulties in the

Table 10.15. *Ab initio* calculations of intrinsic intensity parameters $\bar{A}_t^\lambda$ (in units of $10^{-13}$ cm) for the system $Pr^{3+}-Cl^-$. The calculation is from [RN89], with the sign correction of [XR93]. For a detailed discussion of the contributions see the text.

| Contribution | $\bar{A}_1^2$ | $\bar{A}_3^2$ | $\bar{A}_3^4$ | $\bar{A}_5^4$ | $\bar{A}_5^6$ | $\bar{A}_7^6$ |
|---|---|---|---|---|---|---|
| a: First order | −70 | 53 | 99 | −45 | −107 | 27 |
| b: Crystal field | −2104 | 163 | 369 | −125 | −516 | 115 |
| c: Covalency | 155 | 165 | 280 | −150 | −286 | 124 |
| d: Ligand polarization | 397 | −2927 | 99 | 144 | −68 | 15 |
| e: Ligand polarization exchange | −125 | 93 | 2 | 11 | 27 | −66 |
| a+b+c | −2019 | 381 | 748 | −320 | −909 | 266 |
| d+e | −272 | −2834 | 101 | 155 | −41 | −50 |
| Crude point charge | −4630 | 260 | 610 | −50 | −170 | 20 |
| Crude ligand polarization | 0 | −4276 | 0 | 738 | 0 | −294 |

calculation, discussed further in [RN89]. However, the overall picture that the table gives of the relative size of the various contributions should be correct.

Contributions b and d arise primarily from crystal field and ligand polarization contributions, which are roughly equivalent to the crude point charge and crude ligand polarization calculations. Contribution e is a contribution to ligand polarization arising from the exchange part of the coulomb interaction between the $f$ electrons and ligand electrons. In addition, there is a small first order contribution, a, due to the basis states being molecular orbitals of mixed parity. The other significant contribution, c, is primarily a covalent effect, i.e. due to excitations from ligand orbitals to $4f$ orbitals. These covalent contributions were also calculated in [HFC76, PN84].

In calculations of the phenomenological crystal field in insulating materials all major contributions to the intrinsic parameters (apart from screening) are positive (see Chapter 1). In contrast, the major contributions to the intensity parameters can have different signs, with ligand polarization effects generally being of opposite sign to the other contributions. Comparing these calculations to the phenomenological intrinsic intensity parameters (Table 10.6), it is seen that the calculation does appear to explain the experimental signs of the $\bar{A}_1^2$, $\bar{A}_3^2$, and $\bar{A}_3^4$ parameters. We suspect that the ligand polarization contribution to $\bar{A}_5^4$ is underestimated; if it were larger then we also would have agreement for that parameter. The experimental signs of the other parameters are not well determined.

The cancellation between the ligand polarization and other contributions has implications for the dependence of intensity parameters on the distance to the ligands. In the case of the phenomenological crystal field, as all contributions (apart from screening) have the same sign, the crystal field power law exponents $t_k$ are always positive, corresponding to a monotonic reduction of interaction processes with distance (see Chapter 5). It is possible that, over certain distance ranges, the intensity parameter power law exponents $\tau_t^\lambda$ may not be positive since the different contributions have opposing signs and could have very different distance dependences. Such an effect has been established in the ligand distance dependence of the spin-Hamiltonian intrinsic parameters (see Chapter 7).

It is our hope that routine quantum chemistry calculations will soon be capable of realistic calculations of $4f^N$ transition intensity parameters. An example of progress in this direction is the calculations of [KFR95]. Those calculations were not designed to determine the phenomenological parameters, but a determination of the parameters from such a calculation would not be particularly difficult.

## 10.6 Higher order effects

The calculations described in the previous section only go to first order in the perturbation. If the calculation is extended to higher order, i.e. with two or more of the potential operators in (10.20) appearing in the expansion of (10.26), we have the possibility of two-electron operators (i.e. correlation effects) and spin-dependent operators. In the former case the extra operators include the coulomb interaction, in the latter the spin–orbit interaction. Note that no new spin-independent one-electron parameters are required, though the higher order effects may modify the values of the $A_{tp}^\lambda$ (or $B_{lq}^\lambda$) parameters.

There is not enough information in the available experimental data to test a truly general parametrization. Such a parametrization would modify (10.30), supplementing the $\mathbf{U}^{(\lambda)}$ operators with the two-body operators $\left(\mathbf{U}^{(k_1)}\mathbf{U}^{(k_2)}\right)^{(K)}$, as in the case of the correlation crystal field, and spin-dependent $\mathbf{V}^{(1k)K}$ operators. It is not proposed to discuss the details of these generalized parametrizations here (see, for example, [Wyb68, Rei93, Sme98]). The number of parameters (larger than the number of correlation crystal field parameters) is so great that we cannot hope to determine them experimentally. However, if we only deal with transitions involving a restricted number of multiplets then the extra operators may simply be proportional to the one-electron operators, as in the case of correlation crystal field operators (Chapter 6).

In cases where $\Omega_\lambda$ fits are poor there have sometimes been attempts to extend the parametrization to include odd $\lambda$ parameters. It is believed that these attempts are technically incorrect. Since the effective transition operator should be Hermitian, $\lambda$ must be even (like $k$ in the crystal field calculation) for *one-electron* spin-independent operators. However, as in the case of the correlation crystal field, one can have odd rank *two-electron* operators (but not rank 1 – see Chapter 6). Although adding odd rank one-electron operators is not technically correct it has been attempted for some $Pr^{3+}$ systems [EGRS85]. The success of such fits might be interpreted to demonstrate the importance of two-body (correlation) operators of odd rank.

Extensive calculations of correlation effects have been carried out by Smentek and coworkers (reviewed in [Sme98]). Though these calculations use a point charge approach to the crystal field the evaluation of correlation effects is very sophisticated. These calculations demonstrate that correlation is important, although in many cases it merely modifies the one-electron parameters.

Burdick and coworkers have extended the $\Omega_\lambda$ parametrization to include spin-dependent operators [BDS89]. Their fits are considerably better than the standard ones for cases where the standard operators have very small matrix elements. This is particularly apparent for $Gd^{3+}$. These extensions were first suggested by Judd and Pooler [JP82] for two-photon transitions. The reason why spin-dependent operators are important is that many transitions are 'spin forbidden' ($\Delta S \neq 0$), obtaining intensity because of spin–orbit mixing in the $4f^N$ configuration. If this mixing is small the spin-dependent operators (which take into account spin–orbit mixing in the excited states) provide a means to increase the intensity of transitions that otherwise would be very weak.

There have also been attempts to extend the basis set to include the $4f^{N-1}5d$ states explicitly. Since this configuration is very close to the $4f^N$ configuration it is a major source of correlation and spin–orbit effects [GF92, BRRK95].

## 10.7 Related topics

The techniques discussed may be applied to physical processes other than one-photon transitions. These processes include vibronic transitions, circular dichroism, two-photon absorption, and Raman scattering.

### 10.7.1 Vibronic transitions

The transitions considered so far do not change the vibrational state of the crystal. If a lattice phonon (or local mode quantum) is absorbed or emitted at the same time as the photon then a line offset from the zero-phonon line by the energy of the vibration is observed. References to various approaches to this problem may be found in [GWB98]. In our approach, the extension of the parametrization (10.30) to include vibronic processes is relatively straightforward – the $tp$ combinations, instead of transforming as the identity irreducible representation, must transform as the irreducible representation of the vibration. Therefore the fitting process is the same, but a different set of parameters must be fitted for each vibrational mode. Reid and Richardson [RR84c] and Crooks *et al.* [CRTZ97] have performed such fits for octahedral complexes, where the non-vibronic electric dipole transitions are forbidden.

In rationalizing these fits using the superposition model one needs to differentiate (10.39) with respect to the vibrational modes. In view of our earlier comments that the $\tau_t^\lambda$ power law exponents might, in some cases, be negative, this rationalization is not straightforward.

Since the vibronic intensities are absorbed into the $\Omega_\lambda$ parametrization when we sum over $J$ multiplets it is not possible to separate vibronic from non-vibronic intensity when using that parametrization.

### 10.7.2 Circular dichroism

Molecules or crystals that are chiral (possess a definite handedness) may exhibit differential absorption of left and right circularly polarized light. This is known as *circular dichroism*. Circular dichroism arises from an interference between electric dipole moments and high-order moments (e.g. magnetic dipole, electric quadrupole). The reader is referred to the literature for details (e.g. [PS83]). The oxydiacetate systems exhibit circular dichroism and have been extensively studied by Richardson and coworkers [MRR87a, BSR88, HMR98].

For an axial geometry, the differential absorption (expressed as left−right in the chemists' convention) is related to a rotatory strength given by [MRR87a]

$$R_{FI}(\text{axial}) = -\frac{3}{2}\text{Im}\sum_i\sum_f\sum_{q=\pm1}\langle Ff|-eD_q^{(1)}|Ii\rangle\langle Ff|M_q^{(1)}|Ii\rangle^*, \quad (10.46)$$

where Im means the imaginary part. Note that the negative sign appears because we have rearranged the expression of [MRR87a] to place the initial states to the right of the dipole moment operators. The $D_q^{(1)}$ may be replaced by our $A_{tp}^\lambda$ or $B_{l\pi/\sigma}^\lambda$ parametrized effective operator. Note that this is a signed expression and that the intensity parameters appear linearly. Thus, there is more information in the circular dichroism spectra than in normal absorption spectra, and in principle one could determine the absolute signs of intensity parameters. The parametrized model has been quite successful in reproducing experimental data for the oxydiacetate systems [MRR87a, BSR88].

### 10.7.3 Two-photon absorption and Raman scattering

Processes involving two photons are also possible. A review of work in this area may be found in Downer [Dow89] and recent references in [GWB98].

For two-photon processes, the expression analogous to the electric dipole moment is

$$\sum_k \frac{\langle f|D_{q_2}^{(1)}|k\rangle\langle k|D_{q_1}^{(1)}|i\rangle}{E_i - E_k + \hbar\omega_1} + \sum_k \frac{\langle f|D_{q_1}^{(1)}|k\rangle\langle k|D_{q_2}^{(1)}|i\rangle}{E_i - E_k + \hbar\omega_2}, \quad (10.47)$$

where $q_1$ and $q_2$ and $\hbar\omega_1$, $\hbar\omega_2$ are the polarizations and energies of the two photons. For this formula to be applied to Raman scattering a minus sign must be associated with the $\hbar\omega$ of the outgoing photon. The summation runs over all possible intermediate states (including $4f^N$), so resonance effects are possible [HLC89]. The Fermi golden rule transition rate is proportional to the square of (10.47).

Note that the states and denominators in (10.47) are *exact* and the denominators arise from the time-dependent perturbation theory derivation of Fermi's golden rule. In order to obtain an effective operator we must expand both the states and the denominators using time-independent perturbation theory. For a discussion of these issues see [BR98a].

For two-photon absorption within $4f^N$ with identical photons the lowest order contribution leads to a single $\mathbf{U}^{(2)}$ operator, i.e. there are no free parameters in a calculation of relative intensities [Axe64]. Downer and coworkers [DDNB81] found that this model gave a very poor reproduction of experimental data for $Gd^{3+}$. Judd and Pooler [JP82] were able to explain the measurements by extending the model to include spin–orbit interactions in the excited states. These higher order contributions are important in this case because the lowest order contributions are extremely small. The influence of electron correlation effects has been discussed by Smentek [Sme98] and by Burdick and coworkers [BR93, BKR93]. In their study of $Eu^{2+}$ Burdick *et al.* [BKR93] found that a perturbation expansion failed to converge, since the $4f^6 5d$ configuration substantially overlaps the $4f^7$ configuration, so it was necessary to include the $4f^6 5d$ configuration as part of the model space.

Some two-photon experiments have concentrated on the variation in intensity as the polarization of the incoming light is rotated. These variations are more complex than in the case of one-photon absorption. In some cases the variations fit the model very well [GBMB93]. However, some experiments are not well explained by the theory [MNG97].

Two-photon circular dichroism has been observed by Gunde and Richardson [GR95] and explained theoretically [GBR96]. As in the case of one-photon circular dichroism, these experiments are able to give more information than a normal two-photon absorption measurement.

Raman scattering is also an interesting test of the formalism. In this case the very different denominators in the two terms of (10.47) allow effective operators of rank 1 (time reversal and Hermiticity arguments do not rule out rank 1 operators, since the denominators are not identical), in addition to the rank 2 operators allowed in single beam two-photon absorption, and interesting polarization effects. In some cases the theoretical predictions explain the experimental results [NME+97], whereas in other cases there are difficulties rationalizing the experimental ratios of the rank 1 and 2 operators with calculations [BEW+85].

## 10.8 Future directions

The parametrization of transition intensities between crystal field split levels of the $4f^N$ configuration is a much more complicated problem than the parametrization of crystal field energy levels. Nevertheless, it is becoming possible not only to carry out detailed parametrizations but also to explain the results in terms of *ab initio* calculations. More extensive experimental work is required to gather and analyse data if we are to characterize the transition intensities phenomenologically to the same level of detail and accuracy as has been achieved in crystal field analyses.

# Appendix 1

## Point symmetry

### D. J. NEWMAN

*University of Southampton*

### BETTY NG

*Environment Agency*

This appendix introduces all the relevant point group theoretical concepts and tools which are employed in this book. The aim is to explain the relationships between group theory and crystal field theory. No attempt is made to teach group theory. Readers who are interested in finding out about the mathematical details and understanding the basic theorems of group theory should refer to the standard texts, e.g. [Hei60, Sac63, Fal66, Tin64, LN69, ED79, But81].

## A1.1 Full rotation group $O_3$ and free magnetic ionic states

A free magnetic ion is said to have spherical symmetry (see Chapter 1), which means that its Hamiltonian is invariant under all rotations about axes and reflections through planes which contain its centre of symmetry, i.e. its nucleus. The set of rotation operations through all angles about all axes passing through a given point, together with the inversion, form a group called the *full rotation group* $O_3$. This group also contains, by construction, all the reflection operations through planes containing the given point. The wavefunctions (characterized by the symbols $nlm$) of a single electron in an electrostatic potential of spherical symmetry are obtained by solving the Schrödinger equation. They are given by the products of a radial function characterized by $n$ and a spherical harmonic $Y_{lm}(\theta, \phi)$. The energy of states corresponding to a given wavefunction depends on the value of $n$ and $l$ only. The $N$-electron states of a free magnetic ion can be constructed from determinants, each of which contains $N$ of these single electronic states.

It can be shown mathematically that when the operations of $O_3$ are used to transform any member of a set of spherical harmonics $Y_{lm}$, the resultant state is always a linear combination of members of the set of spherical harmonics with the same $l$ value. In other words, the set of spherical harmonics $Y_{lm}$ are *closed* under the operations of the group $O_3$. As the smallest sets which have this property, they are said to generate *irreducible (matrix) representations* of $O_3$, denoted $D^{(l)}$. Hence, the degenerate solutions of the one-electron Schrödinger equation, namely multiplets with orbital angular momentum $l$, correspond to different irreducible representations $D^{(l)}$ of $O_3$. The transformation properties of these irreducible representations under inversion are distinguished by adding plus or minus signs: $D^{(l)+}$ have even parity and $D^{(l)-}$ have odd parity.

### A1.1.1 Tensor operators and matrix elements

Tensor operators $t_q^{(k)}$ are operators that have the same transformation properties under all the rotations in the group $O_3$ as the irreducible representation $D^{(k)}$ (or the spherical harmonics $Y_{kq}$). $k$ and $q$ are, respectively, the *rank* and *component* labels of the tensor. Allowed values of $q$ run from $-k$ to $k$. Tensor operators can be constructed from sums of products of the angular momentum operators $\mathbf{J}$, which are themselves tensor operators of rank 1, with three components $-1$, 0 and 1. Readers who are interested in angular momentum theory can consult one of the standard texts on this subject, e.g. Brink and Satchler [BS68].

In evaluating a matrix element of a tensor operator $t_q^{(k)}$ between the states $\langle JM_J|$ and $|J'M_{J'}\rangle$, the following factorization (known as the Wigner–Eckart theorem) can be used:

$$\langle \alpha JM_J|t_q^{(k)}|\alpha J'M_J'\rangle = (-1)^{J-M_J} \begin{pmatrix} J & k & J' \\ -M_J & q & M_J' \end{pmatrix} (\alpha J||t^{(k)}||\alpha J').$$

$$(A1.1)$$

In this equation, $J$ and $J'$ can be integral or half-integral. The dependence of the matrix element on $q$, $M_J$ and $M_J'$ is expressed in terms of the so-called $3j$ symbols, which are discussed below.

The symbols $(\alpha J||t^{(k)}||\alpha J')$ are called *reduced matrix elements*. Their values are determined by the normalization of the tensor operators together with the structure of the many-electron states characterized by the symbol $\alpha$. The normalization of tensor operators is arbitrary, but has to be specified in applications.

It is argued in Chapter 2 that crystal field potentials can be expressed in terms of the functional tensor operators $C_q^{(k)}$. In the special case of single-electron matrix elements, where the orbital angular momentum $l = J = J'$, the tensor operators $C_q^{(k)}$ have reduced matrix elements

$$(l||C^{(k)}||l) = (-1)^l (2l+1) \begin{pmatrix} l & k & l \\ 0 & 0 & 0 \end{pmatrix}. \tag{A1.2}$$

These reduced matrix elements determine the normalization of the tensor operators used in this book: many-electron reduced matrix elements can be derived from them.

Another tensor operator normalization which is frequently used in crystal field theory defines the so-called *unit tensor operators*, which are denoted by $u_q^{(k)}$, and have the one-electron reduced matrix elements

$$(l||u^{(k)}||l) = \sqrt{2l+1}. \tag{A1.3}$$

Note that the reduced matrix elements of these operators are non-zero for both even and odd values of $k$. Their many-electron reduced matrix elements for the $LS$ states of the $p^n$, $d^n$ and $f^n$ configurations have been tabulated by Nielson and Koster [NK63].

### A1.1.2 Coupling of angular momentum states

When angular momenta $j_1$ and $j_2$ couple together (e.g. in spin–orbit coupling), the resulting angular momentum eigenstates $|\alpha, j_1, j_2; j, m\rangle$ are related to the eigenstates of $j_1$ and $j_2$ as follows:

$$|\alpha, j_1, j_2; j, m\rangle = \sum_{m_1+m_2=m} |\alpha, j_1, j_2; m_1, m_2\rangle\langle m_1, m_2|j, m\rangle, \tag{A1.4}$$

where $\alpha$ stands for other labels of the eigenstates (see Chapter 2). The Clebsch–Gordan or coupling coefficients $\langle m_1, m_2|j, m\rangle$ are related to the $3j$ symbols through the equation

$$\langle m_1, m_2|j, m\rangle = \langle j, m|m_1, m_2\rangle$$
$$= (-1)^{j_2-j_1-m}\sqrt{2j+1} \begin{pmatrix} j_1 & j_2 & j \\ m_1 & m_2 & -m \end{pmatrix}. \tag{A1.5}$$

### A1.1.3 Properties of the 3j symbols

The $3j$ symbols $\begin{pmatrix} J & k & J' \\ -M_J & q & M'_J \end{pmatrix}$ are non-vanishing only when the following *triangular* relationships among $J$, $J'$ and $k$ are satisfied:

$$J = |k - J'|, |k - J' + 1|, \ldots, k + J'. \tag{A1.6}$$

This triangular relationship is said to be a *selection rule*, because it can be used, in conjunction with the Wigner–Eckart theorem, to determine when the matrix elements of a tensor operator are non-zero. Using the triangular relationship, the possible values for the rank $k$ of crystal field operators for $f$-electrons are restricted to 0, 1, 2, 3, 4, 5 and 6. Hermiticity and time-reversal invariance of the crystal field Hamiltonian ensure that only crystal field parameters of even rank are non-zero [New71]. Hence, the possible ranks $k$ of crystal field operators for $f$-electrons are 0, 2, 4 and 6. The rank $k = 0$ crystal field operator corresponds to an isotropic potential, which does not produce crystal field splittings (see Chapter 2), and hence is omitted from the crystal field. A further selection rule is that the $3j$ symbol $\begin{pmatrix} J_1 & J_2 & J_3 \\ M_1 & M_2 & M_3 \end{pmatrix}$ is zero when the sum $M_1 + M_2 + M_3$ is non-zero. The $3j$ symbols satisfy two important orthogonality relations, the explicit forms of which can be found in any book which discusses angular momentum theory (e.g. see [CO80], p. 180).

The program THREEJ.BAS calculates values of the $3j$ symbol

$$\begin{pmatrix} J1 & J2 & J3 \\ M1 & M2 & M3 \end{pmatrix}.$$

When the program is run, the user is asked to input values of $J1$, $J2$, $J3$, $M1$, $M2$ and $M3$. The value of the required $3j$ symbol is then calculated and printed out on the screen by the program. The code of the program THREEJ.BAS is given in Appendix 2. An example screen display from the running of the program is given below:

```
THIS IS A PROGRAM TO CALCULATE THE 3_J SYMBOL
  RO =    (J1,J2,J3;M1,M2,M3)
CRYSTAL FIELD HANDBOOK, EDITED BY D.J. NEWMAN AND BETTY NG
  CAMBRIDGE UNIVERSITY PRESS


  J1 = ?, J2 = ?, J3 = ? 3,4,3
  M1 = ?, M2 = ?, M3 = ? 0,0,0
THE THREE J SYMBOL IS:          -.1611645928050761


DO YOU WANT TO CALCULATE ANOTHER 3-j SYMBOL?
TYPE IN Y/y (FOR YES), N/n (FOR NO)   n
```

The values for $J$s and $M$s are keyed in by the user.

### *A1.1.4  6j Symbols*

The need to employ $6j$ symbols arose in relating singly reduced and doubly reduced matrix elements of tensor operators using equation (3.3). These symbols can either be expressed as a product of four $3j$ symbols, as in equation (3.61) of [LM86], or as a closed algebraic expression (e.g. see equation (3-7) of [Jud63] or equation (9) of section $9^3$ of [CO80]). A subprogram which generates the $6j$ symbols from the algebraic expression is attached to program REDMAT.BAS, the code of which is listed in Appendix 2.

### *A1.1.5 Orthogonal operators*

Tensor operators are said to be 'orthogonal' over spaces defined by one or more $J$ multiplets (e.g. see [New81]). The orthogonality properties of the $3j$ symbols ensure that the expression

$$\sum_{M_1,M_2} \langle J, M_1 | t_q^{(k)} | J, M_2 \rangle \langle J, M_2 | t_{q'}^{(k')} | J, M_1 \rangle$$

is zero if either, or both, $k \neq k'$, $q \neq q'$. The term 'orthogonal' arises from the simple analogy between this relation and the vanishing of the dot product between orthogonal vectors in elementary vector analysis.

The advantage of using orthogonal operators in linear parametrizations is that parameters associated with each operator can be shown to be independent, in the sense that the values obtained for the fitted parameters do not depend on how many of the other parameters are fitted at the same time. The orthogonality of tensor operators has been used to simplify the crystal field fitting program ENGYFIT.BAS, which is described in Chapter 3 and Appendix 2.

The concept of orthogonal operators can be extended to expressions involving products of two or more tensor operators (see Chapters 3, 6 and 8). In such cases the bases over which the operators are defined necessarily span more than one multiplet. Most theoretical discussions have been concerned with orthogonality defined over whole configurations, although operators defined in this way are not likely to be even approximately orthogonal over the 'reduced' bases that are used in parameter fits. Further work is necessary to introduce sets of standardized tensor product operators which are orthogonal over such reduced bases. Discussions of the advantages of using orthogonal operators in spectroscopy, and their relation to symmetry considerations, can be found in [New81, JHR82, New82, JC84, JS84, DHJL85, HJL87, Lea87, Rei87a, JNN89].

Table A1.1. Commonly occurring site symmetries in crystals.

| Symmetry group | International symbol | Order | •Generators | Type |
|---|---|---|---|---|
| $S_2$ | $\bar{1}$ | 2 | $i$ | Triclinic |
| $C_2$ | 2 | 2 | $C_2$ | Monoclinic |
| $C_{2h}$ | $\frac{2}{m}$ | 4 | $C_2, i$ | Monoclinic |
| $C_{2v}$ | mm2 | 4 | $C_2, \sigma_v, \sigma_v'$ | Orthorhombic |
| $D_{2h}$ | mmm | 8 | $C_2^x, C_2^y, i$ | Orthorhombic |
| $C_{4v}$ | 4mm | 8 | $C_4, \sigma_v, \sigma_v'$ | Tetragonal |
| $D_{2d}$ | $\bar{4}$2m | 8 | $S_4, \sigma_d, i$ | Tetragonal |
| $D_{4h}$ | $\frac{4}{m}$mm | 16 | $C_4, \sigma_h, i$ | Tetragonal |
| $C_{3v}$ | 3m | 6 | $C_3, \sigma_v$ | Trigonal |
| $C_{3h}$ | $\frac{3}{m}$ | 6 | $C_3, \sigma_h$ | Hexagonal |
| $D_{3h}$ | $\frac{3}{m}$m2 | 12 | $C_3, \sigma_v, \sigma_h$ | Hexagonal |
| $D_{6h}$ | $\frac{6}{m}$mm | 24 | $C_6, \sigma_v, \sigma_h, i$ | Hexagonal |
| $T_d$ | $\bar{4}$3m | 24 | $S_4, C_3, \sigma_d$ | Cubic |
| $O_h$ | m3m | 48 | $C_4, C_2', C_3, i$ | Cubic |

## A1.2 Site symmetries and symmetry operators

The first task in crystal field applications is to determine the point symmetry of magnetic ions. In many instances, this information will have been provided by an X-ray or neutron scattering determination of the crystal structure. There are 32 crystallographic point groups (e.g. see table 3.1 of [LN69]). However, only 14 of them (see Table A1.1) are commonly occurring site symmetries for crystal fields. Quasi-crystals and molecules can have sites of icosahedral symmetry, which may also be an approximate site symmetry in a crystal. Details of this symmetry can be found in [But81].

What is meant by a magnetic ion in a crystal field having a point, or site, symmetry group G? It signifies that the crystal field at this ion will be left unchanged (or invariant) by all the point symmetry operations contained in the group G. Point symmetry operations include rotations, reflections and inversion. In Table A1.1, the symbol $C_n$ stands for an anticlockwise rotation about an axis (called the principal axis) through $\frac{360^0}{n}$. (The principal axis is normally identified with the $z$-coordinate direction.) $C_n$ is also known as a *n-fold rotation*. If there is more than one axis of rotation, the different axes are denoted by a superscript. For example, $C_2^x$ represents a rotation through $180^0$ about the $x$ axis. The operation of inversion through the origin is symbolized as $i$. A $C_n$ rotation, followed by an inversion, is denoted $S_n$. Reflections through the diagonal, horizontal and vertical planes are denoted by $\sigma_d$, $\sigma_h$ and $\sigma_v$, respectively. Diagonal and vertical planes contain the principal axis of rotation while the horizontal plane is perpendicular to it.

The groups listed in Table A1.1 all have a finite number of elements, or symmetry operators, which is known as their *order*. They are, therefore, all examples of *finite* point groups. The symmetry operators shown in the fourth column of Table A1.1 can be used to generate all other symmetry operators of the group. For example, applying a rotation $C_n$ $n$ times will produce the identity operation. Applying $C_4$ twice will produce $C_2$.

### A1.2.1 Subgroups

If a group G contains all the symmetry operators (or generators) of another group H, then H is called a *subgroup* of G. For example, $T_d$ is a subgroup of $O_h$ as the generators $C_2$ and $\sigma_d$ of $T_d$ can be obtained, respectively, by applying the $O_h$ group generators $C_4$ twice, and $C_2'$ followed by $i$ (see Table A1.1).

The full rotation group $O_3$ contains the generators of all point symmetry groups. Hence, all the site symmetry groups shown in Table A1.1 are subgroups of $O_3$. The group/subgroup relationships relevant to most crystal field applications are given in Figures A1.1 and A1.2.

### A1.3 Crystal field parameters and site symmetry

The symmetry of the crystal field Hamiltonian (i.e. the *crystal field symmetry*), discussed in Section 2.2, is defined as the point group of symmetry operators which leave the crystal field Hamiltonian invariant. A selection rule of crystal field matrix elements (e.g. see chapter 12 of [LN69]) leads to the general conclusion that the *effective* crystal field always contains the inversion operator. Hence, it may correspond to a higher symmetry than the site symmetry group (the geometrical/physical point symmetry of the crystal at the paramagnetic ion).

The site symmetry at a magnetic ion imposes further restrictions on the possible tensor operators $C_q^{(k)}$ which can appear in the crystal field. Only certain values $q$ of the set $-k$, $-k+1$, ..., $k-1$, $k$ are allowed. They are found by considering the effects of the various symmetry operators on $C_q^{(k)}$ and the fact that the crystal field must be invariant under these operations. Examples of possible $q$ values allowed by symmetry operations are given in

*D. J. Newman and Betty Ng*

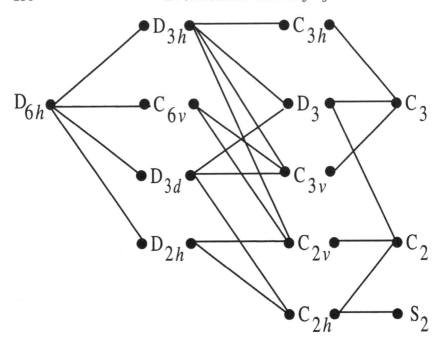

Fig. A1.1 Subgroups of $D_{6h}$.

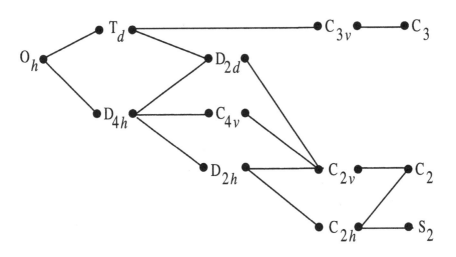

Fig. A1.2 Subgroups of $O_h$.

Table A1.2. $U_2$ is a twofold rotation about an axis perpendicular to the principal axis.

Table A1.2. Values of $q$ allowed by various symmetry operations.

| Symmetry operations | $q$ Values |
|---|---|
| $C_2$, $\sigma_h$ | $0$, $\pm$ (even number) |
| $C_3$ | $0$, $\pm$ (multiples of three) |
| $C_4$ | $0$, $\pm$ (multiples of four) |
| $C_6$ | $0$, $\pm$ (multiples of six) |
| $\sigma_v$, $U_2$ | $0$, any positive number |

## A1.4 Irreducible representations of point symmetry groups and energy levels

Irreducible representations of point symmetry groups are defined in the same way as those for the full rotation group $O_3$. They are the smallest sets of functions which are closed under all operations of the group. Not only does the crystal field symmetry determine the possible crystal field operators in a crystal field Hamiltonian, but its irreducible representation labels also provide identification for the (possibly degenerate) energy levels of a magnetic ion. Suppose $\Phi_\lambda$ is a many-electron state associated with the energy eigenvalue $\lambda$ of the crystal field Hamiltonian $V_{CF}$, i.e.

$$V_{CF}\Phi_\lambda = \lambda\Phi_\lambda. \tag{A1.7}$$

When an operator $\mathbf{R}$ of the crystal field symmetry group is applied to both sides of equation (A1.7), the eigenvalue equation becomes:

$$\mathbf{R}V_{CF}\Phi_\lambda = \mathbf{R}\lambda\Phi_\lambda$$
$$\mathbf{R}V_{CF}\mathbf{R}^{-1}\mathbf{R}\Phi_\lambda = \lambda\mathbf{R}\Phi_\lambda. \tag{A1.8}$$

Since the crystal field is invariant under the operation of $\mathbf{R}$ (by the definition of a crystal field symmetry group), $\mathbf{R}V_{CF}\mathbf{R}^{-1}$ is the same as $V_{CF}$. Hence (A1.8) becomes

$$V_{CF}\mathbf{R}\Phi_\lambda = \lambda\mathbf{R}\Phi_\lambda, \tag{A1.9}$$

showing that all the expressions $\mathbf{R}\Phi_\lambda$, which form an irreducible representation of the crystal symmetry group, are eigenfunctions of $V_{CF}$ with the eigenvalue $\lambda$. The labels of the irreducible representation can therefore be used to identify the different eigenvalues (energy states) of the crystal field operator. It also illustrates that the energy level, corresponding to a single free-ion $J$ multiplet, is broken into a number of energy levels, corresponding to the irreducible representations of its crystal field symmetry group. This

Table A1.3. Reduction/subduction of $O_3$ with respect to $O_h$ and $D_{6h}$.

| $O_3$ | $O_h$ | $D_{6h}$ |
|---|---|---|
| $D^{(0)}$ | $A_{1g}$ | $A_{1g}$ |
| $D^{(1)}$ | $T_{1g}$ | $A_{2g} + E_{1g}$ |
| $D^{(2)}$ | $E_g + T_{2g}$ | $A_{1g} + E_{1g} + E_{2g}$ |
| $D^{(3)}$ | $A_{2g} + T_{1g} + T_{2g}$ | $A_{1g} + B_{1g} + B_{2g} + E_{1g} + E_{2g}$ |
| $D^{(4)}$ | $A_{1g} + E_g + T_{1g} + T_{2g}$ | $A_{1g} + B_{1g} + B_{2g} + E_{1g} + 2E_{2g}$ |
| $D^{(5)}$ | $E_g + 2T_{1g} + T_{2g}$ | $A_{1g} + B_{1g} + B_{2g} + 2E_{1g} + 2E_{2g}$ |
| $D^{(6)}$ | $A_{1g} + A_{2g} + E_g + T_{1g} + 2T_{2g}$ | $2A_{1g} + A_{2g} + B_{1g} + B_{2g} + 2E_{1g} + 2E_{2g}$ |
| $D^{(7)}$ | $A_{2g} + E_g + 2T_{1g} + 2T_{2g}$ | $A_{1g} + 2A_{2g} + B_{1g} + B_{2g} + 3E_{1g} + 2E_{2g}$ |
| $D^{(8)}$ | $A_{1g} + 2E_g + 2T_{1g} + 2T_{2g}$ | $2A_{1g} + A_{2g} + B_{1g} + B_{2g} + 3E_{1g} + 3E_{2g}$ |
| $D^{(5/2)}$ | $\Gamma_{7g} + \Gamma_{8g}$ | $\Gamma_{7g} + \Gamma_{8g} + \Gamma_{9g}$ |
| $D^{(7/2)}$ | $\Gamma_{6g} + \Gamma_{7g} + \Gamma_{8g}$ | $\Gamma_{7g} + \Gamma_{8g} + 2\Gamma_{9g}$ |
| $D^{(9/2)}$ | $\Gamma_{6g} + 2\Gamma_{8g}$ | $\Gamma_{7g} + 2\Gamma_{8g} + 2\Gamma_{9g}$ |
| $D^{(11/2)}$ | $\Gamma_{6g} + \Gamma_{7g} + 2\Gamma_{8g}$ | $2\Gamma_{7g} + 2\Gamma_{8g} + 2\Gamma_{9g}$ |
| $D^{(13/2)}$ | $\Gamma_{6g} + 2\Gamma_{7g} + 2\Gamma_{8g}$ | $3\Gamma_{7g} + 2\Gamma_{8g} + 2\Gamma_{9g}$ |
| $D^{(15/2)}$ | $\Gamma_{6g} + \Gamma_{7g} + 3\Gamma_{8g}$ | $3\Gamma_{7g} + 2\Gamma_{8g} + 3\Gamma_{9g}$ |

idea is fundamental to crystal field applications, as is discussed in Chapter 3.

## A1.5 Subduced and induced representations

The relationship between irreducible representations of a group G and those of its subgroup H is useful in finding the splitting of energy levels of an ion due to the lowering of its symmetry from G to H. Each irreducible representation of G can be expressed as a sum of irreducible representations of H. This process is called *subduction* or *reduction* and the results are obtained by employing the theory of group characters. Some commonly used subduced representations are given in Table A1.3 and Figure A1.3. A definition of group characters and a description of how the reduction is carried out can be found in the standard textbooks, e.g. see [Tin64, LN69, But81, ED79].

Table A1.3 and Figure A1.3 can be used to find the irreducible repesentation labels of energy levels of a magnetic ion in a crystal. For example, in a crystal field of symmetry $C_{4v}$, a $J = 4$ multiplet of an ion will split into $A_1 + A_1 + B_1 + A_2 + E + B_2 + E$, i.e. $2A_1 + B_1 + A_2 + B_2 + 2E$. This

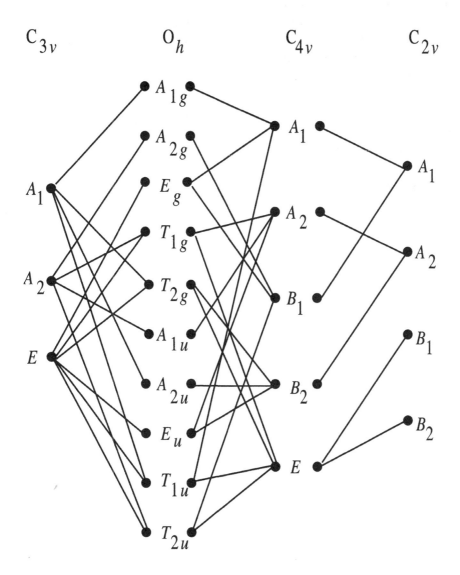

Fig. A1.3 Subduced representations of $O_h$.

is done by first reducing $D^{(4)}$ of $O_3$ into the irreducible representations of $O_h$ (using Table A1.3) and the subsequent reduction of the $O_h$ irreducible representations into those of $C_{4v}$ using Figure A1.3.

The relationships between irreducible representations of a group G and those of its subgroup H, as shown in Figure A1.3, can also be used in the *induction* of an irreducible representation of the subgroup H to those of the higher symmetry group G. For example, the $A_2$ irreducible representation of $C_{4v}$ induces the representation $A_{1g} + E_g + T_{1u}$ of $O_h$. The physical relationship corresponding to induction is important in the semiclassical model, discussed in Chapter 9. Readers who are interested in an explanation of why the same relationship can be used for both subduction and induction processes may refer to [Alt77] or [Led77].

The number of non-zero crystal field parameters for each rank $k$ corresponds to the number of invariant irreducible representations (normally labelled $A_1$ or $A_{1g}$) of the crystal field symmetry group appearing in the reduction of the representation $D^{(k)}$ of $O_3$. For example, in a crystal field of cubic ($O_h$) symmetry, there are only two crystal field parameters, one of rank 4 and one of rank 6 (see Table A1.3).

# Appendix 2

## QBASIC programs

D. J. NEWMAN

*University of Southampton*

BETTY NG

*Environment Agency*

This appendix describes the structure of the QBASIC programs and data files employed in the main text. In a few cases complete printouts of the programs have been provided for the reader's convenience. All programs and data files can be downloaded from the Cambridge University Press web site:

`http://www.cup.cam.ac.uk/physics`

or from the web site maintained by Dr M. F. Reid:

`http://www.phys.canterbury.ac.nz/crystalfield`

The QBASIC programs available from these sites are as follows.

(i) THREEJ.BAS, which calculates values of $3j$ symbols. Its use is described in Section 3.1.1. The subprogram, which forms the core of this program, is listed in Section A2.1.1.

(ii) REDMAT.BAS, which is used to determine $J$-basis reduced matrix elements of the tensor operators $\mathbf{C}^{(k)}$ from $LS$-basis reduced matrix elements, such as those tabulated by Nielson and Koster [NK63]. The use of this program in calculating matrix elements is described in Section 3.1.1. It includes the subprogram SIXj, listed in Section A2.1.2.

(iii) ENGYLVL.BAS, which is used to calculate single $J$ multiplet energy levels and eigenfunctions from crystal field parameters. This program is discussed at length in Chapter 3. A specific example of its use is given in Section 3.1.3.

(iv) ENGYFIT.BAS, which is used to fit crystal field parameters to sets of energy levels of a single $J$ multiplet. This program is discussed at length in Chapter 3. A specific example of its use is given in Section 3.2.

(v) CORFACW.BAS, which determines combined coordination factors (Wybourne normalization) from input coordination angles. An example of the use of this program is given in Section 5.3.2.1.

(vi) CORFACS.BAS, which is the same as CORFACW.BAS, apart from using Stevens normalization. Descriptions of both coordination factor programs, and complete printout of CORFACS.BAS, are given in Section A2.2.

(vii) INTRTOCF.BAS, which converts input intrinsic parameters and coordination factors to crystal field parameters.

(viii) CFTOINTR.BAS, which fits sets of crystal field parameters to intrinsic parameters, using input coordination factors.

A range of data files can also be downloaded from the address given above. Section A2.3 gives the format of data files used as inputs to, and appearing in outputs of, the programs listed above.

In order to overcome the problem of printing lines in the QBASIC programs which exceed the page width it is necessary to introduce a special symbol to indicate line continuations. In the program listings provided in Sections A2.1 and A2.2, a symbol '& ' at the beginning of any line indicates that it should be read as a continuation of the previous line. The symbols '& ' are NOT part of the QBASIC code.

## A2.1 Calculation of $3j$ and $6j$ symbols

### A2.1.1 The $3j$ symbols

The $3j$ symbols (see Section A1.1.3), are necessary in determining energy matrices from crystal field parameters (see Section 3.1.1). They are closely related to the Clebsch–Gordan, or coupling, coefficients defined in Appendix 1.

Although tabulations of the $3j$ symbols exist (e.g. [RBMJ59]), it is usually most convenient to calculate them using using a program such as THREEJ.BAS. This program is based on the subprogram given below, which is used in programs ENGYLVL.BAS and ENGYFIT.BAS. It evaluates a closed algebraic formula expressed, for example, in equations (2.76) and (2.81) of Lindgren and Morrison [LM86]. The evaluation procedure could almost certainly be made a lot more efficient. [LM86] also includes (as appendix C) algebraic expressions for some simple cases and a tabulation for low $j$ values.

```
SUB threej (R0, AJ1, AJ2, AJ3, AM1, AM2, AM3)
        DEFINT J, M-N
        DEFDBL F, R
        DIM F(0 TO 30)

REM DEFINE THE FACTORIAL FUNCTION
        FOR I = 0 TO 30
            F(I) = 1
            FOR J = 1 TO I
               F(I) = F(I) * J
            NEXT J
         NEXT I


REM    SELECTION RULES
        IF AM1 + AM2 + AM3 <> 0 THEN GOTO ZERO:
        IF AJ1 + AJ2 - AJ3 < 0 THEN GOTO ZERO:
        IF AJ3 + AJ1 - AJ2 < 0 THEN GOTO ZERO:
        IF AJ3 + AJ2 - AJ1 < 0 THEN GOTO ZERO:
        IF AJ1 + AJ2 + AJ3 + 1 < 0 THEN GOTO ZERO:
        IF AJ1 - AM1 < 0 THEN GOTO ZERO:
        IF AJ2 - AM2 < 0 THEN GOTO ZERO:
        IF AJ3 - AM3 < 0 THEN GOTO ZERO:
        IF AJ1 + AM1 < 0 THEN GOTO ZERO:
        IF AJ2 + AM2 < 0 THEN GOTO ZERO:
        IF AJ3 + AM3 < 0 THEN GOTO ZERO:


     R4 = F(CINT(AJ1 + AJ2 - AJ3)) * F(CINT(AJ1 - AM1))
&          * F(CINT(AJ2 - AM2)) * F(CINT(AJ3 - AM3))
&          * F(CINT(AJ3 + AM3))
     R5 = F(CINT(AJ1 + AJ2 + AJ3 + 1))
&        * F(CINT(AJ3 + AJ1 - AJ2)) * F(CINT(AJ3 + AJ2 - AJ1))
&        * F(CINT(AJ1 + AM1)) * F(CINT(AJ2 + AM2))
     R6 = 0
     FOR J7 = 0 TO 25
         IF AJ1 + AM1 + J7 < 0 THEN GOTO J7R:
         IF AJ2 + AJ3 - AM1 - J7 < 0 THEN GOTO J7R:
         IF AJ3 + AM3 - J7 < 0 THEN GOTO J7R:
         IF AJ1 - AM1 - J7 < 0 THEN GOTO J7R:
         IF AJ2 - AJ3 + AM1 + J7 < 0 THEN GOTO J7R:
         R8 = F(CINT(AJ1 + AM1 + J7))
```

```
&                    * F(CINT(AJ2 + AJ3 - AM1 - J7))
&                      * (-1) ^ (CINT(AJ1 - AM1 - J7))
         R9 = F(J7) * F(CINT(AJ3 + AM3 - J7))
&                    * F(CINT(AJ1 - AM1 - J7))
&                    * F(CINT(AJ2 - AJ3 + AM1 + J7))
         R6 = R6 + R8 / R9
J7R:   NEXT J7
         R0 = SQR(R4 / R5) * R6 * (-1) ^ CINT((AJ1 - AJ2 - AM3))
         GOTO FIN:
ZERO:   R0 = 0
FIN: END SUB
```

## A2.1.2 *Subprogram for calculating* 6*j* *symbols*

The $6j$ symbols may be calculated from a closed algebraic formula (see equation (3-7) of [Jud63]) or by expressing them in terms of $3j$ symbols (see equation (3.61) of [LM86]). The subprogram given below is based on the first of these methods. It is used in the calculation of reduced matrix elements, see Section 3.1.1, and forms part of the program REDMAT.BAS.

```
SUB SIXj (RJ0, RJ1, RJ2, RJ3, RL1, RL2, RL3)
REM  THIS SUBPROGRAM CALCULATES 6-J SYMBOLS:
REM     R0=(RJ1,RJ2,RJ3;RL1,RL2,RL3)
REM     F(I) = I! IS THE FACTORIAL FUNCTION
REM     I=0 TO 30 IS USUALLY SUFFICIENT
REM     PLEASE NOTE THAT I STARTS FROM ZERO

     DIM F(0 TO 30), RK(1 TO 7)

     FOR I = 0 TO 30
         F(I) = 1
         FOR J = 1 TO I
             F(I) = F(I) * J
         NEXT J
     NEXT I

     RK(7) = RJ3 + RJ1 + RL3 + RL1
     RK(6) = RJ2 + RJ3 + RL2 + RL3
     RK(5) = RJ1 + RJ2 + RL1 + RL2
```

```
     RK(4) = RL1 + RL2 + RJ3
     RK(3) = RL1 + RJ2 + RL3
     RK(2) = RJ1 + RL2 + RL3
     RK(1) = RJ1 + RJ2 + RJ3
     IF RK(5) > RK(6) THEN
         RMAX = RK(6)
         ELSE
         RMAX = RK(5)
     END IF
     IF RMAX < RK(7) THEN
         ELSE
         RMAX = RK(7)
     END IF
     IF RK(1) < RK(2) THEN
         RMIN = RK(2)
         ELSE
         RMIN = RK(1)
     END IF
     FOR I = 3 TO 4
      IF RMIN < RK(I) THEN
        RMIN = RK(I)
        ELSE
      END IF
     NEXT I
     RJ0 = 0

     FOR I = RMIN TO RMAX
      RJ0 = RJ0 + ((-1) ^ I) * F(I + 1) / (F(I - RK(1))
&            * F(I - RK(2)) * F(I - RK(3)) * F(I - RK(4))
&            * F(RK(5) - I) * F(RK(6) - I) * F(RK(7))
     NEXT I

    C1 = SQR(F(RJ1 + RJ2 - RJ3) * F(RJ1 - RJ2 + RJ3)
&        * F(-RJ1 + RJ2 + RJ3) / F(RJ1 + RJ2 + RJ3 + 1))
    C2 = SQR(F(RJ1 + RL2 - RL3) * F(RJ1 - RL2 + RL3)
&        * F(-RJ1 + RL2 + RL3) / F(RJ1 + RL2 + RL3 + 1))
    C3 = SQR(F(RL1 + RJ2 - RL3) * F(RL1 - RJ2 + RL3)
&        * F(-RL1 + RJ2 + RL3) / F(RL1 + RJ2 + RL3 + 1))
    C4 = SQR(F(RL1 + RL2 - RJ3) * F(RL1 - RL2 + RJ3)
&        * F(-RL1 + RL2 + RJ3) / F(RL1 + RL2 + RJ3 + 1))
```

```
RJO = RJO * C1 * C2 * C3 * C4

END SUB
```

## A2.2 Calculation of coordination factors

Separate QBASIC programs are provided to determine the combined coordination factors in Stevens and Wybourne crystal field parameter normalizations. These are in files CORFACS.BAS and CORFACW.BAS, respectively. CORFACS.BAS is listed below.

Both coordination factor programs function in exactly the same way, so that a single description is sufficent for both. In order to provide a standard format for the data files, they output values of the combined coordination factors for all values of $k$ and $q$, even when these are zero. It is advisable to choose output file names which identify both the crystal and the parameter normalization, e.g. S_YVO4.DAT and W_YVO4.DAT for the coordination factors of YVO$_4$ in Stevens and Wybourne normalization, respectively. The structure of these data files is illustrated in Section A2.3.1.

Positions of the ligands can be input by hand or from a file. Separate angular positions of all the ligands are required, together with labels 1, 2, etc. which identify ligands at the same distance without having to specify the actual distances (which may not be known). The contributions of all the ligands at a given distance are included in combined coordination factors, as explained in Section 6.2.1. Input file names can be chosen at will, but it is suggested that they identify the crystal. An example of standard form used in this book is C_YVO.DAT for the coordination angles at the yttrium sites in YVO$_4$. The structure of this file is given in Section A2.3.1.

### A2.2.1 CORFACS.BAS

```
REM    PROGRAM CORFACS
REM    F- AND D- ELECTRON COORDINATION FACTOR PROGRAM FOR
REM    GENERAL SYMMETRY ALLOWING  FOR UP TO 16 LIGANDS AT
REM    ARBITRARY DISTANCES.  GROUPS TOGETHER LIGANDS
REM    WITH SAME DISTANCE OR INTRINSIC PARAMETER AND OUTPUTS
REM    COMBINED COORDINATION FACTORS

       DEFINT I-K, M-N
       DEFSNG A, C-E, P, S-T
```

```
      DIM G2(-2 TO 2, 16), G4(-4 TO 4, 16), G6(-6 TO 6, 16)
      DIM G2C(-2 TO 2, 16), G4C(-4 TO 4, 16), G6C(-6 TO 6, 16)
REM   DEGTORAD = CONVERSION FACTOR FROM DEGREES TO RADIANS
REM   ABSOLUTE VALUE OF ANY NUMBER LESS THAN EPLIN
REM   WILL BE REGARDED AS ZERO
      PI = 3.14156
      DEGTORAD = PI / 180!
      EPLIN = .01
REM   THERE ARE AT MOST 16 LIGANDS, LABELLED 1 TO 16
REM   ID(J) (J=1-16) = DISTANCE LABEL OF LIGAND J
REM   THETA(J) (J=1-16) = THETA OF LIGAND J
REM   PHI(J)   (J=1-16) = PHI OF LIGAND J
REM   SEE CHAPTER 6 FOR DEFINITIONS OF THETA AND PHI
      DIM THETA(1 TO 16), PHI(1 TO 16), ID(1 TO 16)
      DIM JDML(1 TO 16, 1 TO 16), JDMC(1 TO 16), IDM(1 TO 16)
        CLS
      PRINT "THIS PROGRAM ENABLES YOU TO GENERATE"
      PRINT "COORDINATION FACTORS FOR STEVENS PARAMETERS"
      PRINT "FROM INPUT VALUES OF LIGAND DISTANCE LABELS AND"
      PRINT "ANGULAR POSITIONS FOR UP TO 16 LIGANDS"
      PRINT "        "
LINP:     INPUT "NO. OF LIGANDS = ?", NL
      IF NL > 16 THEN
 PRINT "NO. OF LIGANDS TO LARGE. PLEASE TRY AGAIN"
 GOTO LINP
      END IF
      PRINT "    "
      PRINT "TO INPUT LIGAND POSITIONS FROM A FILE"
      INPUT "TYPE IN Y/y (FOR YES), N/n (FOR NO)", QANS$
      IF QANS$ = "Y" OR QANS$ = "y" THEN
 PRINT "        "
 INPUT "NAME OF INPUT FILE =  ", FILE1$
 PRINT "          "
 OPEN FILE1$ FOR INPUT AS #5
 FOR I = 1 TO NL
      INPUT #5, ID(I), THETA(I), PHI(I)
 NEXT I
 CLOSE #5
 ELSE
 FOR I = 1 TO NL
```

```
  PRINT "FOR ", I, "TH LIGAND:"
  INPUT "  DISTANCE LABEL= ?,THETA = ? AND PHI =? ",
&                                    ID(I), THETA(I), PHI(I)
 NEXT I
     END IF
REM     CONVERT THE ANGLES FROM DEGREES TO RADIANS
     FOR I = 1 TO NL
 THETA(I) = THETA(I) * DEGTORAD
 PHI(I) = PHI(I) * DEGTORAD
     NEXT I
     FOR I = 1 TO NL
S = SIN(THETA(I))
C = COS(THETA(I))
C2 = C * C
SP1 = SIN(PHI(I))
CP1 = COS(PHI(I))
SP2 = SIN(2 * PHI(I))
CP2 = COS(2 * PHI(I))
SP3 = SIN(3 * PHI(I))
CP3 = COS(3 * PHI(I))
SP4 = SIN(4 * PHI(I))
CP4 = COS(4 * PHI(I))
SP5 = SIN(5 * PHI(I))
CP5 = COS(5 * PHI(I))
SP6 = SIN(6 * PHI(I))
CP6 = COS(6 * PHI(I))
REM  GK(Q,I)= COORDINATION FACTOR FOR STEVENS FACTOR RANK K
REM  AND COMPONENT Q FOR ITH LIGAND
G2(-2, I) = (3 / 2) * (1 - C2) * SP2
G2(-1, I) = 6 * S * C * SP1
G2(0, I) = .5 * (3 * C2 - 1)
G2(1, I) = 6 * S * C * CP1
G2(2, I) = (3 / 2!) * (1 - C2) * CP2

G4(-4, I) = (35 / 8!) * (1 - C2) * (1 - C2) * SP4
G4(-3, I) = 35 * C * S * (1 - C2) * SP3
G4(-2, I) = 2.5 * (7 * C2 - 1) * (1 - C2) * SP2
G4(-1, I) = 5 * (7 * C2 - 3) * C * S * SP1
G4(0, I) = (35 * C2 * C2 - 30 * C2 + 3) / 8!
G4(1, I) = 5 * (7 * C2 - 3) * C * S * CP1
```

```
G4(2, I) = 2.5 * (7 * C2 - 1) * (1 - C2) * CP2
G4(3, I) = 35 * C * S * (1 - C2) * CP3
G4(4, I) = (35 / 8) * (1 - C2) * (1 - C2) * CP4

G6(-6, I) = 231 * (1 - C2) * (1 - C2) * (1 - C2)
&                     * SP6 / 32
G6(-5, I) = 693 * C * S * (1 - C2) * (1 - C2)
&                     * SP5 / 8
G6(-4, I) = 63 * (11 * C2 - 1) * (1 - C2) * (1 - C2)
&                     * SP4 / 16
G6(-3, I) = 105 * (11 * C2 - 3) * C * S * (1 - C2)
&                     * SP3 / 8
G6(-2, I) = 105 * (33 * C2 * C2 - 18 * C2 + 1)
&                     * (1 - C2) * SP2 / 32
G6(-1, I) = 21 * (33 * C2 * C2 - 30 * C2 + 5) * C
&                     * S * SP1 / 4
G6(0, I) = (231 * C2 * C2 * C2 - 315 * C2 * C2
&                     + 105 * C2 - 5) / 16
G6(1, I) = 21 * (33 * C2 * C2 - 30 * C2 + 5) * C
&                     * S * CP1 / 4
G6(2, I) = 105 * (33 * C2 * C2 - 18 * C2 + 1)
&                     * (1 - C2) * CP2 / 32
G6(3, I) = 105 * (11 * C2 - 3) * C * S * (1 - C2)
&                     * CP3 / 8
G6(4, I) = 63 * (11 * C2 - 1) * (1 - C2) * (1 - C2)
&                     * CP4 / 16
G6(5, I) = 693 * C * S * (1 - C2) * (1 - C2) * CP5 / 8
G6(6, I) = 231 * (1 - C2) * (1 - C2) * (1 - C2)
&                     * CP6 / 32
    NEXT I


REM  MD=NO. OF DIFFERENT DISTANCE LABELS
REM  IDM =THE DIFFERENT DISTANCE LABELS
REM  JDML(I,J)=JTH LIGAND OF DISTINCT DISTANCE LABEL I
REM  JDMC(I)=NO. OF LIGANDS FOR DISTINCT DISTANCE LABEL I
    FOR I = 1 TO NL
        FOR J = 1 TO NL
  JDML(I, J) = 0
        NEXT J
        JDMC(I) = 0
```

```
   NEXT I
   MD = 1
   IDM(1) = ID(1)
   JDML(1, 1) = 1
   JDMC(1) = 1
   FOR I = 1 TO NL
     FOR J = -6 TO 6
G6C(J, I) = 0!
     NEXT J
     FOR J = -4 TO 4
 G4C(J, I) = 0!
     NEXT J
     FOR J = -2 TO 2
 G2C(J, I) = 0!
     NEXT J
     IF I > 1 THEN
 IF ID(I) > IDM(MD) OR ID(I) < IDM(MD) THEN
    MD = MD + 1
    IDM(MD) = ID(I)
 END IF
 JDMC(MD) = JDMC(MD) + 1
 K = JDMC(MD)
 JDML(MD, K) = I
     END IF
     NEXT I


REM    COMBINING COORDINATION FACTORS FOR EQUIDISTANT LIGANDS
REM    (I.E. LIGANDS WITH THE SAME EQUIDISTANT LABELS)
     FOR I = 1 TO MD
       FOR J = 1 TO JDMC(I)
 M = JDML(I, J)
 FOR K = -2 TO 2
    G2C(K, I) = G2C(K, I) + G2(K, M)
 NEXT K
 FOR K = -4 TO 4
    G4C(K, I) = G4C(K, I) + G4(K, M)
 NEXT K
 FOR K = -6 TO 6
   G6C(K, I) = G6C(K, I) + G6(K, M)
  NEXT K
```

```
NEXT J
    NEXT I
REM ************************OUTPUT**************************
    INPUT "NAME OF OUTPUT FILE = ", FILE1$
    PRINT "     "
REM          FILE1$ = "CORFACS.DAT"
 OPEN FILE1$ FOR OUTPUT AS #8
    PRINT #8, "PROGRAM: CORFACS.BAS"
    PRINT #8, "COORDINATION FACTORS FOR STEVENS PARAMETERS"
    PRINT #8, "     "
    FOR I = 1 TO MD
    PRINT #8, "DISTANCE LABEL ="; IDM(I)
    PRINT #8, "RANK 2"
 FOR J = -2 TO 2
 IF ABS(G2C(J, I)) < EPLIN THEN
    G2C(J, I) = 0!
 END IF
 PRINT #8, USING "#####.##"; J; G2C(J, I)
 NEXT J
    PRINT #8, "RANK 4"
 FOR J = -4 TO 4
 IF ABS(G4C(J, I)) < EPLIN THEN
    G4C(J, I) = 0!
 END IF
 PRINT #8, USING "#####.##"; J; G4C(J, I)
NEXT J
    PRINT #8, "RANK 6"
 FOR J = -6 TO 6
 IF ABS(G6C(J, I)) < EPLIN THEN
    G6C(J, I) = 0!
 END IF
 PRINT #8, USING "#####.##"; J; G6C(J, I)
  NEXT J
     NEXT I
 CLOSE #5
 PRINT "COORDINATION FACTORS FOR STEVENS PARAMETERS"
 PRINT "ARE OUTPUT TO FILE:", FILE1$
 PRINT "PROGRAM RUN IS COMPLETED SUCCESSFULLY."
END1:
    END
```

### A2.3  Structure and naming of data files

Various standard formats are used in input data files.

### A2.3.1  *Data files used in superposition model calculations*

The coordination angle file, C_YVO4.DAT, which is used as input to the coordination factor programs CORFACS.BAS and CORFACW.BAS, has the following form:

```
1, 101.9, 0.
1, 101.9, 180.
1, 78.1, 90.
1, 78.1, 270.
2, 32.8, 0.
2, 32.8, 180.
2, 147.2, 90.
2, 147.2, 270.
```

The first column distinguishes the two possible ligand distances. Columns two and three give the $\theta$ and $\phi$ angles, respectively, for each of the eight oxygen ligands at Y sites in $YVO_4$.

Coordination factor files (e.g. W_YVO4.DAT), generated by the program CORFACW.BAS, are required as input into the superposition model programs CFTOINTR.BAS and INTRTOCF.BAS. In order to conserve space, the data for each of the two ligand distances in $YVO_4$ are given side by side, rather than in a single column without a break, as they are in the actual file.

```
PROGRAM: CORFACW.BAS
COORDINATION FACTORS FOR WYBOURNE PARAMETERS
```

| DISTANCE LABEL = 1 | | | DISTANCE LABEL = 2 | |
|---|---|---|---|---|
| RANK 2 | | | RANK 2 | |
| -2 | 0.000 | | -2 | 0.000 |
| -1 | 0.000 | | -1 | 0.000 |
| 0 | -1.745 | | 0 | 2.239 |
| 1 | 0.000 | | 1 | 0.000 |
| 2 | 0.000 | | 2 | 0.000 |
| RANK 4 | | | RANK 4 | |
| -4 | 0.000 | | -4 | 0.000 |
| -3 | 0.000 | | -3 | 0.000 |

| | | | | |
|---|---|---|---|---|
| -2 | 0.000 | | -2 | 0.000 |
| -1 | 0.000 | | -1 | 0.000 |
| 0 | 0.894 | | 0 | -0.362 |
| 1 | 0.000 | | 1 | 0.000 |
| 2 | 0.000 | | 2 | 0.000 |
| 3 | 0.000 | | 3 | 0.000 |
| 4 | 1.918 | | 4 | 0.180 |

| RANK 6 | | | RANK 6 | |
|---|---|---|---|---|
| -6 | 0.000 | | -6 | 0.000 |
| -5 | 0.000 | | -5 | 0.000 |
| -4 | 0.000 | | -4 | 0.000 |
| -3 | 0.000 | | -3 | 0.000 |
| -2 | 0.000 | | -2 | 0.000 |
| -1 | 0.000 | | -1 | 0.000 |
| 0 | -0.272 | | 0 | -1.647 |
| 1 | 0.000 | | 1 | 0.000 |
| 2 | 0.000 | | 2 | 0.000 |
| 3 | 0.000 | | 3 | 0.000 |
| 4 | -0.685 | | 4 | 0.818 |
| 5 | 0.000 | | 5 | 0.000 |
| 6 | 0.000 | | 6 | 0.000 |

(to top of next column)

## A2.3.2 *Crystal field parameter files*

Crystal field parameters for $Er^{3+}$:$YVO_4$ (Wybourne normalization, in $cm^{-1}$) are given in the file er_ yv1.wdt as follows:

```
RANK2   CF PARAMETERS
0,0,-206,0,0
RANK4   CF PARAMETERS
0,0,0,0,364,0,0,0,926
RANK 6 CF PARAMETERS
0,0,0,0,0,0,-688,0,0,0,32,0,0
```

This file structure can be used for any site symmetry. It is required for inputs into programs ENGYLVL.BAS and CFTOINTR.BAS.

### A2.3.3 Energy level files

Energy level files are used as input to program ENGYFIT.BAS, and are determined as output to ENGYLVL.BAS. They consist of simple lists of energies, in arbitrary order. There follows the file ER_ ENGY.DAT, generated by program ENGYLVL.BAS in Section 3.1.3, and used as input to ENGYFIT.BAS in Section 3.2.1.

```
   -17.35
  -153.58
   104.37
   102.43
   -35.87
   -35.87
   102.43
   104.37
  -153.58
   -17.35
```

### A2.3.4 Intrinsic parameter files

The output file from program CFTOINTR.BAS following the calculation described in Chapter 5, takes the form:

```
A2
FOR USING tk = 1
THE  1  ESTIMATES OF AK ARE:
 -564.0
FOR USING tk = 3
THE  1  ESTIMATES OF AK ARE:
-1589.5
FOR USING tk = 5
THE  1  ESTIMATES OF AK ARE:
 2582.8
FOR USING tk = 6
THE  1  ESTIMATES OF AK ARE:
 1173.8
FOR USING tk = 7
THE  1  ESTIMATES OF AK ARE:
  775.2
FOR USING tk = 8
```

```
THE  1  ESTIMATES OF AK ARE:
  587.3
A4
  468.6
  151.6
A6
  377.6
  355.4
```

All parameters are in $cm^{-1}$ (Wybourne normalization). Parameters $\overline{B}_4$ and $\overline{B}_6$ each have two values, corresponding to the nearest neighbour and next nearest neighbour ligands. The derived $\overline{B}_2$ parameters are all for the nearest neighbour ligand, with values corresponding to the assumed power law. Because of the multiple values of $\overline{B}_2$, this file is not in a suitable form to use as an input into INTRTOCF.BAS.

# Appendix 3

## Accessible program packages

Y. Y. YEUNG

*Hong Kong Institute of Education*

M. F. REID

*University of Canterbury*

D. J. NEWMAN

*University of Southampton*

This appendix provides brief descriptions of several special purpose programs and program packages that can be accessed by the reader.

### A3.1 A crystal field analysis computer package for $3d^N$ ions

This computer package was developed by Y. Y. Yeung (see [YR92] and [YR93]) to calculate the energy levels and state vectors for any transition metal ions with the $3d^N$ configuration ($N = 1$–9) doped at orthorhombic or higher symmetry sites. It can be used to predict, analyse and correlate some optical spectroscopic spectra with electron paramagnetic resonance data and magnetic susceptibility data. It involves full diagonalization within the $3d^N$ configuration of a complete crystal field Hamiltonian which consists of the electrostatic repulsion (with the Slater integrals or the Racah parameters) among those $3d$ electrons, the spin–orbit interaction and the Trees correction describing the two body orbit–orbit polarization interaction as well as the crystal field Hamiltonian ($B_q^k$ parameters). Explicit forms of the corresponding Hamiltonians and formulae for evaluating their matrix elements can be found in [YR92]. By incorporating the 'imaginary' crystal field terms (see [CRY94]), this package has been extended to deal with any $3d^N$ ion located at sites with symmetry given by any of the 32 crystallographic point groups.

In essence, the package is composed of four computer programs applicable for the orthorhombic or higher symmetry sites (with real crystal field parameters) and arbitrary low symmetry sites (with complex crystal field parameters) in either the $|d^N \alpha S M_S L M_L\rangle$ or $|d^N \alpha S L J M_J\rangle$ basis. All programs are written in Microsoft QBASIC and are executable in the DOS environment of any 286 (or above) personal computer system.

Two libraries of efficient subroutines have been specifically developed to handle matrix operations and calculations of the $3j$, $6j$ and $9j$ symbols. Reduced matrix elements for all tensor operators (except the Trees correction) have been calculated and their numerical values, in double precision, are stored in data files in order to save computing effort. Some example files of input parameters and output energy levels and state vectors etc. are also included in order to help users gain familiarity with the package.

This computer package is obtainable from Dr Y. Y. Yeung who can be contacted by email at dr.yeung@physics.org or by mail at G.P.O. Box 8594, Central, Hong Kong, China. Further information and a user manual can be found on the following internet website:

http://www.ied.edu.hk/has/phys/apepr/links/software.htm

## A3.2 Crystal field and intensity determinations from optical spectra

The program package used in the correlation crystal field calculations described in Chapter 6 (e.g. see [Rei87a, BJRR94, BR98c]) and the transition intensity calculations described in Chapter 10 (e.g. see [DRR84, MRR87a, BJRR94, BCR99]) was developed during the 1980s at the University of Virginia and The University of Hong Kong.

The details of the operation of the program suite, the programs, and some documentation are available from Dr M. F. Reid, who may be contacted via email at: M.Reid@phys.canterbury.ac.nz. The programs are available on the web site http://www.phys.canterbury.ac.nz/crystalfield, but support is limited. Only the major features are described here.

### A3.2.1 Matrix elements

Matrix elements and coefficients of fractional parentage for the $f^N$ configurations were originally supplied by Hannah Crosswhite. Matrix elements for $d^N$ systems are also available. Correlation crystal field matrix elements may be calculated, as well as the transformations between the $G^k_{iq}$ and $B^k_q(k_1 k_2)$ parameters described in Chapter 6.

Intermediate coupled reduced matrix elements may be generated. These are useful for multiplet–multiplet transition intensity calculations, and for assessing the relative effect of various energy level and transition intensity operators. The same intermediate coupling truncation scheme as in the Crosswhite programs (described in Chapter 4) may be used. In this approach the free-ion Hamiltonian is diagonalized and then matrix elements of both free-ion and crystal field operators are generated for a truncated basis.

### A3.2.2 *Crystal field calculations*

Crystal field and correlation crystal field calculations for any symmetry (with complex values allowed for the crystal field parameters) are possible, and parameters may be fitted. Zeeman calculations for arbitrary orientation choices are also possible.

### A3.2.3 *Transition intensities*

Transition intensity calculations using the $A_{tp}^{\lambda}$ and $B_{lq}^{\lambda}$ parameters are implemented. Circular dichroism and two-photon transition intensities may also be calculated. However, the two-photon calculations are currently restricted in scope.

### A3.2.4 *Parameters*

Superposition model and simple point charge and ligand polarization calculations for crystal field and transition intensity parameters for arbitrary symmetries are implemented. These calculations are often useful in exploring the symmetries involved in the calculations. For example, how the parameters change under certain inversions and rotations.

### A3.2.5 *Problems and difficulties*

There are some shortcomings of the current program package.

(i) The code is written in the Pascal language, which is not as widely available as FORTRAN or C. However, it is available on many Unix systems (including Linux), as well as VMS. The programs will run under DOS on PCs using the Turbo Pascal compiler, but only for small basis sets.

(ii) General transition intensities for low symmetry are not properly implemented, i.e. only $\sigma$, $\pi$, axial and isotropic polarizations are supported.

(iii) Only $JM_J$ basis calculations are possible.

### A3.2.6 Future plans

Current work is concentrated on producing a new package, based on the RACAH software developed by Butler and coworkers at the University of Canterbury [But81, RMSB96]. This software will allow point group basis calculations, and much more flexibility in choice of coupling schemes and configurations. It is hoped to make this software freely available in 2000.

## A3.3 Crystal field determinations from inelastic neutron scattering

Neutron scattering is widely used to determine energy spectra of materials, such as metals, which are not transparent, and cannot therefore be investigated by optical spectroscopy. This technique is capable of providing information about magnetic dipole transition intensities, magnetic interactions and electron–phonon coupling as well as the crystal field. Its use to investigate lanthanide ions in various environments has been reviewed by Fulde and Loewenhaupt [FL86].

Neutron scattering spectra are continuous functions of energy. Even in the absence of strong coupling between the open-shell electron and other types of excitation, they are dependent on temperature and include information about magnetic dipole transition intensities and line broadening due to both instrument resolution and coupling between the open-shell electrons and phonons. It is useful to include the observed magnetic dipole intensities in fitting crystal field parameters as they provide constraints on the form of the open-shell wavefunctions.

The Rutherford Appleton Laboratory has published an 'interactive crystal electric field parameter fitting computer package using neutron scattering data', called FOCUS†. This package was developed by Dr Peter Fabi in 1995. Some new commands have now been added to the original version described in the FOCUS manual.

In order to use the FOCUS package it is necessary to have access to the ISISE-Network on the ISIS Facility of the Rutherford Appleton Laboratory. It is presumed that readers who might need to use this package would have, or could easily obtain, such access. The aim of this section is not to describe how to use the package, which is covered by the FOCUS manual, but to identify its main features, and to relate them to the simple approach used in program ENGYFIT.BAS which is described in Chapter 3.

---

† The FOCUS manual is obtainable from Library and Information Services, Rutherford Appleton Laboratory, Chilton, Didcot, Oxfordshire OX11 0QX, UK.

As is the case in all fitting packages it is necessary to provide estimated starting values for the crystal field parameters. However, FOCUS can generate its own starting values using a constrained Monte-Carlo procedure. This provides a useful check on the possible existence of multiple solutions and false minima.

Because the relative magnetic dipole intensities can be calculated precisely from the wavefunctions, they provide useful additional information about the crystal field parameters. The amount of this additional information is dependent on whether a single crystal or polycrystalline sample has been used in the experiment. In principle, it should be possible to determine crystal field parameters in situations where there are fewer energy levels than parameters. However, the editors are not aware of any theoretical work that would make this statement more precise.

### A3.4 Mathematica program to calculate energy level diagrams for cubic symmetry sites

Dr K. S. Chan contributed the Mathematica program LLWDIAG, which determines the Lea, Leask and Wolf [LLW62] energy diagrams for the ground multiplets of lanthanide and actinide ions in cubic symmetry sites. This program is particularly relevant to the discussions in Chapter 9. The program listing is available, along with the QBASIC programs that accompany this book, at the website addresses given in Appendix 2.

# Appendix 4

## Computer package CST

CZ. RUDOWICZ

*City University of Hong Kong*

The intrinsic physical properties of the crystal field Hamiltonian have led over the years to various formats and conventions being used in the literature to express crystal field parameters. Hence comparison of crystal field parameter sets from various sources often requires several manipulations and conversions. To facilitate the computations involved in this process, the user-friendly computer package 'CST' (Conversion, Standardization and Transformation) has been developed [RAM98, RAM97]. The package is useful for various general manipulations of the format of the experimental zero-field splitting parameters as well as the crystal field ones. Its capabilities include CONVERSIONS: *unit conversions* – between several units most often used for crystal field and zero-field splitting parameters; *normalization (or notation) conversions* – between several major normalizations for crystal field and zero-field splitting parameters; STANDARDIZATION of orthorhombic, monoclinic and triclinic crystal field and zero-field splitting parameters; and TRANSFORMATIONS of crystal field and zero-field splitting parameters into an arbitrary axis system, including the rotation invariants.

In this appendix the structure and capabilities of the CST package are briefly presented. In keeping with the overall approach of a book designed as a 'Do-It-Yourself' text, the emphasis is put on the practical implementations of the package, illustrated using examples of experimental data taken from recent crystal field literature. A detailed description of the CST modules is provided in the Manual [RAM97], whereas examples of applications to zero-field splitting parameters are dealt with in [RAM97, Rud97, RM99]. The reader can obtain the CST package, i.e. the program and the Manual, from the author upon request.

## A4.1  Properties of the crystal field and zero-field splitting Hamiltonians

### A4.1.1  Parameter normalizations

Two major types of operators have been used to express the parametric crystal field Hamiltonian (see Chapter 2): (i) the spherical tensor operators [Rud87a], of which the Wybourne operators $C_q^{(k)}$ [Wyb65a, Hüf78, RAM97] are the most commonly used; and (ii) the tesseral tensor operators [Rud87a], of which the Stevens operators [Rud85b, NU75] were first introduced and are still widely used. In the CST package we adopt as the *reference* normalization the *extended* Stevens operators $O_k^q$ [Rud85b], for which $q$ can be negative as well as positive. Explicit listings of the extended Stevens operators and references are provided in the CST Manual[RAM97]. Further details of the different normalizations for crystal field and zero-field splitting parameters that appear in the literature will also be found in [RAM97]. For meaningful comparison of crystal field data for different systems it is sometimes useful to consider the crystal field invariants, or 'strength' parameters, that are discussed in Chapter 8 (also see [Rud86]).

### A4.1.2  Parameter standardization

The coordinate system which is implicit in a set of crystal field parameters can be standardized using a process [RB85, Rud85a] which can be applied to either the crystal field parameters or the zero-field splitting parameters. The following discussion is limited to zero-field splitting parameters $b_k^q$ (Stevens normalization, as defined in Section 7.1) and crystal field parameters (Wybourne normalization) in orthorhombic symmetry. Standardization of parameters for monoclinic symmetry is discussed in the Manual [RAM97].

The standardization idea is, for systems of orthorhombic or lower symmetry, to limit the zero-field splitting parameter ratio

$$\lambda' = b_2^2 / b_2^0, \tag{A4.1}$$

to the 'standard' range $(0, 1)$ by an appropriate choice of axis system. This idea has been extended to the fourth and sixth order zero-field splitting (or crystal field) terms for orthorhombic [RB85, Rud91] and monoclinic symmetry [Rud86].

The standardization transformations $S_i$ $(i = 1–6)$ have been defined as follows [RB85, Rud85a]: $S_1$[original]$(x, y, z)$, $S_2(x, z, -y)$, $S_3(y, x, -z)$, $S_4(z, x, y)$, $S_5(y, z, x)$, $S_6(z, y, -x)$. If the original value of $\lambda'$ is in the range $(-\infty, -3)$ use $S_6$; $(-3, -1)$ use $S_4$; $(-1, 0)$ use $S_3$; $(1, 3)$ use $S_2$; $(3, +\infty)$

use $S_5$. The transformed $\lambda'$ is limited to $(0, 1)$ by using the transformation $S_i$ as indicated.

As described in Chapters 3 and 4, the crystal field parameters $B_q^k$ are determined by fitting calculated levels to the observed spectra. In general, the ratios $B_2^2/B_0^2 = \kappa$ can be anywhere between $-\infty$ and $+\infty$ [RB85, Rud91, ML82]. The ratio $\kappa$ may fall in any of the six different regions: $(-\infty, -3/\sqrt{6})$, $(-3/\sqrt{6}, -1/\sqrt{6})$, $(-1/\sqrt{6}, 0)$, $(0, 1/\sqrt{6})$, $(1/\sqrt{6}, 3/\sqrt{6})$ and $(3/\sqrt{6}, +\infty)$. Hence for any set of crystal field parameters there exist five other sets related by simple rotations [RB85, Rud86] and yielding an identical energy level structure. On transforming to a proper axis system any "non-standard" ratio $B_2^2/B_0^2$ can be brought within the "standard" range $(0, 1/\sqrt{6})$ for the Wybourne parameters [RB85, Rud86].

The intrinsic properties of $\kappa$ and $\lambda'$ have been used to unify different orthorhombic and lower symmetry crystal field parameter sets, which facilitates inter-comparison of experimental results [RB85, Rud86, Rud91]. Alternative choices of standarized crystal field parameter sets have also been used in the literature (e.g. see [ML82] p. 632).

The major physical implication of the standardization concerns site structure. As discussed in [RM99] the ratio of the zero-field splitting parameters: $E/D \equiv \lambda$ defines a 'rhombicity' parameter, which measures the deviation from axial symmetry, and its value may be restricted to the range $0 < \lambda < 1/3$ [RB85]. A number of authors have noted that $\lambda = 0$ corresponds to axial symmetry, whereas the maximum possible rhombicity is characterized by $\lambda = 1/3$. However, since we deal with the *effective* Hamiltonian $H_{ZFS}$ [Rud87a], it may appear that the ratio $\lambda$ ($\lambda'$ or $\kappa$) describes the rhombicity in the *effective* and not crystallographic sense. Using the superposition model (Chapter 5), which provides direct relationships between the zero-field splitting (or crystal field) parameters and the structural ones, we have shown [RM99] that the 'maximum rhombicity' limit: $E/D = 1/3$ or $A_{22}\langle r^2 \rangle / A_{20}\langle r^2 \rangle = b_2^2/b_2^0 = 1$, is valid not only in the *effective* spin-Hamiltonian sense but also in the crystallographic sense. The same conclusion applies for the crystal field parameter sets.

## A4.2 Structure and capabilities of the package

The CST package is written in standard FORTRAN 77 and can run either on a mainframe computer or PC. The input form: KEYBOARD or FILE; and the output form: SCREEN, FILE or BOTH, are provided. The user selects options and/or suboptions from a MENU and can control the flow of the

Table A4.1. Main menu options.

| | |
|---|---|
| 1. | FREE UNIT CONVERSION |
| 2. | UNIT CONVERSION |
| 3. | NOTATION CONVERSION |
| 4. | STANDARDIZATION |
| 5. | TRANSFORMATION |
| Q. | EXIT |

Table A4.2. Symmetry menu options.

| | | |
|---|---|---|
| 1. | CUBIC I | $(O, T_d, O_h)$ |
| 2. | CUBIC II | $(T, T_h)$ |
| 3. | HEXAGONAL I | $(D_6, C_{6v}, D_{3h}, D_{6h})$ |
| 4. | HEXAGONAL II | $(C_6, C_{3h}, C_{6h})$ |
| 5. | TRIGONAL I | $(D_3, C_{3v}, D_{3d})$ |
| 6. | TRIGONAL II | $(C_3, C_{3i} \equiv S_6 )$ |
| 7. | TETRAGONAL I | $(D_4, C_{4v}, D_{2d}, D_{4h})$ |
| 8. | TETRAGONAL II | $(C_4, S_4, C_{4h})$ |
| 9. | ORTHORHOMBIC | $(D_2, C_{2v}, D_{2h})$ |
| 10. | MONOCLINIC | $(C_2, C_s \equiv C_{1h}, C_{2h})$ |
| 11. | TRICLINIC | $(C_1, C_i \equiv S_2 )$ |
| M. | EXIT TO MAIN MENU | |

calculations according to the prompts. The general structure of all modules and the list of subroutine files are provided in appendix 2 of [RAM97].

The opening screen MAIN MENU (Table A4.1) presents the major options. Two lower level menus: the SYMMETRY MENU (Table A4.2) followed by the RANK MENU (Table A4.3) will be activated if one of the options 2–5 in Table A4.1 is selected. The package can handle crystal field parameters (zero-field splitting parameters) of arbitrary symmetry since it includes all possible point symmetry groups (Table A4.2). A separate *GTRANS* module for transformations of the Zeeman **g**-matrix is included.

In order to avoid input of zero values for non-existent crystal field parameters (zero-field splitting parameters) of an irrelevant higher rank $k = 4$ or 6, as the case may be, three options are provided (Table A4.3). The package can process several consecutive data sets.

Table A4.3. Rank menu options.

| | |
|---|---|
| 1. | ONLY 2ND ORDER TERMS |
| 2. | 2ND AND 4TH ORDER TERMS |
| 3. | 2ND, 4TH AND 6TH ORDER TERMS |
| M. | EXIT TO MAIN MENU |

### *A4.2.1 Unit conversion*

Two options are provided in the MAIN MENU (Table A4.1) for conversion between various units: (1) FREE UNIT CONVERSION, which is not related to any specific symmetry; and (2) UNIT CONVERSION, which uses a specified symmetry (Table A4.2) and normalization (see Section A4.1.1) for zero-field splitting or crystal field parameters. The following factors [Rud87a] are used to convert to the standard unit $[10^{-4}$ cm$^{-1}]$ a value $U$ in a given unit: $8065.54077 \times 10^4 \times U$ [eV], $0.5035 \times 10^{20} \times U$ [erg], g $\times$ $0.466856 \times U$ [Gauss] (where the $g$-factor must be provided), $6950.38605 \times U[K]$, $0.33356 \times U$ [MHz]. Options (1) and (2) convert an input original unit A, selected from the above six options, to the output format: (i) 'A to B' – conversion from units A to any other units B, whereas for option (2) a second output format is available: (ii) 'A to all' – automatic conversion to all other five unit systems.

### *A4.2.2 Normalization conversion*

The NOTATION MENU (Table A4.4) controls conversions between the major normalizations used for zero-field splitting parameters and crystal field parameters. Abbreviations employed in this MENU of the CST program are fully defined in appendix 1 of the Manual [RAM97]. A brief summary is given below.

(i)  ES: extended Stevens, $Bkq \equiv A_{kq}\langle r^k \rangle \theta_k$, $bkq \equiv b_k^q$.
(ii)  NS: normalized Stevens.
(iii)  NCST: normalized combinations of spherical tensor operator.
(iv)  BST: Buckmaster, Smith and Thornley operator.
(v)  Ph.M.BST: phase modified BST.
(vi)  KS BCS: Koster–Statz/Buckmaster–Chatterjee–Shing, or Wybourne normalization.

Table A4.4. Notation menu.

| | |
|----|----------------|
| 1. | ES (bkq) |
| 2. | ES (Bkq) |
| 3. | NS |
| 4. | NCST |
| 5. | BST |
| 6. | Ph.M.BST |
| 7. | KS BCS |
| 8. | CONVENTIONAL |
| M. | EXIT TO MAIN MENU |

(vii) CONVENTIONAL: zero–field splitting parameters when operators
are expressed in terms of spin operators.

The $b_k^q$ parameters are related to the extended Stevens parameters $B_k^q$ in Section 7.1.

Selection of one of the options activates the prompts regarding the symmetry (Table A4.2) and rank (Table A4.3). The input of the original crystal field parameters or zero-field splitting parameters must then be done according to the chosen symmetry. Two output formats are provided: (i) 'A to B' – conversion from A to any other normalization B selected from the eight options, and (ii) 'A to all' – automatic conversion to all other tensorial normalizations (options 1–7). Option (8) is applicable only for some symmetry cases and requires special input format [RAM97].

To facilitate calculations within the standardization and transformation modules of the CST package, two separate options are provided for input of the crystal field parameters: (i) extended Stevens; and (ii) Wybourne normalization. The equivalence between the Wybourne and the Buckmaster, Smith and Thornley normalizations has been used for internal conversions within the program, where all expressions within the standardization and transformation modules are given in the extended Stevens normalization. However, sometimes a direct conversion between the crystal field parameters in the extended Stevens normalization and those in the Wybourne normalization may be required. This can be achieved by selecting the extended Stevens normalization for $A_{kq}\langle r^k \rangle \theta_k$ and the equivalent Buckmaster, Smith and Thornley operator normalization for the compact (i.e. complex) parameters $\hat{B}_q^k$, defined in Section 2.2, as input or output. Further details of the input procedures are given in the CST Manual [RAM97].

Table A4.5. Sample input file, see [RJR94].

```
Rukmini et al. J. Phys CM 6, 5919 (1994)
114., -172.,
1192., -125., 6.,
1487., 235., -358., 870.
```

Table A4.6. Sample output file for the notation (i.e. normalization) conversion option using input in Table A4.5.

```
                       NOTATION CONVERSION

            [ORIGINAL]        (BST)

  B20        .1140D+03
  B22       -.1720D+03     B22M        .0000D+00
  B40        .1192D+04
  B42       -.1250D+03     B42M        .0000D+00
  B44        .6000D+01     B44M        .0000D+00
  B60        .1487D+04
  B62        .2350D+03     B62M        .0000D+00
  B64       -.3580D+03     B64M        .0000D+00
  B66        .8700D+03     B66M        .0000D+00

            [FINAL]          ( ES(Bkq) )

    B20 = .57000D+02   B22 =-.21066D+03
    B40 = .14900D+03   B42 =-.98821D+02   B44 = .62750D+01
    B60 = .92938D+02   B62 = .15050D+03   B64 =-.25116D+03
    B66 = .82643D+03
```

An example of this application is provided by inputting the parameter values given in Table A4.5 (Wybourne normalization) for $Nd^{3+}$ in $NdF_3$ (see table 2 of [RJR94]). The program converts these to the extended Stevens $(A_{kq}\langle r^k \rangle \theta_k)$ parameters shown in Table A4.6. All parameter values are in $cm^{-1}$.

Note that if, after the normalization conversion, the user requests the program to continue the calculations, selects the STANDARIZATION MENU (see Section A4.2.3) and specifies the original parameters in the Wybourne normalization, the program uses the FINAL extended Stevens $(B_q^k)$ values in Table A4.6. This yields the standardized crystal field parameters in the latter normalization and not in the Wybourne normalization. In order to avoid misinterpretation of the results, which may lead to incorrect values, it is better to perform the normalization conversion calculations and the standardization ones in separate runs, although the same input file as in Table A4.5 can be used.

### A4.2.3 Standardization

Selection of the main option (4) (see Table A4.1) activates the STANDARD-IZATION MENU, which provides two choices: zero-field splitting STAN-DARDIZATION and crystal field STANDARDIZATION. For each choice the SYMMETRY and RANK MENU options must then be specified. Note that the standardization procedure deals only with the orthorhombic, monoclinic and triclinic symmetry. An input normalization (Table A4.4) must be specified since the standardization expressions [RB85, Rud86] apply directly only to the crystal field parameters (zero-field splitting parameters) in the extended Stevens ($A_{kq}\langle r^k \rangle \theta_k$ and $b_k^q$) normalization. For normalizations other than the extended Stevens one the conversion to the extended Stevens normalization and reconversion is carried out automatically. The crystal field (zero-field splitting) STANDARDIZATION MENU provides three options:

(i) AUTOMATIC STANDARDIZATION – depending on the value of $\lambda'$.

(ii) STANDARDIZATION TRANSFORMATION – using a particular transformation $S_i$.

(iii) STANDARDIZATION ERRORS – for details, see [RAM97].

For illustration the orthorhombic crystal field standardization is applied to the crystal field parameter data set for $Nd^{3+}$ in $NdF_3$ [RJR94] in Table A4.5. The crystal field standardization, Wybourne normalization, orthorhombic type standardization, and automatic standardization options are selected. The transformation $S_6$ is automatically chosen to standardize crystal field parameters. The original ($S_1$) and standardized ($S_6$) crystal field parameters (in $cm^{-1}$) for $NdF_3$ are given in Table A4.7 (output file only).

### A4.2.4 Transformation of the crystal field (zero-field splitting) parameters

The transformation relations [Rud85b, Rud85a] apply directly to the crystal field parameters $A_{kq}\langle r^k \rangle \theta_k$ in extended Stevens normalization or the

Table A4.7. Sample output file for the standarization option using input
in Table A4.5.

---

ORTHORHOMBIC TYPE CRYSTAL FIELD STANDARDIZATION

ORIGINAL PARAMETERS (WYBOURNE NOTATION) ARE :

| B20 |      | .1140D+03  |     |           |
|-----|------|------------|-----|-----------|
| B22 | Re:  | -.1720D+03 | Im: | .0000D+00 |
| B40 |      | .1192D+04  |     |           |
| B42 | Re:  | -.1250D+03 | Im: | .0000D+00 |
| B44 | Re:  | .6000D+01  | Im: | .0000D+00 |
| B60 |      | .1487D+04  |     |           |
| B62 | Re:  | .2350D+03  | Im: | .0000D+00 |
| B64 | Re:  | -.3580D+03 | Im: | .0000D+00 |
| B66 | Re:  | .8700D+03  | Im: | .0000D+00 |

FINAL (S6) PARAMETERS (WYBOURNE NOTATION) ARE :

| B20 |      | -.2677D+03 |     |           |
|-----|------|------------|-----|-----------|
| B22 | Re:  | -.1619D+02 | Im: | .0000D+00 |
| B40 |      | .5521D+03  |     |           |
| B42 | Re:  | -.5297D+03 | Im: | .0000D+00 |
| B44 | Re:  | .5414D+03  | Im: | .0000D+00 |
| B60 |      | .7634D+03  |     |           |
| B62 | Re:  | .8949D+03  | Im: | .0000D+00 |
| B64 | Re:  | -.5514D+03 | Im: | .0000D+00 |
| B66 | Re:  | .8060D+03  | Im: | .0000D+00 |

---

zero-field splitting parameters $b_k^q$. To facilitate applications for crystal field
parameters, the TRANSFORMATION module provides directly two op-
tions: the extended Stevens normalization and the Wybourne normalization [Wyb65a, Rud86, RAM97] for the input parameter normalization. The
SYMMETRY and RANK MENU options must be specified, and 'PHI' ($\phi$-
rotation about the original $z$-axis) and 'THETA' ($\theta$-rotation about the new
$y$-axis) must be provided. Input of $\phi$ and $\theta$ may be in degrees or radi-
ans. The rotational invariants $s_k$ (discussed in Chapter 8) are calculated
automatically within the transformation module.

## A4.3  Summary and conclusions

A summary of the structure and capabilities of the computer package CST
has been presented. This package enables various general manipulations of
the zero-field splitting parameters and crystal field parameters to be carried

out. The Manual [RAM97] provides several other examples of applications of the CST package to zero-field splitting and crystal field parameters.

The menu-driven structure and the operating instructions displayed on screen during execution of each subprogram make the package easy and convenient to use. The CST package facilitates interpretation and comparison of experimental data derived from various sources. The effort to convert the zero-field splitting and crystal field parameters expressed in different normalizations and formats can be substantially reduced by efficient usage of the package. Application of the package may also help in clarifying confusion arising from the inadvertent usage by some authors of identical symbols having sometimes different meanings (see, e.g. [RAM97, Rud87a, RM99]). The present package provides an efficient way for bringing various data available in the literature for a given magnetic ion and crystal system into a unified form.

It is worth mentioning a recent (unpublished) application of the CST package for standardization of crystal field parameters for several rare earth compounds as well as to facilitate the computations required for the multiple correlated fitting technique [Rud86]. This technique is based on independent fitting of crystal field parameter sets lying in distinct regions of the multi-parameter space and being correlated with each other via the standardization transformations $S_i$ [RB85, Rud86]. Having several independently fitted yet correlated crystal field parameter sets may increase the reliability of the final crystal field parameter set.

The author may be contacted by e-mail at `apceslaw@cityu.edu.hk`

## Acknowledgements

Help with the development of the program CST by Drs I. Akhmadulline and S. B. Madhu is gratefully acknowledged. This work was partly supported by the UGC and the City University of Hong Kong Strategic Research Grant.

# Bibliography

[AB70]  A. Abragam and B. Bleaney. *Electron paramagnetic resonance of transition ions*. Clarendon Press, Oxford, 1970.

[AD62]  J. D. Axe and G. H. Dieke. Calculation of crystal-field splittings of $Sm^{3+}$ and $Dy^{3+}$ levels in $LaCl_3$ with inclusion of $J$ mixing. *J. Chem. Phys.*, **37**: 2364–2371, 1962.

[AFB$^+$89]  P. Allenspach, A. Furrer, P. Brüesch, R. Marsolais, and P. Unternährer. A neutron spectroscopic comparison of the crystalline electric field in tetragonal $HoBa_2Cu_3O_{6.2}$ and orthorhombic $HoBa_2Cu_3O_{6.8}$. *Physica C*, **157**:58–64, 1989.

[AH94]  S. L. Altmann and P. Herzig. *Point-group theory tables*. Clarendon Press, Oxford, 1994.

[Alt77]  S. L. Altmann. *Induced representations in crystals and molecules*. Academic Press, London, 1977.

[AM83]  F. Auzel and O. L. Malta. A scalar crystal field strength parameter for rare-earth ions: meaning and usefulness. *J. Phys. C: Solid State Phys.*, **44**:201–206, 1983.

[AN78]  S. Ahmad and D. J. Newman. Finite ligand size and Sternheimer antiscreening in lanthanide ions. *Austral. J. Phys.*, **31**:421–426, 1978.

[AN80]  S. Ahmad and D. J. Newman. Finite ligand size and the Sternheimer anti-shielding factor $\gamma_\infty$. *Austral. J. Phys.*, **33**:303–306, 1980.

[Auz84]  F. Auzel. A scalar crystal field strength parameter for rare-earth ions. In B. Di Bartolo, editor, *Energy transfer processes in condensed matter*, pages 511–520. Plenum Press, London, 1984.

[Axe63]  J. D. Axe. Radiative transition probabilities within $4f^n$ configurations: the fluorescence spectrum of europium ethylsulphate. *J. Chem. Phys.*, **39**:1154–1160, 1963.

[Axe64]  J. D. Axe. Two-photon processes in complex atoms. *Phys. Rev.*, **136A**:42–45, 1964.

[Bal62]  C. J. Ballhausen. *Introduction to ligand field theory*. McGraw-Hill, New York, 1962.

[BCR99]  G. W. Burdick, S. M. Crooks, and M. F. Reid. Ambiguities in the parametrization of $4f^N - 4f^N$ electric-dipole transition intensities. *Phys. Rev. B*, **59**:7789–7793, 1999.

[BDS89]  G. W. Burdick, M. C. Downer, and D. K. Sardar. A new contribution to spin-forbidden rare earth optical transition: intensities: analysis of all trivalent lanthanides. *J. Chem. Phys.*, **91**:1511–1520, 1989.

[BEW$^+$85]  P. C. Becker, N. Edelstein, G. M. Williams, J. J. Bucher, R. E. Russo, J. A. Koningstein, L. A. Boatner, and M. M. Abraham. Intensities and asymmetries of electronic Raman scattering in ErPO$_4$ and TmPO$_4$. *Phys. Rev. B*, **31**:8102–8110, 1985.

[BHS78]  R. Biederbick, A. Hofstaetter, and A. Scharmann. Temperature-dependent Mn$^{2+}$-$^6S$ state splitting in scheelites. *Phys. Stat. Sol. (b)*, **89**:449–458, 1978.

[BHSB80]  R. Biederbick, A. Hofstaetter, A. Scharmann, and G. Born. Zero-field splitting of Mn$^{3+}$ in scheelites. *Phys. Rev. B*, **21**:3833–3838, 1980.

[BJRR94]  G. W. Burdick, C. K. Jayasankar, F. S. Richardson, and M. F. Reid. Energy–level and line–strength analysis of optical transitions between stark levels in Nd$^{3+}$:Y$_3$Al$_5$O$_{12}$. *Phys. Rev. B*, **50**:16 309–16 325, 1994.

[BKR93]  G. W. Burdick, H. J. Kooy, and M. F. Reid. Correlation contributions to two-photon lanthanide absorption intensities: direct calculations for Eu$^{2+}$ ions. *J. Phys.: Condens. Matter*, **5**:L323–L328, 1993.

[BR93]  G. W. Burdick and M. F. Reid. Many-body perturbation theory calculations of two-photon absorption in lanthanide compounds. *Phys. Rev. Lett.*, **70**:2491–2494, 1993.

[BR98a]  A. R. Bryson and M. F. Reid. Transition amplitude calculations for one- and two-photon absorption. *J. Alloys Compd.*, **275–277**:284–287, 1998.

[BR98b]  G. W. Burdick and F. S. Richardson. Application of the correlation-crystal-field delta-function model in analyses of Pr$^{3+}$ ($4f^2$) energy-level structures in crystalline hosts. *Chem. Phys.*, **228**:81–101, 1998.

[BR98c]  G. W. Burdick and F. S. Richardson. Correlation-crystal-field delta-function analysis of $4f^2$ (Pr$^{3+}$) energy-level structure. *J. Alloys Compd.*, **275–277**:379–383, 1998.

[Bra67]  B. H. Brandow. Linked-cluster expansions for the nuclear many-body problem. *Rev. Mod. Phys.*, **39**:771–828, 1967.

[BRRK95]  G. W. Burdick, F. S. Richardson, M. F. Reid, and H. J. Kooy. Direct calculation of lanthanide optical transition intensities: Nd$^{3+}$:YAG. *J. Alloys Compd.*, **225**:115–119, 1995.

[BS68]  D. M. Brink and G. R. Satchler. *Angular momentum*. Clarendon Press, Oxford, 1968.

[BSR88]  M. T. Berry, C. Schweiters, and F. S. Richardson. Optical absorption spectra, crystal-field analysis, and electric dipole intensity parameters for europium in Na$_3$[Eu(ODA)$_3$].2NaClO$_4$.6H$_2$O. *Chem. Phys.*, **122**:105–124, 1988.

[But81]  P. H. Butler. *Point group symmetry applications*. Plenum Press, London, 1981.

[Car92]  W. T. Carnall. A systematic analysis of the spectra of trivalent actinide chlorides in D$_{3h}$ site symmetry. *J. Chem. Phys.*, **96**:8713–8726, 1992.

[CBC$^+$83]  W. T. Carnall, J. V. Beitz, H. Crosswhite, K. Rajnak, and J. B. Mann. Spectroscopic properties of the f-elements in compounds and solutions. In S. P. Sinha, editor, *Systematics and the properties of the lanthanides*, pages 389–450. D. Reidel, Amsterdam, 1983.

[CC83]  W. T. Carnall and H. Crosswhite. Further interpretation of the spectra of Pr$^{3+}$:LaF$_3$ and Tm$^{3+}$:LaF$_3$. *J. Less Common Met.*, **93**:127–135, 1983.

[CC84] H. Crosswhite and H. M. Crosswhite. Parametric model for $f$-shell configurations. I. The effective-operator hamiltonian. *J. Opt. Soc. Am. B*, 1:246–254, 1984.

[CCC77] W. T. Carnall, H. Crosswhite, and H. M. Crosswhite. Energy level structure and transition probabilities in the spectra of the trivalent lanthanides in $LaF_3$. Technical report, Argonne National Laboratory, 1977.

[CCER77] H. M. Crosswhite, H. Crosswhite, N. Edelstein, and K. Rajnak. Parametric energy level analysis of $Ho^{3+}$: $LaCl_3$. *J. Chem. Phys.*, 67:3002–3010, 1977.

[CCJ68] H. Crosswhite, H. M. Crosswhite, and B. R. Judd. Magnetic parameters for the configuration $f^3$. *Phys. Rev.*, 174:89–94, 1968.

[CFFW76] A. K. Cheetham, B. E. F. Fender, H. Fuess, and A. F. Wright. A powder neutron diffraction study of lanthanum and cerium trifluorides. *Acta Cryst. B*, 32:94–97, 1976.

[CFR68] W. T. Carnall, P. R. Fields, and K. Rajnak. Electronic energy levels in the trivalent lanthanide aquo ions. I. $Pr^{3+}$, $Nd^{3+}$, $Pm^{3+}$, $Sm^{3+}$, $Dy^{3+}$, $Ho^{3+}$, $Er^{3+}$, and $Tm^{3+}$. *J. Chem. Phys.*, 49:4424–4442, 1968.

[CGRR88] W. T. Carnall, G. L. Goodman, K. Rajnak, and R. S. Rana. A systematic analysis of the spectra of the lanthanides doped into single crystal $LaF_3$. Technical Report ANL-88-8, Argonne National Laboratory, 1988.

[CGRR89] W. T. Carnall, G. L. Goodman, K. Rajnak, and R. S. Rana. A systematic analysis of the spectra of the lanthanides doped into single crystal $LaF_3$. *J. Chem. Phys.*, 90:3443–3457, 1989.

[CLWR91] W. T. Carnall, G. K. Liu, C. W. Williams, and M. F. Reid. Analysis of the crystal-field spectra of the actinide tetrafluorides. I. $UF_4$, $NpF_4$ and $PuF_4$. *J. Chem. Phys.*, 95:7194–7203, 1991.

[CM67] H. M. Crosswhite and H. W. Moos. *Optical properties of ions in crystals.* Interscience, New York, 1967.

[CN70] M. M. Curtis and D. J. Newman. Crystal field in rare-earth trichlorides. V. Estimation of ligand-ligand overlap effects. *J. Chem. Phys.*, 52:1340–1344, 1970.

[CN81] S. C. Chen and D. J. Newman. Dynamic crystal field contributions of next-nearest neighbour ions in octahedral symmetry. *Physica B*, 107:365–366, 1981.

[CN82] S. C. Chen and D. J. Newman. The orbit–lattice interaction in lanthanide ions III: superposition model analysis for arbitrary symmetry. *Austral. J. Phys.*, 35:133–145, 1982.

[CN83] S. C. Chen and D. J. Newman. Superposition model of the orbit–lattice interaction I: analysis of strain results for $Dy^{3+}$:$CaF_2$. *J. Phys. C: Solid State Phys.*, 16:5031–5038, 1983.

[CN84a] S. C. Chen and D. J. Newman. Orbit–lattice coupling parameters for the lanthanides in $CaF_2$. *Phys. Lett.*, 102A:251–252, 1984.

[CN84b] S. C. Chen and D. J. Newman. Superposition model of the orbit–lattice interaction II: analysis of strain results for $Er^{3+}$:$MgO$. *J. Phys. C: Solid State Phys.*, 17:3045–3048, 1984.

[CN84c] H. Crosswhite and D. J. Newman. Spin-correlated crystal field parameters for lanthanide ions substituted into $LaCl_3$. *J. Chem. Phys.*, 81:4959–4962, 1984.

[CNT71] G. M. Copland, D. J. Newman, and C. D. Taylor. Configuration interaction in rare earth ions: II. Magnetic interactions. *J. Phys. B: At. Mol. Phys.*, **4**:1388–1392, 1971.

[CNT73] R. Chatterjee, D. J. Newman, and C. D. Taylor. The relativistic crystal field. *J. Phys. C: Solid State Phys.*, **6**:706–714, 1973.

[CO80] E. U. Condon and H. Odabaşi. *Atomic structure.* Cambridge University Press, 1980.

[CR89] D. K. T. Chan and M. F. Reid. Intensity parameters for $Eu^{3+}$ luminescence – tests of the superposition model. *J. Less Common Metals*, **148**:207–212, 1989.

[Cro77] H. M. Crosswhite. Systematic atomic and crystal-field parameters for lanthanides in $LaCl_3$ and $LaF_3$. *Colloques Internationaux du Centre National de la Recherche Scientifique*, **255**:65–69, 1977.

[CRTZ97] S. M. Crooks, M. F. Reid, P. A. Tanner, and Y. Y. Zhao. Vibronic intensity parameters for $Er^{3+}$ in $Cs_2NaErCl_6$. *J. Alloys Compd.*, **250**:297–301, 1997.

[CRY94] Y. M. Chang, Cz. Rudowicz, and Y. Y. Yeung. Crystal field analysis of the $3d^N$ ions at low symmetry sites including the 'imaginary' terms. *Computers Phys.*, **8**:583–588, 1994.

[CTDRG89] C. Cohen-Tannoudji, J. Duport-Roc, and G. Grynberg. *Photons and atoms.* Wiley, New York, 1989.

[CYYR93] Y. M. Chang, T. H. Yeom, Y. Y. Yeung, and C. Rudowicz. Superposition model and crystal-field analysis of the $^4A_2$ and $^2_aE$ states of $Cr^{3+}$ ions in $LiNbO_3$. *J. Phys.: Condens. Matter*, **5**:6221–6230, 1993.

[DBM98] R. G. Denning, A. J. Berry, and C. S. McCaw. Ligand dependence of the correlation crystal field. *Phys. Rev. B*, **57**:R2021–R2024, 1998.

[DD63] L. G. DeShazer and G. H. Dieke. Spectra and energy levels of $Eu^{3+}$ in $LaCl_3$. *J. Chem. Phys.*, **38**:2190–2199, 1963.

[DDNB81] M. Dagenais, M. Downer, R. Neumann, and N. Bloembergen. Two-photon absorption as a new test of the Judd–Ofelt theory. *Phys. Rev. Lett.*, **46**:561–565, 1981.

[DEC93] H. Donnerberg, M. Exner, and C. R. A. Catlow. Local geometry of $Fe^{3+}$ ions on the potassium sites in $KTaO_3$. *Phys. Rev. B*, **47**:14–19, 1993.

[DHJL85] H. Dothe, J. E. Hansen, B. R. Judd, and G. M. S. Lister. Orthogonal scalar operators for $p^n d$ and $pd^n$. *J. Phys. B: At. Mol. Phys.*, **18**:1061–1080, 1985.

[Die68] G. H. Dieke. *Spectra and energy levels of rare earth ions in crystals.* Interscience, New York, 1968.

[Div91] M. Diviš. Crystal fields in some rare earth intermetallics analysed in terms of the superposition model. *Phys. Stat. Sol. (b)*, **164**:227–234, 1991.

[Don94] H. Donnerberg. Geometrical microstructure of $Fe_{Nb}^{3+} - V_0$ defects in $KNbO_3$. *Phys. Rev. B*, **50**:9053–9062, 1994.

[Dow89] M. C. Downer. The puzzle of two-photon rare earth spectra in solids. In W. M. Yen, editor, *Laser spectroscopy of solids II*, pages 29–75. Springer, Berlin, 1989.

[DRR84] J. J. Dallara, M. F. Reid, and F. S. Richardson. Anisotropic ligand polarizability contributions to intensity parameters for the trigonal $Eu(ODA)_3^{3-}$ and $Eu(DBM)_3H_2O$ systems. *J. Phys. Chem.*, **88**:3587–3594, 1984.

[EBA+79] L. Esterowitz, F. J. Bartoli, R. E. Allen, D. E. Wortman, C. A.

Morrison, and R. P. Leavitt. Energy levels and line intensities of $Pr^{3+}$ in $LiYF_4$. *Phys. Rev. B*, **19**:6442–6455, 1979.

[ED79] J. P. Elliott and P. G. Dawber. *Symmetry in physics, Volume 1: principles and simple applications*. Macmillan Press, London, 1979.

[Ede95] N. M. Edelstein. Comparison of the electronic structure of the lanthanides and actinides. *J. Alloys Compd.*, **223**:197–203, 1995.

[EGRS85] M. Eyal, E. Greenberg, R. Reisfeld, and N. Spector. Spectroscopy of praeseodymium (III) in zirconium fluoride crystals. *Chem. Phys. Lett.*, **117**:108–114, 1985.

[EHHS72] R. J. Elliott, R. T. Harley, W. Hayes, and S. R. P. Smith. Raman scattering and theoretical studies of Jahn–Teller induced phase transitions in some rare-earths compounds. *Proc. Roy. Soc.*, **A328**:217–266, 1972.

[Eis63a] J. C. Eisenstein. Spectrum of $Er^{3+}$ in $LaCl_3$. *J. Chem. Phys.*, **39**:2128–2133, 1963.

[Eis63b] J. C. Eisenstein. Spectrum of $Nd^{3+}$ in $LaCl_3$. *J. Chem. Phys.*, **39**:2134–2140, 1963.

[Eis64] J. C. Eisenstein. Erratum: spectrum of $Nd^{3+}$ in $LaCl_3$. *J. Chem. Phys.*, **40**:2044, 1964.

[EN75] A. Edgar and D. J. Newman. Local distortion effects on the spin-Hamiltonian parameters of $Gd^{3+}$ substituted into the fluorites. *J. Phys. C: Solid State Phys.*, **8**:4023–4030, 1975.

[Fal66] L. M. Falicov. *Group theory and its physical applications*. University of Chicago Press, Chicago, 1966.

[FBU88a] A. Furrer, P. Brüesch, and P. Unternährer. Crystalline electric field in $HoBa_2Cu_3O_{7-\delta}$ determined by inelastic neutron scattering. *Solid State Commun.*, **67**:69–73, 1988.

[FBU88b] A. Furrer, P. Brüesch, and P. Unternährer. Neutron spectroscopic determination of the crystalline electric field in $HoBa_2Cu_3O_{7-\delta}$. *Phys. Rev. B*, **38**:4616–4623, 1988.

[FGC⁺89] M. Faucher, D. Garcia, P. Caro, J. Derouet, and P. Porcher. The anomalous crystal field splittings of $^2H_{11/2}(Nd^{3+}, 4f^3)$. *J. Phys. (Paris)*, **50**:219–243, 1989.

[FGP89] M. Faucher, D. Garcia, and P. Porcher. Empirically corrected crystal field calculations within the $^2H(2)_{11/2}$ level of $Nd^{3+}$. *C. R. Acad. Sci. Ser II*, **308**:603–608, 1989.

[FKV72] A. Furrer, J. Kjems, and O. Vogt. Crystalline electric field levels in the neodymium monopnictides determined by neutron spectroscopy. *J. Phys. C: Solid State Phys.*, **5**:2246–2258, 1972.

[FL86] P. Fulde and M. Loewenhaupt. Magnetic excitations in crystal-field split $4f$ systems. *Adv. Phys.*, **34**:589–661, 1986.

[FM97] M. Faucher and O. K. Moune. $4f^2/4f6p$ configuration interaction in $LiYF_4:Pr^{3+}$. *Phys. Rev. A*, **55**:4150–4154, 1997.

[FW62] A. J. Freeman and R. E. Watson. Theoretical investigation of some magnetic and spectroscopic properties of rare earth ions. *Phys. Rev.*, **127**:2058–2075, 1962.

[GBMB93] J. C. Gâcon, G. W. Burdick, B. Moine, and H. Bill. $^7F_0 \rightarrow {}^5D_0$ two-photon absorption transitions of $Sm^{2+}$ in $SrF_2$. *Phys. Rev. B*, **47**:11 712–11 716, 1993.

[GBR96] K. E. Gunde, G. W. Burdick, and F. S. Richardson. Chirality-dependent two-photon absorption probabilities and circular dichroic line strengths: Theory, calculation and measurement. *Chem. Phys.*, **208**:195–219, 1996.

[GdSH89] T. Gregorian, H. d'Amour Sturm, and W. B. Holzapfel. Effect of pressure and crystal structure on energy levels of $Pr^{3+}$ in $LaCl_3$. *Phys. Rev. B*, **39**:12 497–12 519, 1989.

[GF89] D. Garcia and M. Faucher. An explanation of the $^1D_2$ anomalous crystal field splitting in $PrCl_3$. *J. Chem. Phys.*, **90**:5280–5283, 1989.

[GF92] D. Garcia and M. Faucher. First direct calculation of $4f \rightarrow 4f$ oscillator strengths for dipolar electric transitions in $PrCl_3$. *J. Alloys Compd.*, **180**:239–242, 1992.

[GHOS69] P. Grünberg, S. Hüfner, E. Orlich, and J. Schmitt. Crystal field in dysprosium garnets. *Phys. Rev.*, **184**:285–293, 1969.

[GLS91] G. L. Goodman, C.-K. Loong, and L. Soderholm. Crystal-field properties of $f$-electron states in $RBa_2Cu_3O_7$ for R = Ho, Nd and Pr. *J. Phys.: Condens. Matter*, **3**:49–67, 1991.

[GMO92] E. A. Goremychkin, A. Yu. Muzychka, and R. Osborn. Crystal field potential of $NdCu_2Si_2$: A comparison with $CeCu_2Si_2$. *Physica B*, **179**:184–190, 1992.

[GNKHV+98] O. Guillot-Noël, A. Kahn-Harari, B. Viana, D. Vivien, E. Antic-Fidancev, and P. Porcher. Optical spectra and crystal field calculations of $Nd^{3+}$ doped zircon-type $YMO_4$ laser hosts M = V, P, As. *J. Phys.: Condens. Matter*, **6**:6491–6503, 1998.

[GO93] E. A. Goremychkin and R. Osborn. Crystal-field excitations in $CeCu_2Si_2$. *Phys. Rev. B*, **47**:14 280–14 290, 1993.

[GOM94] E. A. Goremychkin, R. Osborn, and A. Yu. Muzychka. Crystal-field effects in $PrCu_2Si_2$: an evaluation of evidence for heavy-fermion behavior. *Phys. Rev. B*, **50**:13 863–13 866, 1994.

[GR95] K. E. Gunde and F. S. Richardson. Fluorescence-detected two-photon circular dichroism of $Gd^{3+}$ in trigonal $Na_3[Gd(C_4H_4O_5)_3] \cdot 2NaCl_4 \cdot 6H_2O$. *Chem. Phys.*, **194**:195–206, 1995.

[Gri61] J. S. Griffith. *The theory of transition metal ions*. Cambridge University Press, 1961.

[GS73] M. Gerloch and R. C. Slade. *Ligand-field parameters*. Cambridge University Press, 1973.

[GWB96] C. Görller-Walrand and K. Binnemans. Rationalization of crystal–field parameterization. In K. A. Gschneidner Jr and L. Eyring, editors, *Handbook on the Physics and Chemistry of Rare Earths*, volume **23**, pages 121–283. North-Holland, Amsterdam, 1996.

[GWB98] C. Görller-Walrand and K. Binnemans. Spectral intensities of $f - f$ transitions. In K. A. Gschneidner Jr and L. Eyring, editors, *Handbook on the Physics and Chemistry of the Rare Earths*, volume **25**, pages 101–264. North Holland, Amsterdam, 1998.

[GWBP+85] C. Görller-Walrand, M. Behets, P. Porcher, O. K. Moune-Minn, and I. Laursen. Analysis of the fluorescence spectrum of $LiYF_4$:$Eu^{3+}$. *Inorganica Chimica Acta*, **109**:83–90, 1985.

[Har66] W. A. Harrison. *Pseudopotentials in the theory of metals*. W. A. Benjamin, Reading, Mass., 1966.

[Hei60] Volker Heine. *Group theory in quantum mechanics*. Pergamon Press, Oxford, 1960.

[HF93] V. Hurtubise and K. F. Freed. The algebra of effective hamiltonians and operators: exact operators. *Adv. Chem. Phys.*, **83**:465–541, 1993.

[HF94] V. Hurtubise and K. F. Freed. Perturbative and complete model space linked diagrammatic expansions for the canonical effective operator. *J. Chem. Phys.*, **100**:4955–4968, 1994.

[HF96a] K. Hummler and M. Fähnle. Full-potential linear-muffin-tin-orbital calculations of the magnetic properties of rare-earth-transition-metal intermetallics. I. Description of the formalism and application to the series $RCo_5$ (R = rare-earth atom). *Phys. Rev. B*, **53**:3272–3289, 1996.

[HF96b] K. Hummler and M. Fähnle. Full-potential linear-muffin-tin-orbital calculations of the magnetic properties of rare-earth-transition-metal intermetallics. II. $Nd_2Fe_{14}B$. *Phys. Rev. B*, **53**:3272–3289, 1996.

[HF98] W. Henggeler and A. Furrer. Magnetic excitations in rare-earth-based high-temperature superconductors. *J. Phys.: Condens. Matter*, **10**:2579–2596, 1998.

[HFC76] D. E. Henri, R. L. Fellows, and G. R. Choppin. Hypersensitivity in the electronic transitions of lanthanide and actinide complexes. *Coord. Chem. Rev.*, **18**:199–224, 1976.

[HI89] B. Henderson and G. F. Imbusch. *Optical spectroscopy of inorganic solids*. Clarendon Press, Oxford, 1989.

[HJC96] J. E. Hansen, B. R. Judd, and H. Crosswhite. Matrix elements of scalar three-electron operators for the atomic $f$ shell. *Atomic Data and Nuclear Data Tables*, **62**:1–49, 1996.

[HJL87] J. E. Hansen, B. R. Judd, and G. M. S. Lister. Parametric fitting to $2p^n 3d$ configurations using orthogonal operators. *J. Phys. B: At. Mol. Phys.*, **20**:5291–5324, 1987.

[HLC89] J. Huang, G. K. Liu, and R. L. Cone. Resonant enhancement of direct two-photon absorption in $Tb^{3+}$:$LiYF_4$. *Phys. Rev. B*, **39**:6348–6354, 1989.

[HMR98] T. A. Hopkins, D. H. Metcalf, and F. S. Richardson. Electronic state structure and optical properties of $Tb(ODA)_3^{3-}$ complexes in trigonal $Na_3[Tb(ODA)_3].2NaClO_4.6H_2O$ crystals. *Inorg. Chem.*, **37**:1401–1412, 1998.

[HP79] W. G. Harter and C. W. Patterson. Asymptotic eigensolution of fourth and sixth rank octahedral tensor operators. *J. Math. Phys.*, **20**:1452–1459, 1979.

[HR63] M. T. Hutchings and D. K. Ray. Investigation into the origin of crystalline electric field effects on rare earth ions I. Contribution from neighbouring induced moments. *Proc. Phys. Soc.*, **81**:663–676, 1963.

[HSE+81] T. Hayhurst, G. Shalimoff, N. Edelstein, L. A. Boatner, and M. M. Abraham. Optical spectra and Zeeman effect for $Er^{3+}$ in $LuPO_4$ and $HfSiO_4$. *J. Chem. Phys.*, **74**:5449–5452, 1981.

[Hüf78] S. Hüfner. *Optical spectra of transparent rare earth compounds*. Academic Press, London, 1978.

[Hut64] M. T. Hutchings. Point-charge calculations of energy levels of magnetic ions in crystalline electric fields. *Solid State Phys.*, **16**:227–273, 1964.

[IMEK97] M. Illemassene, K. M. Murdoch, N. M. Edelstein, and J. C. Krupa. Optical spectroscopy and crystal field analysis of $Cm^{3+}$ in $LaCl_3$. *J. Luminescence*, **75**:77–87, 1997.

[JC84] B. R. Judd and H. Crosswhite. Orthogonalized operators for the $f$ shell. *J. Opt. Soc. Amer. B*, **1**:255–260, 1984.

[JCC68] B. R. Judd, H. M. Crosswhite, and H. Crosswhite. Intra-atomic magnetic interactions for $f$ electrons. *Phys. Rev.*, **169**:130–138, 1968.

[JDR69] L. F. Johnson, J. F. Dillon, and J. P. Remeika. Optical studies of $Ho^{3+}$ ions in YGaG and YIG. *J. Appl. Phys.*, **40**:1499–1500, 1969.

[JHR82] B. R. Judd, J. E. Hansen, and A. J. J. Raassen. Parametric fits in the atomic d shell. *J. Phys. B: At. Mol. Phys.*, **15**:1457–1472, 1982.

[JJ64] C. K. Jørgensen and B. R. Judd. Hypersensitive pseudo-quadrupole transitions in lanthanides. *Mol. Phys.*, **8**:281–290, 1964.

[JL96] B. R. Judd and E. Lo. Factorization of the matrix elements of three-electron operators used in configuration-interaction studies of the atomic $f$ shell. *Atomic Data and Nuclear Data Tables*, **62**:51–75, 1996.

[JNN89] B. R. Judd, D. J. Newman, and B. Ng. Properties of orthogonal operators. In B. Gruber and F. Iachello, editors, *Symmetries in science III*, pages 215–224. Plenum Press, New York, 1989.

[Jor62] C. K. Jorgensen. *Orbitals in atoms and molecules*. Academic Press, London, 1962.

[JP82] B. R. Judd and D. R. Pooler. Two-photon transitions in gadolinium ions. *J. Phys. C*, **15**:591–598, 1982.

[JRTH93] C. K. Jayasankar, M. F. Reid, Th. Tröster, and W. B. Holzapfel. Analysis of correlation effects in the crystal-field splitting of $Nd^{3+}:LaCl_3$ under pressure. *Phys. Rev. B*, **48**:5919–5921, 1993.

[JS84] B. R. Judd and M. A. Suskin. Complete set of orthogonal scalar operators for the configuration $f^3$. *J. Opt. Soc. Amer. B*, **1**:261–265, 1984.

[Jud62] B. R. Judd. Optical absorption intensities of rare-earth ions. *Phys. Rev.*, **127**:750–761, 1962.

[Jud63] B. R. Judd. *Operator techniques in atomic spectroscopy*. McGraw-Hill, New York, 1963. A corrected reprint was published by Princeton University Press, Princeton, NJ, in 1998.

[Jud66] B. R. Judd. Hypersensitive transitions in rare earth ions. *J. Chem. Phys.*, **44**:839–840, 1966.

[Jud77a] B. R. Judd. Correlation crystal fields for lanthanide ions. *Phys. Rev. Lett.*, **39**:242–244, 1977.

[Jud77b] B. R. Judd. Ligand field theory for actinides. *J. Chem. Phys.*, **66**:3163–3170, 1977.

[Jud78] B. R. Judd. Ligand polarizations and lanthanide ion spectra. In P. Kramer and A. Rieckers, editors, *Group theoretical methods in physics, Proceedings 1977*, volume 79, pages 417–419. Springer, Berlin, 1978.

[Jud79] B. R. Judd. Ionic transitions hypersensitive to environment. *J. Chem. Phys*, **70**:4830–4833, 1979.

[Jud88] B. R. Judd. Atomic theory and optical spectroscopy. In K. A. Gschneidner Jr and L. Eyring, editors, *Handbook on the physics and chemistry of rare earths*, volume **11**, pages 81–195. North-Holland, Amsterdam, 1988.

[Kam95] A. A. Kaminskii. Today and tomorrow of laser-crystal physics. *Phys. Stat. Sol. (a)*, **148**:9–79, 1995.

[Kam96] A. A. Kaminskii. *Crystalline lasers: physical processes and operating schemes*. CRC Press, Boca Raton, Florida, 1996.

[KD66] N. H. Kiess and G. H. Dieke. Energy levels of $Er^{3+}$ and $Pr^{3+}$ in hexagonal $LaBr_3$. *J. Chem. Phys.*, **45**:2729–2734, 1966.

[KDM$^+$97] M. Karbowiak, J. Drozdzynski, K. M. Murdoch, N. M. Edelstein, and

S. Hubert. Spectroscopic studies and crystal-field analysis of $U^{3+}$ ions in $RbY_2Cl_7$ single crystals. *J. Chem. Phys.*, 106:3067–3077, 1997.

[KDWS63] G. F. Koster, J. O. Dimmock, R. G. Wheeler, and H. Statz. *Properties of the thirty-two point groups.* MIT Press, Cambridge, MA, 1963.

[KEAB93] K. Kot, N. M. Edelstein, M. M. Abraham, and L. A. Boatner. Zero-field splitting of $Cm^{3+}$:$LuPO_4$ single crystals. *Phys. Rev. B*, 48:12704–12712, 1993.

[KFR95] M. Kotzian, T. Fox, and N. Rösch. The calculation of electronic spectra of hydrated Ln(III) ions within the INDO/S-CI approach. *J. Phys. Chem.*, 99:600–605, 1995.

[KG89] M. Kibler and J.-C. Gâcon. Energy levels of paramagnetic ions: algebra VI. Transition intensity calculations. *Croatica Chemica Acta*, 62:783–797, 1989.

[KMR80] R. Kuroda, S. F. Mason, and C. Rosini. Anistropic contributions to the ligand polarization model for the $f$–$f$ transition probabilities of Eu(III) complexes. *Chem. Phys. Lett.*, 70:11–16, 1980.

[Kru66] W. F. Krupke. Optical absorption and fluorescence intensities in several rare-earth doped $Y_2O_3$ and $LaF_3$ crystals. *Phys. Rev.*, 145:325–337, 1966.

[Kru71] W. F. Krupke. Radiative transition probabilities within the $4f^n$ ground configuration of Nd:YAG. *IEEE J. Quantum Electron.*, 7:153–159, 1971.

[Kru87] J. C. Krupa. Spectroscopic properties of tetravalent actinide ions in solids. *Inorg. Chi. Acta*, 139:223–241, 1987.

[Kus67] D. Kuse. Optische Absorbtionsspektra und Kristallfeld auspaltungen des $Er^{3+}$ Ions in $YPO_4$ und $YVO_4$. *Z. Phys.*, 203:49–58, 1967.

[LBH93] G. K. Liu, J. V. Beitz, and J. Huang. Ground-state splitting of S-state ion $Cm^{3+}$ in $LaCl_3$. *J. Chem. Phys.*, 99:3304–3311, 1993.

[LC83a] L. I. Levin and V. I. Cherpanov. Metal-ligand exchange effects and crystal-field screening for rare-earth ions. *Soviet Phys.: Solid State*, 25:394–399, 1983.

[LC83b] L. I. Levin and V. I. Cherpanov. Superposition-exchange model of the second-rank crystal field for rare-earth ions. *Soviet Phys.: Solid State*, 25:399–403, 1983.

[LCJ+94] G. K. Liu, W. T. Carnall, R. C. Jones, R. L. Cone, and J. Huang. Electronic energy level structure of $Tb^{3+}$ in $LiYF_4$. *J. Alloys Compd.*, 207–208:69–73, 1994.

[LCJW94] G. K. Liu, W. T. Carnall, G. Jursich, and C. W. Williams. Analysis of the crystal-field spectra of the actinide tetrafluorides. II. $AmF_4$, $CmF_4$, $Cm^{4+}$:$CeF_4$ and $Bk^{4+}$:$CeF_4$. *J. Chem. Phys.*, 101:8277–8289, 1994.

[LE87] L. I. Levin and K. M. Eriksonas. Characteristic parameters of the $Eu^{2+}$ S-state splitting in low-symmetric centres with $F^-$ and $Cl^-$ ligands. *J. Phys. C: Solid State Phys.*, 20:2081–2088, 1987.

[Lea82] R. P. Leavitt. On the role of certain rotational invariants in crystal-field theory. *J. Chem. Phys.*, 77:1661–1663, 1982.

[Lea87] R. C. Leavitt. A complete set of $f$-electron scalar operators. *J. Phys. A: Math. Gen.*, 20:3171–3183, 1987.

[Led77] W. Ledermann. *Introduction to group characters.* Cambridge University Press, 1977.

[Leś90] K. Leśniak. Crystal fields and dopant–ligand separations in cubic centres of rare-earth ions in fluorites. *J. Phys.: Condens. Matter*, 2:5563–5574, 1990.

[LG92] L. I. Levin and A. D. Gorlov. $Gd^{3+}$ crystal-field effects in low-symmetric centres. *J. Phys.: Condens. Matter*, 4:1981–1992, 1992.

[LL75] C. Linares and A. Louat. Interpretation of the crystal field parameters by the superposition and angular overlap models. Application to some lanthanum compounds. *J. Physique*, 36:717–725, 1975.

[LLW62] K. R. Lea, M. J. M. Leask, and W. P. Wolf. The raising of angular momentum degeneracy of $f$-electron terms by cubic crystal fields. *J. Phys. Chem. Solids*, 23:1381–1405, 1962.

[LLZ+98] G. K. Liu, S. T. Li, V. V. Zhorin, C. K. Loong, M. M. Abraham, and L. A. Boatner. Crystal-field splitting, magnetic interaction and vibronic excitations of $^{244}Cm^{3+}$ in $YPO_4$ and $LuPO_4$. *J. Chem. Phys.*, 109:6800–6808, 1998.

[LM86] I. Lindgren and J. Morrison. *Atomic many-body theory*. Springer, Berlin, 1986.

[LN69] J. W. Leech and D. J. Newman. *How to use groups*. Methuen, London, 1969.

[LN73] B. F. Lau and D. J. Newman. Crystal field and exchange parameters in NiO and MnO. *J. Phys. C: Solid State Phys.*, 6:3245–3254, 1973.

[LR90] C. L. Li and M. F. Reid. Correlation crystal field analysis of the $^{2}H(2)_{11/2}$ multiplet of $Nd^{3+}$. *Phys. Rev. B*, 42:1903–1909, 1990.

[LR93] T. S. Lo and M. F. Reid. Group-theoretical analysis of correlation crystal-field models. *J. Alloys Compd.*, 193:180–182, 1993.

[LSA+93] C.-K. Loong, L. Soderholm, M. M. Abraham, L. A. Boatner, and N. M. Edelstein. Crystal-field excitations and magnetic properties of $TmPO_4$. *J. Chem. Phys.*, 98:4214–4222, 1993.

[LSD+73] G. Löhmuller, G. Schmidt, B. Deppisch, V. Gramlich, and C. Scheringer. Die Kristallstruckturen von Yttrium–Vanadat, Lutetium–Phosphat und Lutetium–Arsenat. *Acta Crystallogr. B*, 29:141–142, 1973.

[LSH+93] C.-K. Loong, L. Soderholm, J. P. Hammonds, M. M. Abraham, L. A. Boatner, and N. M. Edelstein. Rare-earth energy levels and magnetic properties of $HoPO_4$ and $ErPO_4$. *J. Phys.: Condens. Matter*, 5:5121–5140, 1993.

[Mar47] H. H. Marvin. Mutual magnetic interactions of electrons. *Phys. Rev.*, 71:102–110, 1947.

[MB82] V. M. Malhotra and H. A. Buckmaster. A study of the host lattice effect in the lanthanide hydroxides. 34 GHz $Gd^{3+}$ impurity ion EPR spectra at 77 and 294K. *Canad. J. Phys.*, 60:1573–1588, 1982.

[MEBA96] K. M. Murdoch, N. M. Edelstein, L. A. Boatner, and M. M. Abraham. Excited state absorption and fluorescence line narrowing studies of $Cm^{3+}$ and $Gd^{3+}$ in $LuPO_4$. *J. Chem. Phys.*, 105:2539–2546, 1996.

[ML79] C. A. Morrison and R. P. Leavitt. Crystal field analysis of triply ionized rare earth ions in lanthanum trifluoride. *J. Chem. Phys.*, 71:2366–2374, 1979.

[ML82] C. A. Morrison and R. P. Leavitt. Spectroscopic properties of triply ionized lanthanides in transparent host crystals. In K. A. Gschneider Jr and L. Eyring, editors, *Handbook on the physics and chemistry of rare earths*, volume 5, chapter 46, pages 461–692. North-Holland, Amsterdam, 1982.

[MN73] B. Morosin and D. J. Newman. $La_2O_2S$ structure refinement and crystal field. *Acta Cryst.*, B29:2647–2648, 1973.

[MN87] P. K. MacKeown and D. J. Newman. *Computational techniques in physics.* Adam Hilger, Bristol, 1987.

[MNG97] K. M. Murdoch, A. D. Nguyen, and J. C. Gâcon. Two-photon absorption spectroscopy of $Cm^{3+}$ in $LuPO_4$. *Phys. Rev. B*, **56**:3038–3045, 1997.

[MPS74] S. F. Mason, R. D. Peacock, and B. Stewart. Dynamic coupling contributions to the intensity of hypersensitive lanthanide transitions. *Chem. Phys. Lett.*, **29**:149–153, 1974.

[MRB96] L. F. McAven, M. F. Reid, and P. H. Butler. Transformation properties of the delta function model of correlation crystal fields. *J. Phys. B*, **29**:1421–1431, 1996. Corrigenda: 4319.

[MRR87a] P. S. May, M. F. Reid, and F. S. Richardson. Circular dichroism and electronic rotatory strengths of the samarium $4f$–$4f$ transitions in $Na_3[Sm(oxydiacetate)_3] \cdot 2NaClO_4 \cdot 6H_2O$. *Mol. Phys.*, **62**:341–364, 1987.

[MRR87b] P. S. May, M. F. Reid, and F. S. Richardson. Electric dipole intensity parameters for the samarium $4f$–$4f$ transitions in $Na_3[Sm(oxydiacetate)_3] \cdot 2NaClO_4 \cdot 6H_2O$. *Mol. Phys.*, **61**:1471–1485, 1987.

[MWK76] C. A. Morrison, D. E. Wortmann, and N. Karayianis. Crystal field parameters for triply-ionized lanthanides in yttrium aluminium garnet. *J. Phys. C: Solid State Phys.*, **9**:L191–194, 1976.

[NB75] D. J. Newman and G. Balasubramanian. Parametrization of rare-earth ion transition intensities. *J. Phys. C: Solid State Phys.*, **8**:37–44, 1975. In this paper the sense of $\sigma$ and $\pi$ polarization is reversed from the convention used in Chapter 10.

[NBCT71] D. J. Newman, S. S. Bishton, M. M. Curtis, and C. D. Taylor. Configuration interaction and lanthanide crystal fields. *J. Phys. C: Solid State Phys.*, **4**:3234–3248, 1971.

[NC69] D. J. Newman and M. M. Curtis. Crystal field in rare-earth fluorides. I. Molecular orbital calculation of $PrF_3$ parameters. *J. Phys. Chem. Solids*, **30**:2731–2737, 1969.

[NE76] D. J. Newman and A. Edgar. Interpretation of $Gd^{3+}$ spin-Hamiltonian parameters in garnet host crystals. *J. Phys. C: Solid State Phys.*, **9**:103–109, 1976.

[New70] D. J. Newman. Origin of the ground state splitting of $Gd^{3+}$ in crystals. *Chem. Phys. Lett.*, **6**:288–290, 1970.

[New71] D. J. Newman. Theory of lanthanide crystal fields. *Adv. Phys.*, **20**:197–256, 1971.

[New73a] D. J. Newman. Band structure and the crystal field. *J. Phys. C: Solid State Phys.*, **6**:458–466, 1973.

[New73b] D. J. Newman. Crystal field and exchange parameters in $KNiF_3$. *J. Phys. C: Solid State Phys.*, **6**:2203–2208, 1973.

[New73c] D. J. Newman. Slater parameter shifts in substituted lanthanide ions. *J. Phys. Chem. Solids*, **34**:541–545, 1973.

[New74] D. J. Newman. Quasi-localized representation of electronic band structures in cubic crystals. *J. Phys. Chem. Solids*, **35**:1187–1199, 1974.

[New75] D. J. Newman. Interpretation of $Gd^{3+}$ spin-Hamiltonian parameters. *J. Phys. C: Solid State Phys.*, **8**:1862–1868, 1975.

[New77a] D. J. Newman. Ligand ordering parameters. *Austral. J. Phys.*, **30**:315–323, 1977.

[New77b] D. J. Newman. On the g-shift of S-state ions. *J. Phys. C: Solid State Phys.*, **29**:L315–L318, 1977.

[New77c] D. J. Newman. Parametrization of crystal induced correlation between $f$–electrons. *J. Phys. C: Solid State Phys.*, **10**:4753–4764, 1977.

[New78] D. J. Newman. Parametrization schemes in solid state physics. *Austral. J. Phys.*, **31**:489–531, 1978.

[New80] D. J. Newman. The orbit-lattice interaction for lanthanide ions. II. strain and relaxation time predictions for cubic systems. *Austral. J. Phys.*, **33**:733–743, 1980.

[New81] D. J. Newman. Matrix mutual orthogonality and parameter independence. *J. Phys. A: Math. Gen.*, **14**:L429–L431, 1981.

[New82] D. J. Newman. Operator orthogonality and parameter uncertainty. *Phys. Lett.*, **92A**:167–169, 1982.

[New83a] D. J. Newman. Models of lanthanide crystal fields in metals. *J. Phys. F: Met. Phys.*, **13**:1511–1518, 1983.

[New83b] D. J. Newman. Unique labelling of $J$-states in octahedral symmetry. *Phys. Lett.*, **97A**:153–154, 1983.

[New85] D. J. Newman. Lanthanide and actinide crystal field intrinsic parameter variations. *Lanthanide Actinide Res.*, **1**:95–102, 1985.

[NK63] C. W. Nielson and G. F. Koster. *Spectroscopic coefficients for the $p^n$, $d^n$ and $f^n$ configurations.* MIT Press, Cambridge, Mass., 1963.

[NME$^+$97] A. D. Nguyen, K. Murdoch, N. Edelstein, L. A. Boatner, and M. M. Abraham. Polarization dependence of phonon and electronic raman intensities in $PrVO_4$ and $NdVO_4$. *Phys. Rev. B*, **56**:7974–7987, 1997.

[NN84] B. Ng and D. J. Newman. Models of the correlation crystal field for octahedral $3d^n$ systems. *J. Phys. C: Solid State Phys.*, **17**:5585–5594, 1984.

[NN86a] B. Ng and D. J. Newman. *Ab-initio* calculation of crystal field correlation effects in $Pr^{3+}$–$Cl^-$. *J. Phys. C: Solid State Phys.*, **19**:L585–588, 1986.

[NN86b] B. Ng and D. J. Newman. A linear model of crystal field correlation effects in $Mn^{2+}$. *J. Chem. Phys.*, **84**:3291–3296, 1986.

[NN87a] B. Ng and D. J. Newman. Many-body crystal field calculations I: Methods of computation and perturbation expansion. *J. Chem. Phys.*, **87**:7096–7109, 1987.

[NN87b] B. Ng and D. J. Newman. Many-body crystal field calculations II: Results for the system $Pr^{3+}$–$Cl^-$. *J. Chem. Phys.*, **87**:7110–7117, 1987.

[NN88] B. Ng and D. J. Newman. Spin-correlated crystal field parameters for trivalent actinides. *J. Phys. C: Solid State Phys.*, **21**:3273–3276, 1988.

[NN89a] D. J. Newman and B. Ng. Crystal field superposition model analyses for tetravalent actinides. *J.Phys.:Condens. Matter*, **1**:1613–1619, 1989.

[NN89b] D. J. Newman and B. Ng. The superposition model of crystal fields. *Rep. Prog. Phys.*, **52**:699–763, 1989.

[NNP84] D. J. Newman, B. Ng, and Y. M. Poon. Parametrization and interpretation of paramagnetic ion spectra. *J. Phys. C: Solid State Phys.*, **17**:5577–5584, 1984.

[NP75] D. J. Newman and D. C. Price. Determination of the electrostatic contributions to lanthanide quadrupolar crystal fields. *J. Phys. C: Solid State Phys.*, **8**:2985–2991, 1975.

[NPR78] D. J. Newman, D. C. Price, and W. A. Runciman. Superposition model analysis of the near infrared spectrum of $Fe^{3+}$ in pyrope-almandine garnets. *Amer. Mineral.*, **63**:1278–1288, 1978.

[NS69] D. J. Newman and G. E. Stedman. Interpretation of crystal field parameters in the rare-earth-substituted garnets. *J. Chem. Phys.*, **51**:3013–3023, 1969.

[NS71] D. J. Newman and G. E. Stedman. Analysis of the crystal field in rare-earth substituted oxysulphide and vanadate systems. *J. Phys. Chem. Solids*, **32**:535–542, 1971.

[NS76] D. J. Newman and E. Siegel. Superposition model analysis of $Fe^{3+}$ and $Mn^{2+}$ spin-Hamiltonian parameters. *J. Phys. C: Solid State Phys.*, **9**:4285–4292, 1976.

[NSC70] D. J. Newman, G. E. Stedman, and M. M. Curtis. The use of simplified models in crystal field theory. *Colloques Internationaux, CNRS*, **180**:505–512, 1970.

[NSF82] D. J. Newman, G. G. Siu, and W. Y. P. Fung. Effect of spin-polarization on the crystal field of lanthanide ions. *J. Phys. C: Solid State Phys.*, **15**:3113–3125, 1982.

[NU72] D. J. Newman and W. Urban. A new interpretation of the $Gd^{3+}$ ground state splitting. *J. Phys. C: Solid State Phys.*, **5**:3101–3109, 1972.

[NU75] D. J. Newman and W. Urban. Interpretation of S-state ion E.P.R. spectra. *Adv. Phys.*, **24**:793–843, 1975.

[NV76] P. Novák and I. Veltruský. Overlap and covalency contributions to the zero field splitting of S-state ions. *Phys. Stat. Sol. (b)*, **73**:575–586, 1976.

[Ofe62] G. S. Ofelt. Intensities of crystal spectra of rare-earth ions. *J. Chem. Phys.*, **37**:511–520, 1962.

[OH69] E. Orlich and S. Hüfner. Optical measurements in erbium iron and erbium gallium garnet. *J. Appl. Phys.*, **40**:1503–1504, 1969.

[PC78] P. Porcher and P. Caro. Crystal field parameters for $Eu^{3+}$ in $KY_3F_{10}$ II. Intensity parameters. *J. Chem. Phys.*, **68**:4176–4187, 1978.

[Pea75] R. D. Peacock. The intensities of lanthanide $f$–$f$ transitions. *Struct. Bonding*, **22**:83–122, 1975.

[PFTV86] W. H. Press, B. P. Flannery, S. A. Teukolsky, and W. T. Vetterling. *Numerical recipes: The art of scientific computing*. Cambridge University Press, 1986.

[Pil91a] B. Pilawa. Electron correlation crystal-field splittings of $Ho^{3+}$: $Ho^{3+}$ in $LaCl_3$ and $Y(OH)_3$. *J. Phys.: Condens. Matter*, **2**:667–673, 1991.

[Pil91b] B. Pilawa. Electron correlation crystal-field splittings of $Ho^{3+}$: $Ho^{3+}$ in $YVO_4$, $YAsO_4$ and $HoPO_4$. *J. Phys.: Condens. Matter*, **2**:655–666, 1991.

[PN84] Y. M. Poon and D. J. Newman. Overlap and covalency contributions to lanthanide ion spectral intensity parameters. *J. Phys. C: Solid State Phys.*, **17**:4319–4325, 1984.

[PS71] E. Pytte and K. W. H. Stevens. Tunneling model of phase changes in tetragonal rare-earth crystals. *Phys. Rev. Lett.*, **27**:862–865, 1971.

[PS83] S. B. Piepho and P. N. Schatz. *Group theory in spectroscopy, with applications to magnetic circular dichroism*. Wiley, New York, 1983.

[QBGFR95] J. R. Quagliano, G. W. Burdick, D. P. Glover-Fischer, and F. S. Richardson. Electronic absorption spectra, optical line strengths, and crystal-field energy-level structure of $Nd^{3+}$ in hexagonal $[Nd(H_2O)_9]CF_3(SO_3)_3$. *Chem. Phys.*, **201**:321–342, 1995.

[RAM97] C. Rudowicz, I. Akhmadoulline, and S. B. Madhu. Manual for the computer package CST for Conversions, Standardization and Transformations of the spin Hamiltonian and the crystal field Hamiltonian,

Research report AP-97-12. Technical report, Department of Physics and Materials Science, City University of Hong Kong, 1997.

[RAM98] C. Rudowicz, I. Akhmadoulline, and S. B. Madhu. Conversions, standardization and transformations of the spin Hamiltonian and the crystal field Hamiltonian – computer package 'CST'. In C. Rudowicz, K. N. Yu, and H. Hiraoka, editors, *Modern applications of EPR/ESR: from bio-physics to materials science: Proceedings of the First Asia-Pacific EPR/ESR Symposium, Hong Kong, 20–24 January 1997*, pages 437–444. Springer, Singapore, 1998.

[RB85] Cz. Rudowicz and R. Bramley. On standardization of the spin Hamiltonian and the ligand field Hamiltonian for orthorhombic symmetry. *J. Chem. Phys.*, **83**:5192–5197, 1985.

[RB92] J. R. Ryan and R. Beach. Optical absorption and stimulated emission of neodymium in yttrium lithium fluoride. *J. Opt. Soc. Am. B*, **9**:1883–1887, 1992.

[RBMJ59] M. Rotenberg, R. Bivins, N. Metropolis, and J. K. Wooten Jr. *The 3-j and 6-j symbols*. MIT Press, Cambridge, Mass., 1959.

[RDR83] M. F. Reid, J. J. Dallara, and F. S. Richardson. Comparison of calculated and experimental $4f$–$4f$ intensity parameters for lanthanide complexes with isotropic ligands. *J. Chem. Phys.*, **79**:5743–5751, 1983.

[RDYZ93] Cz. Rudowicz, M. Du, Y. Y. Yeung, and Y. Y. Zhou. Crystal field levels and zero-field splitting parameters of $Cr^{2+}$ $Rb_2Mn_xCr_{1-x}Cl_4$. *Physica B*, **191**:323–333, 1993.

[Rei87a] M. F. Reid. Correlation crystal field analyses with orthogonal operators. *J. Chem. Phys.*, **87**:2875–2884, 1987.

[Rei87b] M. F. Reid. Superposition-model analysis of intensity parameters for $Eu^{3+}$ luminescence. *J. Chem. Phys.*, **87**:6388–6392, 1987.

[Rei88] M. F. Reid. On the use of $\mathbf{E} \cdot \mathbf{r}$ and $\mathbf{A} \cdot \mathbf{p}$ in perturbation calculations of transition intensities for paramagnetic ions in solids. *J. Phys. Chem. Solids*, **49**:185–189, 1988.

[Rei93] M. F. Reid. Additional operators for crystal-field and transition-intensity models. *J. Alloys Compd.*, **193**:160–164, 1993.

[RJR94] E. Rukmini, C. K. Jayasankar, and M. F. Reid. Correlation-crystal-field analysis of the $Nd^{3+}(4f^3)$ energy-level structures in various crystal hosts. *J. Phys.: Condens. Matter*, **6**:5919–5936, 1994.

[RK67] K. Rajnak and W. F. Krupke. Energy levels of $Ho^{3+}$ in $LaCl_3$. *J. Chem. Phys.*, **46**:3532–3542, 1967.

[RK68] K. Rajnak and W.F. Krupke. Erratum: Energy levels of $Ho^{3+}$ in $LaCl_3$. *J. Chem. Phys.*, **48**:3343–3344, 1968.

[RM99] C. Rudowicz and S. B. Madhu. Orthorhombic standardization of spin-Hamiltonian parameters for transition-metal paramagnetic centres in various crystals. *J. Phys.: Condens. Matter*, **11**:273–287, 1999.

[RMSB96] H. J. Ross, L. F. McAven, K. Shinagawa, and P. H. Butler. Calculating spin-orbit matrix elements with RACAH. *J. Comp. Phys.*, **128**:331–340, 1996.

[RN89] M. F. Reid and B. Ng. Complete second-order calculations of intensity parameters for one-photon and two-photon transitions of rare-earth ions in solids. *Mol. Phys.*, **67**:407–415, 1989.

[RR83] M. F. Reid and F. S. Richardson. Anisotropic ligand polarizability contributions to lanthanide $4f$–$4f$ intensity parameters. *Chem. Phys. Lett.*, **95**:501–507, 1983.

[RR84a] M. F. Reid and F. S. Richardson. Electric dipole intensity parameters for $Pr^{3+}$ in $LiYF_4$. *J. Luminescence*, **31/32**:207–209, 1984.

[RR84b] M. F. Reid and F. S. Richardson. Lanthanide $4f$–$4f$ electric dipole intensity theory. *J. Phys. Chem.*, **88**:3579–3586, 1984.

[RR84c] M. F. Reid and F. S. Richardson. Parametrization of electric dipole intensities in the vibronic spectra of rare-earth complexes. *Mol. Phys.*, **51**:1077–1094, 1984.

[RR85] M. F. Reid and F. S. Richardson. Free–ion, crystal–field, and spin–correlated crystal–field parameters for lanthanide ions in the cubic $Cs_2NaLnCl_6$ and $Cs_2NaYCl_6$:$Ln^{3+}$ (doped) systems. *J. Chem. Phys.*, **83**:3831–3836, 1985.

[Rud85a] Cz. Rudowicz. Relations between arbitrary symmetry spin-Hamiltonian parameters $B_k^q$ and $b_k^q$ in various axis systems. *J. Magn. Reson.*, **63**:95–106, 1985.

[Rud85b] Cz. Rudowicz. Transformation relations for the conventional $O_k^q$ and normalised $O_k'^q$ Stevens operator equivalents with $k = 1$ to 6 and $-k \le q \le k$. *J. Phys. C: Solid State Phys.*, **18**:1415–1430, 3837, 1985.

[Rud86] Cz. Rudowicz. On standardization and algebraic symmetry of the ligand field Hamiltonian for rare earth ions at monoclinic symmmetry sites. *J. Chem. Phys.*, **84**:5045–5058, 1986.

[Rud87a] Cz. Rudowicz. Concept of spin Hamiltonian, form of zero field splitting and electronic Zeeman Hamiltonians and relations between parameters used in EPR. A critical review. *Magn. Reson. Rev.*, **13**:1–89, 1987.

[Rud87b] Cz. Rudowicz. On the derivation of superposition-model formulae using the transformation relations for the Stevens operators. *J. Phys. C: Solid State Phys.*, **20**:6033–6037, 1987.

[Rud91] Cz. Rudowicz. Correlations between orthorhombic crystal field parameters for rare-earth $(f^n)$ and transition-metal $(d^n)$ ions in crystals: $REBa_2Cu_3O_{7-x}$,$RE_2F_{14}B$, RE-garnets, RE:$LaF_3$ and $MnF_2$. *Mol. Phys.*, **74**:1159–1170, 1991.

[Rud97] Cz. Rudowicz. On the analysis of EPR data for monclinic symmetry sites. In A. I. Bahtin, editor, *Proceedings of the International Conference Spectroscopy, X-ray and Crystal Chemistry of Minerals, Kazan, 3 September – 20 October*, pages 31–41, Kazan University Press, Kazan, 1997.

[Sac63] M. Sachs. *Solid state theory*. McGraw-Hill, New York, 1963.

[SBP68] R. M. Sternheimer, M. Blume, and R. F. Peierls. Shielding of crystal field at rare-earth ions. *Phys. Rev.*, **173**:376–389, 1968.

[SDO66] R. R. Sharma, T. P. Das, and R. Orbach. Zero field splitting in S-state ions I. Point multipole model. *Phys. Rev.*, **149**:257–269, 1966.

[SDO67] R. R. Sharma, T. P. Das, and R. Orbach. Zero field splitting in S-state ions II. Overlap and covalency model. *Phys. Rev.*, **155**:338–352, 1967.

[SDO68] R. R. Sharma, T. P. Das, and R. Orbach. Zero field splitting in S-state ions III. Corrections to Parts I and II and application to distorted cubic crystals. *Phys. Rev.*, **171**:378–388, 1968.

[SHF+87] B. Schmid, B. Häig, A. Furrer, W. Urland, and R. Kremer. Structure and crystal fields of $PrBr_3$ and $PrCl_3$: a neutron study. *J. Appl. Phys.*, **61**:3426–3428, 1987.

[SIB89] J. Sytsma, G. F. Imbusch, and G. Blaase. The spectroscopy of $Gd^{3+}$ in yttrium oxychloride: Judd–Ofelt parameters from emission data. *J. Chem. Phys.*, **91**:1456–1461, 1989.

[Sla65] J. C. Slater. *Quantum theory of molecules and solids, Vol. 2: symmetry and energy bands in crystals.* McGraw-Hill, New York, 1965.

[SLGD91] L. Soderholm, C.-K. Loong, G. L. Goodman, and B. D. Dabrowski. Crystal-field and magnetic properties of $Pr^{3+}$ and $Nd^{3+}$ in $RBa_2Cu_3O_7$. *Phys. Rev. B,* **43**:7923–7935, 1991.

[SLK92] L. Soderholm, C.-K. Loong, and S. Kern. Inelastic-neutron-scattering study of the $Er^{3+}$ energy levels in $ErBa_2Cu_3O_7$. *Phys. Rev. B,* **45**:10062–10070, 1992.

[SM79a] E. Siegel and K. A. Müller. Local position of $Fe^{3+}$ in ferroelectric $BaTiO_3$. *Phys. Rev. B,* **20**:3587–3595, 1979.

[SM79b] E. Siegel and K. A. Müller. Structure of transition-metal-oxygen-vacancy pair centres. *Phys. Rev. B,* **19**:109–120, 1979.

[SME$^+$95] J. Sytsma, K. M. Murdoch, N. M. Edelstein, L. A. Boatner, and M. M. Abraham. Spectroscopic studies and crystal-field analysis of $Cm^{3+}$ and $Gd^{3+}$ in $LuPO_4$. *Phys. Rev. B,* **52**:12 668–12 676, 1995.

[Sme98] L. Smentek. Theoretical description of the spectroscopic properties of rare earth ions in crystals. *Phys. Rep.,* **297**:155–237, 1998.

[SN71a] G. E. Stedman and D. J. Newman. Crystal field in rare-earth fluorides. II. Parameters for $Er^{3+}$ and $Nd^{3+}$ in $LaF_3$. *J. Phys. Chem. Solids,* **32**:109–114, 1971.

[SN71b] G. E. Stedman and D. J. Newman. Crystal field in rare-earth fluorides. III. Analysis of experimental data for the alkaline earth fluorides. *J. Phys. Chem. Solids,* **32**:2001–2006, 1971.

[SN74] G. E. Stedman and D. J. Newman. Analysis of the spin-lattice parameters for $Gd^{3+}$ an $Eu^{2+}$ in cubic crystals. *J. Phys. C: Solid State Phys.,* **7**:2347–2352, 1974.

[SN83] G. G. Siu and D. J. Newman. Spin-correlation effects in lanthanide ion spectroscopy. *J. Phys. C: Solid State Phys.,* **16**:5031–5038, 1983.

[Ste52] K. W. H. Stevens. Matrix elements and operator equivalents connected with the magnetic properties of rare-earth ions. *Proc. Phys. Soc.,* **A65**:209–215, 1952.

[Ste85] G. E. Stedman. Polarization dependence of natural and field-induced one-photon and multiphoton interactions. *Adv. Phys,* **34**:513–587, 1985.

[Ste90] G. E. Stedman. *Diagram techniques in group theory.* Cambridge University Press, 1990.

[TGH93] Th. Tröster, T. Gregorian, and W. B. Holzapfel. Energy levels of $Nd^{3+}$ and $Pr^{3+}$ in $RCl_3$ under pressure. *Phys. Rev. B,* **48**:2960–2967, 1993.

[THE94] P. Thouvenot, S. Hubert, and N. Edelstein. Spectroscopic study and crystal-field analysis of $Cm^{3+}$ in the cubic-symmetry site of $ThO_2$. *Phys. Rev. B,* **50**:9715–9720, 1994.

[Tin64] M. Tinkham. *Group theory and quantum mechanics.* McGraw-Hill, New York, 1964.

[Tra63] G. T. Trammell. Magnetic ordering properties of rare-earth ions in strong cubic crystal fields. *Phys. Rev.,* **131**:932–948, 1963.

[TS92] P. A. Tanner and G. G. Siu. Electric quadrupole allowed transitions of lanthanide ions in octahedral symmetry. *Mol. Phys.,* **75**:233–242, 1992.

[Url78] W. Urland. The interpretation of the crystal field parameters for $f^n$ electron systems by the angular overlap model. Rare-earth ions in $LaCl_3$. *Chem. Phys. Lett.,* **53**:296–299, 1978.

[USH74] W. Urban, E. Siegel, and W. Hillmer. Trigonal centre of $Gd^{3+}$ in CdS investigated by variable frequency EPR. *Phys. Stat. Sol. (b)*, **62**:73–81, 1974.

[VBG83] Y. Vails, J. Y. Buzaré, and J. Y. Gesland. Zero-field splitting of $Gd^{3+}$ in $LiYF_4$ determined by EPR. *Solid State Commun.*, **45**:1093–1098, 1983.

[VP74] Vishwamittar and S. P. Puri. Investigation of the crystal field in rare-earth doped scheelites. *J. Chem. Phys.*, **61**:3720–3727, 1974.

[WA99] M. Wildner and M. Andrut. Crystal structure, electronic absorption spectra, and crystal field superposition model analysis of $Li_2Co_3(SeO_3)_4$. *Z. Krist.*, **214**:216–222, 1999.

[WDM+97] R. T. Wegh, H. Donker, A. Meijerink, R. J. Lamminmaki, and J. Hölsa. Vacuum-ultraviolet spectroscopy and quantum cutting for $Gd^{3+}$ in $LiYF_4$. *Phys. Rev. B*, **56**:13 841–13 848, 1997.

[Wei78] M. Weissbluth. *Atoms and molecules*. Academic Press, New York, 1978.

[Wil96] M. Wildner. Polarized electronic absorption spectra of $Co^{2+}$ ions in the kieserite-type compounds $CoSO_4.H_2O$ and $CoSeO_4.H_2O$. *Phys. Chem. Minerals*, **23**:489–496, 1996.

[WL73] W. Wüchner and J. Laugsch. Observation of induced magnetism and magnetic ordering in $TbAsO_4$ by optical spectroscopy. *Int. J. Magn*, **5**:181–185, 1973.

[WS93] Q. S. Wang and G. E. Stedman. Spin-assisted matter-field coupling and lanthanide transition intensities. *J. Phys. B*, **26**:1415–1423, 1993.

[WS94] Q. Wang and G. E. Stedman. Time reversal symmetry and fermion many-body operators. *J. Phys. B*, **27**:3829–3847, 1994.

[Wyb65a] B. G. Wybourne. *Spectroscopic properties of rare earths*. Interscience, New York, 1965.

[Wyb65b] B. G. Wybourne. Use of relativistic wave functions in crystal field theory. *J. Chem. Phys.*, **43**:4506–4507, 1965.

[Wyb66] B. G. Wybourne. Energy levels of trivalent gadolinium and ionic contribution to the ground state splitting. *Phys. Rev.*, **148**:317–327, 1966.

[Wyb68] B. G. Wybourne. Effective operators and spectroscopic properties. *J. Chem. Phys.*, **48**:2596–2611, 1968.

[XR93] S. D. Xia and M. F. Reid. Comment: theoretical intensities of $4f - 4f$ transitions between Stark levels of the $Eu^{3+}$ ion in crystals. *J. Phys. Chem. Sol.*, **54**:777–778, 1993.

[YN85a] Y. Y. Yeung and D. J. Newman. Crystal field invariants and parameters for low symmetry sites. *J. Chem. Phys.*, **82**:3747–3752, 1985.

[YN85b] Y. Y. Yeung and D. J. Newman. Unique labelling of $J$-states in finite symmetry. *J. Chem. Phys.*, **83**:4691–4696, 1985.

[YN86a] Y. Y. Yeung and D. J. Newman. High order crystal field invariants and the determination of lanthanide crystal field parameters. *J. Chem. Phys.*, **84**:4470–4473, 1986.

[YN86b] Y. Y. Yeung and D. J. Newman. A new approach to the determination of lanthanide spin-correlated crystal field parameters. *J. Phys. C: Solid State Phys.*, **19**:3877–3884, 1986.

[YN87] Y. Y. Yeung and D. J. Newman. Orbitally correlated crystal field parametrization for lanthanide ions. *J. Chem. Phys.*, **86**:6717–6721, 1987.

[YR89] Y. Y. Yeung and M. F. Reid. Crystal-field and superposition-model analyses of $Pr^{3+}$:$LaF_3$ in $C_2$ symmetry. *J. Less Common Metals*, **148**:213–217, 1989.

[YR92] Y. Y. Yeung and Cz. Rudowicz. Ligand field analysis of the $3d^N$ ions at orthorhombic or higher symmetry sites. *Computers Chem.*, **16**:207–216, 1992.

[YR93] Y. Y. Yeung and Cz. Rudowicz. Crystal field energy levels and state vectors for the $3d^N$ ions at orthorhombic or higher symmetry sites. *J. Comput. Phys.*, **109**:150–152, 1993.

[ZY94] L. Zundu and H. Yidong. Crystal-field analysis of the energy levels and spectroscopic characteristics of $Nd^{3+}$ in $YVO_4$. *J. Phys.: Condens. Matter*, **6**:3737–3748, 1994.

# Index

287